Water Services Management

Water Services Management

David Stephenson

University of Botswana, Gaborone, Botswana

Publishing

Published by IWA Publishing, Alliance House, 12 Caxton Street, London SW1H 0QS, UK
Telephone: +44 (0) 20 7654 5500; Fax: +44 (0) 20 7654 5555; Email: publications@iwap.co.uk
Web: www.iwapublishing.com

First published 2005
© 2005 IWA Publishing

Printed by TJ International (Ltd), Padstow, Cornwall, UK.

Disclaimer

British Library Cataloguing in Publication Data
A CIP catalogue record for this book is available from the British Library

Library of Congress Cataloging- in-Publication Data
A catalog record for this book is available from the Library of Congress

ISBN 1843390809

Contents

[v]

Preface

Water supply and sewerage systems in cities are critical from health and planning points of view, but many are aging and deteriorating. This may be because engineers have tended to focus on development rather than management. As world populations stabilize, less hydraulic design and better management of existing facilities is required.

Modern practices are needed to manage the systems. Business practices, customer satisfaction, and economic planning are all as important as engineering design and can no longer be disregarded by water service providers.

The subjects covered in this book include potable water supply, sewerage and stormwater drainage. Basic design methods are reviewed followed by ways of improving designs and management of those facilities. The book forms a comprehensive guide for design and operation of water services and managing the associated infrastructure.

Hydraulic management implies optimum storage, peak flow attenuation, pollution control and effluent discharge. Infrastructure management includes rehabilitation, reconstruction, upgrading and maintenance.

Topics relevant to economic efficiency are asset management, privatization, and risk analysis. Efficient use of energy and construction project management are also ways of improving economic viability.

The particular problems of developing countries are covered in a special chapter, but a number of other chapters have ideas of relevance, viz. low cost sanitation, staged water supply expansion and off-grid energy sources. Capacity building and appropriate technologies are particularly appropriate.

Students and practitioners are becoming aware of the changes in emphasis of water engineering and university courses and continuing education courses are now orientated to the subjects covered here. The book is a condensation of graduate courses give by the author at a number of universities and to a number of water utilities organizations worldwide. It also serves as a reference book for planners, designers and operators of water services. The broader field of engineering is likely to expand over the coming years, i.e. technical calculations will be relegated to computers and the engineer will be able to broaden his scope. New relevant topics, for example in the IT area, may therefore emerge in future years. The topics here can form the springboard for new focus and terminology.

The manuscript of the book was set out and typed by April Thompson.

About the author

David Stephenson is Professor of Water Engineering at the University of Botswana. He is also Professor Emeritus at the University of the Witwatersrand and a Visiting Professor at the University of Stuttgart. He teaches Water Supply Management and Water Resources Management at graduate level at these Institutes. He is the author of ten books and many papers on the subjects. He started his career in practice and consults internationally.

He lives on a game farm in Africa.

1

Water sources and quality

1.1 AVAILABILITY OF FRESH WATER

Of the 1.4×10^9 km^3 of water in the world, 97% is saline (in oceans), 2% is locked in icebergs, and the balance (2×10^7 km^3) is relatively fresh (potable). Of this fresh water volume, 10^7 km^3 is in the ground, 10^4 km^3 is in the atmosphere and 10^5 km^3 is in rivers and lakes. The latter are renewable resources but sustainability is an important factor in exploiting these resources. In the case of ground water, mining of the water can cause lowering of the water table, since recharge rates are low compared with surface resources.

The concept of 'safe yield' used to be applied in sizing reservoirs or assessing boreholes. That is, a figure is quoted which is taken as the amount that may be drawn from the reservoir. However, there is always a risk associated with sourcing a river or an aquifer. Because past hydrological records showed the source is guaranteed does not imply there could be a worse drought than so far recorded. So the concept of risk analysis is conceived. Indeed taking risks can reduce the cost of the source, e.g. dam size.

There is also risk of pollution or danger to health associated with accessible water sources. Surface sources are generally turbid and contain suspended matter and possibly chemical or bacteriological pollutants from upstream. Groundwater can likewise be contaminated by human activities but this is more difficult to detect.

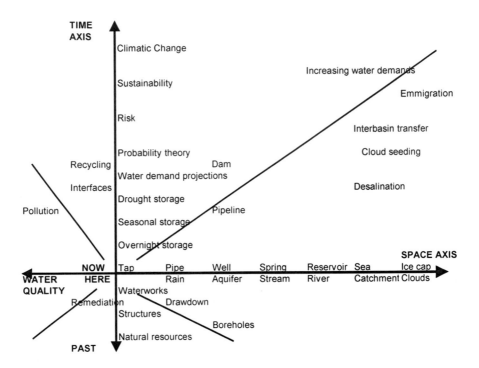

Figure 1.1. Management of water in space and time (Stephenson, 2003)

1.1.1 Surface water

Many bulk supplies of water are drawn from rivers or surface reservoirs. Large scale pumping out of a river is generally cheaper than from an aquifer.

To safeguard water source, water authorities may control or manage the catchment from which the water runs off. This includes pollution control and should fall within the ambit of the water authority in the catchment to ensure good quality potable water and to safeguard the runoff volume. Catchment management may thus extend to control of abstractions from streams or the ground. There may be legal problems associated with abstraction from rivers (UN, 1975). Water may be 'owned' by the state or riparian landowners. It may be that water cannot be transported or directed out of the catchment or province.

Table 1.1. Worldwide stable runoff, by continent (Source: Lvovitch, 1973)

	Stable runoff, km³ p.a.				Total river runoff, incl. flood	Total stable runoff as per cent of total river runoff
	Of under-ground origin	Regulated by lakes	Regulated by water reservoirs	Total		
Europe	1 065	60	300	1 325	3 110	43
Asia	3 410	35	560	4 005	13 190	30
Africa	1 465	40	400	1 905	4 225	45
N. America	1 740	150	490	2 380	5 960	40
S. America	3 740	-	160	3 900	10 380	38
Australasia	465	-	30	495	1 965	25
Total land area except polar zones	11 885	285	1 840	14 010	38 830	36

To overcome drought or seasonal variation in flow, storage is required. This may be by means of dams or lakes. The reservoir could be in-channel and water may be released from an intake tower. A control as to the level of drawoff could avoid drawing in warm water or polluted water. Intake towers may be dry (with pipes running under the dam wall to a pumpstation) or wet (with a tunnel under the dam wall).

If the intake is in the river, some form of sediment rejection is generally necessary (e.g. Petersen, 1984). This may include screens for debris, and sediment traps or ejectors. Intake should be on the outside of a river bend to deflect sediment which tends to be carried to the inside bank by the transverse circulation. Water level fluctuations due to flow variations can make intakes expensive. They have to be high enough to minimize the risk of flooding pumps and motors, and even washing away the pumpstation. The pump sump has to be anchored to avoid floating in high waters. A weir or sump has to be provided in small streams.

The treatment of surface waters is generally confined to removal of suspended solids, followed by disinfection. Sedimentation and filtration are the dominant methods of removing pollutants.

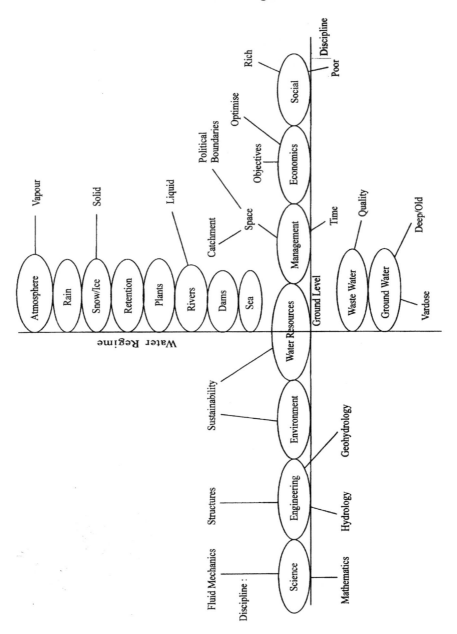

Figure 1.2. Multi-disciplinarity of water resources planning

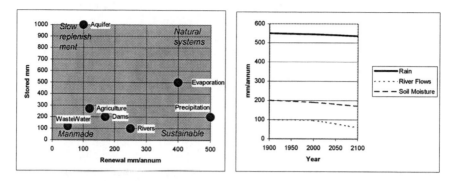

Figure 1.3. Sustainability

1.1.2 Groundwater

Precipitation is separated into runoff to surface water and infiltration into the ground. Most of this subsequently evaporates or transpires into the atmosphere and subsequently falls as precipitation and the cycle is repeated. Water occurs in the ground as follows:

- The water is contained in the soil above an impermeable layer, e.g. bedrock. The distribution of water above and within a porous aquifer is as follows:
  (1) The aeration zone which is the upper layer of water generally only partly saturated;
  (2) The capillary fringe
  (3) The lower layer below the water table referred to as the zone of saturation with water filling the pores. The water table may be recharged by infiltration, but many aquifers are ancient (hardly replenished for centuries). In abstracting groundwater, the rate of recharge needs to be addressed.

There may also be perched water tables, and there is movement between layers due to gravity, capillarity and vapour.

- The water-bearing layer is referred to as an aquifer, which is confined or unconfined depending on whether overlain by an impermeable layer of clay or rock. The type of aquifer is important to note for controlling pollution of the aquifers:
  (1) An unconfined aquifer is open to infiltration from the ground surface.
  (2) A confined aquifer is contained within an impermeable layer and is often under pressure due to the elevation of the recharge area that feeds the aquifer.

Groundwater occurs in primary, dolomitic and secondary aquifers.

- Dolomitic aquifers occur mainly in karstified dolomite. The caverns, the partly decomposed rock and smaller fissures are able to store large quantities of water, which can be abstracted at high rates.
- Primary aquifers consist of sand, gravel and pebbles, which may be consolidated or unconsolidated deposits.
- Ground water is contained in secondary aquifers consisting of weathered fractured rock at depths less than 50 m.

1.1.2.1 Springs

A spring is a visible outlet from a natural underground water system. Management and protection of the aquifer and surface are necessary for quality control. Generally springs fall into three broad categories (Dept. Natl. Housing, 1994). These are:

Open springs: (Occurring as pools). Some form of sump or central collection point, from which an outlet pipe can be led is all that is required.

Closed springs: (The more common form of spring found in rolling or steep topography). In this case, a "spring chamber" is constructed around the eye of the spring, completely enclosing it. It should not be the function of the spring chamber to store water since a rise in the chamber's water level above the eye of the spring can result in the underground flow finding other outlets. A course filter layer is desirable around the spring chamber.

Seepage field: (Where the spring has several eyes or is seeping out over a large area). In this case infiltration trenches are dug and sub-soil drains constructed. The drains feed the spring water to a central collector pipe. Sub-soil drains can be made of stone, gravel, brushwood, tiles, river sand, slotted pipes, filter material or a combination of the above.

1.1.2.2 Wells and boreholes

Where the underground water does not emerge above the natural surface of the ground, digging a well in the case of shallow water tables, or drilling a borehole when the water level is deep can access this water.

1.1.2.3 Hand-dug wells

A well is a shaft which is excavated vertically to a suitable depth below the free-standing surface of the underground water. It is generally necessary to provide some form of lining to prevent the walls of the shaft collapsing, both during and after construction.

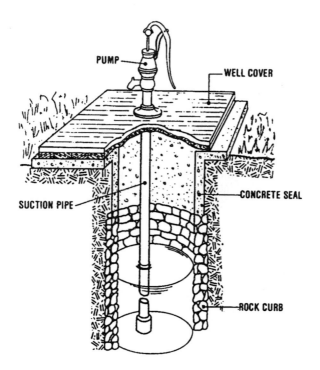

PUMP

WELL COVER

CONCRETE SEAL

SUCTION PIPE

ROCK CURB

Figure 1.4. Hand-dug well (IRC, 1981)

Types of linings used include:
- reinforced concrete rings (caissons)
- curved concrete blocks
- masonry (bricks, blocks or stone)
- cast in situ ferro-cement
- curved galvanised iron sections
- wicker works (saplings, reeds, bamboo, etc.)

The well must be sunk sufficiently deep below the freestanding surface of the ground water to provide adequate water storage, to increase the filtration capacity into the well, and to accommodate seasonal fluctuations in the depth of the water table. It is advisable to cover the bottom of the well with a gravel or stone layer to prevent silt from being stirred up as the water percolates upwards, or as the water is disturbed by the bucket or pump used for abstraction (see Fig. 1.4).

The well shaft should extend above the ground surface to prevent contaminated surface water from running down into the well. An apron is advisable. Joints can be sealed with mortar or bitumen above the water table, but left open below it. The well should be covered with a slab and equipped with a suitable pump or bucket and raising mechanism.

1.1.2.4 Tube wells

In sandy soils, the hand digging of wells is difficult since loose sands tend to collapse. Therefore, hand digging in sandy soils is not recommended as cheaper, more efficient, methods are available. These methods include jetting, hand-drilling and augering of small diameter holes (50 to 500 mm). The holes are lined using PVC or mild steel casings to prevent collapse. The section below the water table is fitted with some form of well screen to allow for filtration of the ground water but to prevent the ingress of silt.

1.1.2.5 Boreholes

Generally, underground water is of a better quality, in terms of bacteria and suspended solids, than surface sources, and its supply is often more reliable. For these reasons human settlements throughout history have shown a preference for underground water, when available, for domestic water supplies.

When the water table is at a deep level (30-100m), or when the subsurface formations are of hard rock or of a material unsuitable for hand-dug or tube wells, a relatively small hole is drilled using mechanical equipment.

The diameter of the hole should suit the size of the casing to be installed, plus any temporary casing required to keep the hole open during drilling and gravel packing. For most hand-pump installations a casing diameter of 100mm is adequate, while submersible pumps normally require a minimum diameter of 120mm.

1.1.3 Siting of wells and boreholes

The presence, amount and depth of underground water cannot normally be predicted beforehand with a high degree of accuracy. Boreholes and wells previously sunk in the area could give valuable information as to the depth and amount of water available. Trained hydrogeologists or geophysicists are able to estimate the most likely sites and even the approximate depth of the water table using techniques such as the study of aerial photographs for the intersection of faults, sonar, magnetic and resistivity measurement. Care must be taken not to locate a well or borehole close to an on-site latrine or soak pit (see Fig. 1.5). (See also Cairncross and Feachem, 1978).

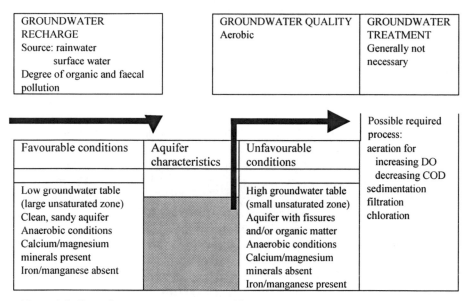

Figure 1.5. Groundwater parameters to consider

1.1.4 Determination of yield

Once a hole has been sunk to below the water table, tests should be carried out to estimate the safe yield from that borehole. The sustained yield of a borehole or well is the flow rate of water into the hole when the hole is pumped continuously for a period of 48 hours to cause maximum drawdown (approximately 1m minimum head on the pump suction). The safe yield is estimated at 30 to 50% of the sustained yield.

Initially a step drawdown test, and then a maximum drawdown or constant rate endurance test should be carried out to determine the maximum yield. The period of the test shall be determined by the duty of the borehole in normal use, as indicated in the following table:

Table 1.2. Yield test duration criteria

Production –demand period (hours per 24-hour period)	Minimum test period (hours)
up to 2 hrs	4
2 – 4 hrs	6
5 – 11 hrs	24
12 – 17 hrs	48
18 – 24 hrs	168

1.2 RAINWATER

Rainwater is a source of good quality water that can be collected and stored. The harvesting of rainwater from roof runoff can supplement domestic supplies, even in semi-arid areas. In particular, rainwater can be harvested not only for domestic use, but also for providing drinking water at remote public institutions like schools and clinics, as well as resorts. Often the limit is not the amount of rainfall that can be collected, but the size of the storage tank that will provide a sustained supply during periods of little or no rainfall.

Rainwater collection from roofs constructed from corrugated iron, asbestos sheeting or tiles is not difficult. Guttering is available in asbestos cement, galvanised iron, uPVC plastic or aluminium. The guttering and downpipes can be attached directly to the ends of rafters or trusses, and to facia boards.

Because the first water to run off a roof can contain a significant amount of debris and dirt that has accumulated on the roof, some mechanism to discard the first flush is desirable. In addition, the inlet to the storage tank should be protected with gauze screen to keep out leaves, etc., as well as mosquitoes and other insects or rodents.

The cost of diversion, storage and purification of rainwater are significant, especially for poorer communities. Therefore, urban rainwater collection, e.g. retention basins, offers economy of scale and catches considerably more water (as well as acting as stormwater control). In regions subject to drought, alternative sources of water are required, and groundwater sometimes can be used as a backup source.

The average quantity of water available from a rainwater catchment area is found by multiplying the area (in plan) with the mean annual rainfall in that area, and adjusting by an efficiency factor (average rainwater (litres) = catchment area (m) x mean rainfall (mm) x efficiency where efficiency has a value between 0 and 1.0). For roofs an efficiency of 0.8 is acceptable.

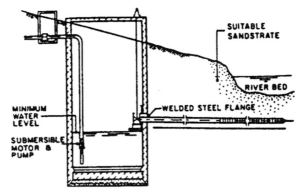

Figure 1.6. Small bank river intake using infiltration drains

Rainfall enhancement by means of cloud seeding does not strictly fall in this category, since the majority of the precipitation is on uncovered land. It is probably of greater benefit to farmers, but problems such as the higher costs, hail damage and diversion of precipitation from others who would have got it, result in the process not being all that acceptable.

1.3 WASTEWATER AND RECYCLING

Since less than 25% of urban water supply is actually consumed, reclamation of wastewater is a large potential source. Water used in the urban environment is primarily for washing and it is polluted. The proportion used for drinking and returned via sewers is the most polluted and care is needed in re-using it. Wastewater treatment works are as a rule planned to catch all this sewage. Basic sewage treatment is designed to remove suspended and organic matter. As a rule dissolved chemicals are not removed, as the processes involved are expensive.

Reclaimed wastewater needs special attention and often-expensive treatment to render it potable. The alternatives could be to re-use it only for designated uses, e.g. irrigation, or street washing. Separate reticulation systems could be constructed for this water. There is still a danger of mistaken drinking of this water.

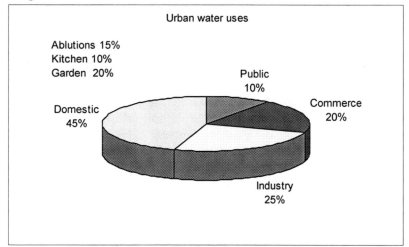

Figure 1.7. Urban water allocation

Table 1.3. Water sources, in order of abundance

Source	Locality	Special features
Sea		Desalinate
Ice	Sea	Tow
	Land	Melt
Snow	Land	Melt
Rain	Enhanced	Cloud seed
	Roofs and paving	Purify
Surface	Rivers – Mountain	Natural storage
	– Storm runoff	Purify, store
	Lakes and Dams	Environmental resistance
Groundwater	River wellpoint	Polluted river
	Springs	Protect
	Wells	Hand draw
	Boreholes	Pump
	Mines	Pump, demineralize
Wastewater	Discharged to rivers	Purify
	Internal recirculation	Demineralize
	Separate systems	Disinfect
Mist, clouds		Condense
Vegetation		Extract
Fauna		Devour

The other alternative is to provide drinking water in containers, as the volume is relatively small (e.g. 3ℓ/person/day) compared with the total per capita usage (up to 300ℓ/day).

1.4 SEAWATER OR SALINE WATER

Water with high dissolved salt concentrations is neither drinkable nor usable for the majority of purposes. Desalination is an expensive process and seldom warranted for general municipal use. Even demineralisation of brackish water (less than 5000 mg/ℓ salt concentration) is too expensive in comparison with alternative sources.

Where desalination is unavoidable, e.g. the Middle East or desert countries, then there may be some alleviating circumstances, such as cheap energy in oil producing countries.

1.5 WATER POLLUTION

Pollutants in water can be classified as:
- Suspended solids - floating debris, settleable sediment or buoyant algae.

- A quality unrelated necessarily to solids, e.g. temperature, acidity, colour, odour or taste. The latter are often associated with matter in the water, however.
- Dissolved chemicals - many cations, anions, with different effects, e.g. on health, equipment or toxicity.
- Bacteria, pathogens, viruses, organic micropollutants.

There are further sub classifications, e.g. nutrients, which include phosphates and nitrates. Sometimes organic matter is measured by TOC, or related to the oxygen requirements of the water, measured in terms of BOD or COD.

1.5.1 Protection of public health

The probability of contracting a water-related disease is statistically related to the concentration of pathogens or viruses. Improved laboratory techniques have resulted in the identification of harmful organic and inorganic matter. These may occur in water sources naturally but also from industry and agriculture. Of concern to health are also heavy metals such as mercury, cadmium and lead as well as pesticides such as DDT and carcinogenic compounds.

Diseases may also be transmitted by water-based parasites, e.g. snails and vectors, e.g. mosquitoes, especially in the tropics.

Pathogenic bacterial related diseases include cholera, typhoid and dysentery, while viruses, which are smaller and require a host and include hepatitis. It is difficult to detect bacteria in water, so evidence of excreta is sought and indices such as E. coli are used.

1.5.2 Pollution risk assessment and management

Risk assessment is the determination of the chance of harm that would beset an individual consuming water of a particular quality or lack thereof. The following four essential steps have been identified in risk assessment (Cortruvo, 1989).
(1) Hazard identification, which is the qualitative evaluation of the health danger.
(2) Exposure assessment, which is the evaluation of the number of people exposed, the type, magnitude and length of exposure.
(3) Hazard assessment, which is the attempt to correlate the health danger to dosage levels.
(4) Characterisation of the risk.

Risk management is the public process of:
(1) Determining public acceptability of the established risk level
(2) Deciding on action to be taken

1.6 INTERNATIONAL STANDARDS AND CRITERIA

A comparison of the international approaches to establishing water quality shows that there are essentially two approaches, namely enforceable standards and guidelines. The United States of America enacted a Safe Water Drinking Act in 1974 (amended in 1986) that required the USEPA to establish national primary drinking water regulations along with the identification of enforceable maximum contaminant levels and non-enforceable maximum contaminant level goals. Other significant features of the act are as follows (Sayre, 1988):

- All surface water supplies must be filtered and disinfected to prevent viral infections.
- Pipe solder or flux used on pipes must be lead-free.
  (1) Each state must submit a detailed programme for protecting ground water used for public water supplies.
  (2) Every public water authority must monitor for contaminants at least every five years.

The European Economic Community (EEC) in 1980 adopted a set of standards that address maximum admissible concentrations of water contaminants. The WHO recognised that uniform water quality standards could not practically be applied throughout the world. However, it noted the need for guidance to regulatory agencies on water quality to ensure maintenance of good health. It decided on and in 1984 published drinking-water quality guidelines to be used as a basis for the development of standards in each country. The WHO also stated that the judgement of acceptable risk levels is undertaken by society as a whole. Therefore the adoption of the proposed guidelines is for each country to decide. The guidelines were developed assuming lifelong consumption and that specific geographic, socio-economic, dietary and industrial conditions would also have to be considered.

A guideline value represents the level (a concentration or a count) of a constituent that ensures aesthetically acceptable water, and does not result in significant risk to the health of the consumer. In the case of supplies to small communities, the parameters used to assess and measure the quality of water must necessarily be limited in number. Similarly, the guideline values may be considered long-term goals, rather than rigid standards to be complied with at all times and in all supply systems.

1.6.1 Microbiological determinants

Emphasis should be placed on the microbiological and biological safety of drinking water supplies. Drinking water should not contain any micro-organisms known to be pathogenic. Since the most common sources of micro-organisms are from faecal material, drinking water should be free from faecal pollution. The primary bacterial indicator recommended for this purpose is the coliform group of organisms. Although, as a group, they are not exclusively of faecal origin, they are universally present in large numbers in the faeces of man and other warm-blooded animals. A sub-group of these coliform organisms, the faecal (thermo-tolerant) coliforms (or, in particular, *Escherichia coli*) provide definite evidence of faecal pollution.

1.6.2 Chemical and physical determinants

There are a number of chemical and physical parameters, which should be assessed when determining the quality of a water source. These include turbidity in surface waters and fluorides, nitrates, iron hardness and total salts in ground waters.

Turbidity is the ability to pass light, but is used as a measure of suspended matter. Turbidity affects the aesthetic quality of the water, but its primary importance is in relation to water disinfection. The amount of chlorine needed for disinfection often increases as the turbidity increases. Certain suspended solids, which are organic in nature, may also impart an undesirable taste to the water. The suspended solids are usually clay and silt particles (often increasing as a result of rainfall), algae as a result of nutrient enrichment of the surface watercourses, and organic matter from rotting vegetation or industrial and domestic effluent.

Fluorides. Excess fluorides in the diet result in mottling of teeth and bone fluorosis. Ground waters contain fluoride levels as high as 20 mg/litres. These levels may be a result of the natural geological formations. Some industrial effluents contain high levels of fluoride.

Nitrates. While being relatively non-toxic to adults, nitrates are potentially lethal to infants. Fatal *methaemoglobinaemia* can occur at nitrate concentrations in excess of 10 mg/litre. Nitrates occur in underground water from natural formations, from excessive use of fertilizers by farmers, and from leachates from pit latrines and other sanitation systems or waste disposal sites.

Iron. While iron in high concentrations is potentially toxic, its aesthetic undesirability manifests well below potentially toxic concentrations. Iron concentrations above 0.3 mg/ℓ in water give rise to discoloration, staining and offensive taste. Many ground waters contain some iron from the natural geological formations. Iron is also used as a flocculant in water purification, or it can arise from corrosion in the distribution system.

Hardness. Hardness, or the presence of multivalent cations (especially calcium and magnesium), is not a problem as far as health is concerned. In fact, some hardness in the water is desirable as a protective factor against heart disease. Excess hardness, however, can result in the scaling of pipes and hot water systems, as well as in excessive consumption of soap for laundry. Groundwater may contain high levels of hardness, particularly in dolomitic zones.

Total salts. Due to the osmotic effects of high salt levels, saline water is generally unpalatable. Certain underground waters in arid and semi-arid areas, as well as in certain geological formations, are saline and thus not fit for human consumption. The total dissolved salts are usually measured in terms of the electrical conductivity of the water.

Guideline values for the above physical/chemical water quality parameters are given in Table 1.4.

Table 1.4. Guideline values for physical/chemical water quality parameters

Determinant	Unit	Recommended	Maximum level for insignificant risk	Maximum level for low risk
Turbidity	NTU	1	5	10
Fluoride	mg F/ℓ	1	1.5	2
Nitrate	mg N/ℓ	6	10	20
Iron	mg Fe/ℓ	0.1	1.0	2.0
Hardness	mg CaCo$_3$/ℓ	20 – 300	650	1 300
Conductivity	mS/m 25°C	70	300	400

Table 1.5. Comparison of US regulations with EC and WHO guidelines (Sayre, 1988; Twort et al, 1994)

Substance	US maximum contaminant level *	EC maximum admissible +	WHO guideline
Inorganics: Arsenic	0.05 mg/ℓ	0.05 mg/ℓ	0.01 mg/ℓ
Barium	2.0 mg/ℓ	0.1 mg/ℓ	0.7 mg/ℓ
Cadmium	0.005 mg/ℓ	0.005 mg/ℓ	0.003 mg/ℓ
Chromium	0.01 mg/ℓ	0.05 mg/ℓ	0.05 mg/ℓ
Fluoride	4.0 mg/ℓ	NS	1.5 mg/ℓ
Lead	0.015 mg/ℓ	0.05 mg/ℓ	0.01 mg/ℓ
Mercury	0.002 mg/ℓ	0.001 mg/ℓ	0.001 mg/ℓ
Nitrate	10.0 (N) mg/ℓ	50.0 (NO$_3$) mg/ℓ	10.0 (N) mg/ℓ
Organics: 2,4-D	100 μg/ℓ	NS	1 μg/ℓ
Endrin	0.2 μg/ℓ	NS	NS
Hydrocarbons		10 μg/ℓ	
Lindane	0.4 μg/ℓ	NS	NS
Methoxychlor	100 μg/ℓ	NS	1 μg/ℓ
Pesticides (total)	NS	0.5 μg/ℓ	NS
Trihalomethanes	100 μg/ℓ	1 μg/ℓ	30 (CHCl$_3$ only) μg/ℓ

Substance	US maximum contaminant level *	EC maximum admissible +	WHO guideline
Volatiles: Benzene	5 µg/ℓ	NS	10 µg/ℓ
Carbon tetra chloride	5 µg/ℓ	3 µg/ℓ	2 µg/ℓ
1,1-Dichloro-ethylene	7 µg/ℓ	NS	3 µg/ℓ
1,2-Dochloro-ethane	5 µg/ℓ	NS	30 µg/ℓ
para-Dichloro-benzene	75 µg/ℓ	NS	NS
1,1,1-Trichloro-ethane	200 µg/ℓ	NS	2000 µg/ℓ
Trichloroethylene	5 µg/ℓ	30 µg/ℓ	30 µg/ℓ
Vinyl chloride	2 µg/ℓ	NS	5 µg/ℓ
Coloforms-organisms/100 mℓ	< 1	0	0
Turbidity – NTU	1 – 5	0 – 4	< 1
Bta particle and photon	4 mrem	NS	1.0 Bq/ℓ
Gross alpha particle activity	15 pCi/ℓ (0.56 Bq/ℓ)	NS	0.1 Bq/ℓ
Radium-226 + radium-228	5 pCi/ℓ (0.19 Bq/ℓ)	NS	NS
Chloride	250 mg/ℓ	NS	250 mg/ℓ
Colour	15 CU	20 mg Pt-Co/ℓ	15 CU
Copper	1 mg/ℓ	NS	2.0 mg/ℓ
Fluoride	4 mg/ℓ	1.5 (8° – 10°C)	1.5 mg/ℓ
Foam agents	0.5 mg/ℓ	NS	NS
Iron	0.3 mg/ℓ	0.2 mg/ℓ	0.3 mg/ℓ
Manganese	0.05 mg/ℓ	0.05 mg/ℓ	0.5 mg/ℓ
Odour	3 TON	2 dilution at 12°C	
pH	6.5 – 8.5	NS	6.5 – 8.5
Sulphate	250 mg/ℓ	250 mg/ℓ	400 mg/ℓ
Total solids	500 mg/ℓ	1 500 mg/ℓ	1 000 mg/ℓ
Zinc	5 mg/ℓ	NS	5.0 mg/ℓ
Hardness		60 mg/ℓ	
Phenols		0.0005	

* = Enforceable; + = Non-enforceable; NTU = Nephelometric turbidity units
NS = Not specified; CU = Colour units; TON = Threshold odour number

1.6.3 Organic pollution

In addition to the above, there may be certain organic contaminants which affect water quality. These include the following:

(1) Pesticides (e.g. chlorinated hydrocarbons) used in farming and subsequently appearing in the runoff into the water courses;
(2) Oil (especially crude petroleum or mineral oil) from leakage or dumping, which enters the water source by seepage or runoff;
(3) Various toxic substances from industrial waste-dump sites; and
(4) Organic oxygen-demanding wastes, which can result in depletion of the oxygen levels in the water source.

These contaminants can usually be detected by the taste or odour they impart to the water. Should such contamination be suspected, the water source should not be used until the nature and source of the contamination has been determined and a decision made regarding its safety.

1.6.4 Stability of water supplies

Water that is put into distribution pipelines should be neither corrosive nor scale forming. Corrosive waters may lead to corrosion of the pipeline, fittings, storage tanks, etc., resulting in costly maintenance and/or corrosion products in the final water being delivered to the consumer.

Water which is scale forming could result in reduced capacity of the distribution network and/or malfunctioning of water meters, taps and valves, and so forth. Ideally water should be just sufficiently scale forming to maintain a thin protective coating in the distribution system.

The stability of a water is expressed as the degree of calcium carbonate saturation of the water. It is a function of its calcium and magnesium concentrations, its pH, alkalinity, ionic strength and temperature. The pH at which the water is just saturated, under the given conditions, is called the saturation pH (pHs). If the actual pH is less than pHs, the water is corrosive; if the actual pH is higher than pHs, the water is scale-forming. The stability of a water is often expressed in terms of the Langelier Index (LI) where:

Langelier Index (LI) = pH – pHs

If the LI is negative, the water is corrosive; if the LI is positive, the water is scale-forming.

1.7 WATER TREATMENT

In many cases, water obtained from a particular source will require some treatment before being distributed for domestic use. Water obtained from boreholes, protected wells, protected springs and harvested rainfall will often require little or no treatment. However, as a precautionary measure and to minimise biological activity in the storage reservoirs and pipelines, even such waters should be disinfected before distribution.

Okun and Schulz (1983) suggest that under all circumstances groundwater is the preferred choice for community supplies, as it generally does not require treatment. When treatment is required the treatment will be determined by the extent of contamination and by the raw water characteristics. Most surface waters will require treatment, both to remove turbidity and to disinfect them. Certain surface- and ground waters will require additional treatment for the removal of organic and/or inorganic contaminants.

A simple approach for the selection of a treatment system for small poor communities is given by Thanh and Hettiaratchi (1982) (see Table 1.6).

The emphasis on slow sand filtration is valid for areas where skilled personnel may not be permanently available to operate the plant, where chemical shortages may occur, where space is available at low cost, and where supervision may be irregular. Marx and Johannes (1988) found slow sand filtration to be an economical and successful option for water treatment plants in developing areas.

Table 1.6. Treatment selection criteria for rural communities (Thanh and Hettiaratchi, 1982)

Raw water quality	Treatment suggested
Turbidity: 0 – 5 NTU Faecal coliform 0.100 mℓ Guinea worm or schistosomiasis Not endemic	No treatment
Turbidity: 0 – 5 NTU Faecal coliform 0.100 mℓ Guinea worm or schistosomiasis Endemic	Slow sand filtration
Turbidity: 0 – 10 NTU Faecal coliform 1 – 500/100 mℓ	Slow sand filtration Chlorination if possible
Turbidity: 20 – 30 NTU Up to 30 NTU for a few days only Faecal coliform 1 – 500/100 mℓ	Pretreatment advantageous Slow sand filtration Chlorination if possible
Turbidity: 20 – 30 NTU Up to 30 NTU for a several weeks Faecal coliform 1 – 500/100 mℓ	Pretreatment advisable Slow sand filtration Chlorination if possible
Turbidity: 20 – 150 NTU Faecal coliforms 500 – 5000/100 mℓ	Pretreatment Slow sand filtration Chlorination if possible
Turbidity: 30 – 150 NTU Faecal coliforms > 5000/100 mℓ	Pretreatment Slow sand filtration Chlorination if possible
Turbidity: > 150 NTU	Detailed investigation (and possible pilot-plant study)

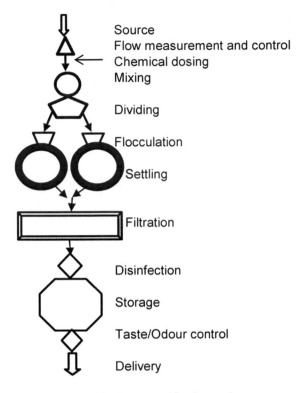

Figure 1.8. Typical layout of a municipal water purification works

Where sufficient money and skilled operators are available standard water treatment plants (e.g. chemical flocculation, settler, rapid sand filtration and chlorination) have worked well. In other situations a more positive, or a more basic treatment may be warranted.

Water treatment may be classified in four main processes:
- removal of suspended solids (turbidity);
- removal of undesirable dissolved substances;
- water stabilisation; and
- disinfection.

The normal processes in treatment of surface water to bring it to potable standards (see Fig. 1.8) are:
- Screening
- Grit settling
- Addition of coagulant

- Mixing
- Flocculation
- Sedimentation
- Filtration
- Disinfection

Other more specialized processes may be added, e.g. pH correction, stabilization, demineralization, taste or odour removal.

Raw water from surface sources is usually treated as above, but ground water is frequently only disinfected. Wastewater on the other hand receives more rigorous treatment with organics being removed by biochemical means. The distinction between various types of water is likely to disappear as different sources are utilized and as recycling becomes more imperative to relieve shortages.

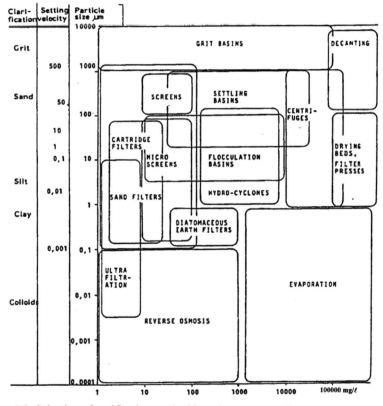

Figure 1.9. Selection of purification method based on water quality

1.8 REFERENCES

Cairncross, G. and Feachem, R. (1978) *Small Water Supplies*. Bulletin No. 12, Ross Inst. Tropical Hygiene, London 78p.
Cortruvo, J.A. (1989) Drinking water standards and risk assessment. *Jnl. Wat. and Environ. Mngmnt.* **3**, Feb, 6-12.
Dept. Natl. Housing (1994) Guidelines for the Provision of Engineering Services and Amenities in Residential Township Development. Ch.8, *Water Supply*. CSIR, Pretoria.
Doxiadis, C.A. (1967) Water and the environment. *Int. Conf. on Water for Peace*. Washington D.C.
I.R.C. (Intl. Ref. Centre for Community Water Supply) (1981). *Small community water supplies.* The Hague, Tech paper 14, 91p.
Lvovitch, M.I. (1973) The global water balance. *E.O.S.* **54**(1), Trans. American Geophysical Union, Jan.
Marx, C.J. and Johannes, W.G. (1988) Comparison between a rapid gravity filtration and slow sand filtration water treatment plant. *Proc. Quinquennial Convention of the SAICE*, Univ. Pretoria, Pretoria.
Okun, D.A. and Schulz, C.R. (1983) Practical water treatment for communities in developing countries. *Aqua*, no. **1**, 23-26.
Petersen, M.S. (1984) *Water Resource Planning and Development*. Prentice Hall, Englewood Cliffs.
Sayre, I.M. (1988) International standards for drinking water. *J. Am Wat. Wks. Ass.* **80**(1), 53-60.
Stephenson, D. (2003) *Water Resources Management*, Balkema.
Thanh, N. C. and Hettiaratchi, J. P. A. (1982) *Surface water filtration for rural areas: guidelines for design, construction operation and maintenance*. Environmental Sanitation Information Centre, Bangkok, Thailand, 78pp.
Twort, A.C., Low, F.M., Crowley, F.W. and Ratnayaka, D.D. (1994) *Water Supply*. 4th Edn. Edward Arnold, London.
United Nations (1975) *Proc. Interregional Seminar on River Basin and Interbasin Development*, Budapest.
Van der Leeden, F. (1975) *Water Resources of the World.* Water Information Centre, N.Y.
WHO (1984) *Guidelines for Drinking Water Quality.* World Health Organization, Geneva.

2

Urban water supply

2.1 WATER USE

The total volume of water required for domestic use in the world is some 200 km^3/a, which is only 0.5% of the average total river runoff of 40000 km^3/a. The total urban water consumption is some 2000 km^3/a, which includes industry. However, the largest user of water is agriculture, accounting for 4000 km^3/a.

Theoretically then there is sufficient water to go around, but the problems of distribution in space, time and affordability, result in water shortages. However, with growing world populations and increasing per capital consumption, the total consumption could rapidly reach the limits of availability (Fig. 2.1). The problem is likely to get worse in rural developing communities in particular.

The majority (4 billion) of the world population have inadequate water for healthy living. Even in developed economies there are often water shortages, due to droughts, inadequate local resources, pollution, finance, or simply poor planning. This book assists in planning water supply systems. Ways of ameliorating the other effects are suggested, but many other factors come into the equation, such as water resources available, sustainability, politics and ownership.

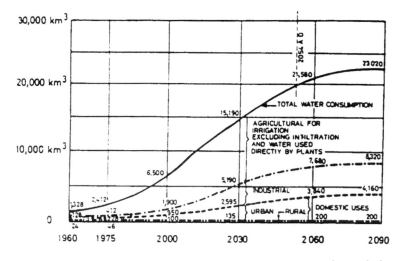

Figure 2.1. Projection of world-wide total water demand (in $km^3 = 10^9 m^3$ per year) (Doxiadis, 1967)

Concern has been expressed as to the carrying capacity of the earth primarily from the point of view of food and water (Meadows et al, 1972). Populations continue to grow and the world population in 2000 was estimated to be 6500 million. The biggest concentrations are in China (1000 million) and India (700 million) followed by the USSR and USA. However, population densities indicate there is little danger of crowding out in our lifetime. That of China (100 per sq km) may be compared with Bangladesh (600 per sq km). There is no clear distinction between developed and developing countries.

The availability of resources for larger populations is only a problem if the countries are to develop to the consumption levels of the first world. The populations of the USA (250 million), Europe (400 million) and Japan (120 million) total only 12% of the world population, so one could expect usage of metals, timber and oil to increase manyfold if the entire world reached the advanced levels of these countries.

2.1.1 Volumes required

Water requirements vary with type of user and various other factors. The following urban users need to be considered:

- Domestic - drinking, washing, gardening, pools.
- Commercial - shops, offices.
- Industrial - mining, heavy users, washing.
- Municipal - institutions, education, washing, firefighting, power, transport.

- Losses - leaks, misuse, testing, overflow.
- Irrigation - parks, recreational, agriculture, stock.

Some users consume water lost by evaporation, e.g. agriculture, but most used urban water returns to sewers or stormwater drains. Very little (10-20%) is probably lost but the balance returned to the system is generally polluted by the user. Some is difficult to retrieve, e.g. discharged to groundwater. The amount of water used by each type of consumer can be influenced by:

- Income
- Standard and type of living, e.g. bungalows versus apartments
- Size of dwelling unit and stand
- Climate
- Type of supply, e.g. communal or multiple household connections
- Price
- Whether it is metered
- Who pays, e.g. public water is not paid for by the user
- Pressure
- Type of sanitation, e.g. waterborne
- Social customs
- Availability from the supplier, e.g. intermittent

2.1.2 Planning basis

The steps for estimating water demand for design purposes are as follows:
(1) Decide on time horizon to plan for.
(2) Estimate average water consumption for that time (or population and per capita consumption plus other demands).
(3) Calculate peak supply rate.
(4) Size pipes, reservoirs.

The possibility of variations in demands due to changes in population, etc., should be considered and risk may affect the time horizon or type of project installed. There may also be a technical upper limit to which a project can be installed for example limited availability of water, or size of pipes available. In planning to meet supply demand, it is necessary to add losses, which can add 10-25% to requirements (see Anis et al, 1985).

The actual demand can vary widely according to circumstances of the population. In affluent societies, people may demand water for drinking, ablutions, cooking, washing in kitchen, laundry or garage, gardening or recreation. They may be relatively unconcerned about wastage. The draw off could be over a few hours a day in total, so that the distribution system has to be sized for peak rates.

Table 2.1. Domestic water usage in warm climates in litres per capita per day – according to standard of housing

	Type of housing	Water usage
(i)	Areas of high quality housing	200 – 250 ℓ/c/d
(ii)	City residential areas including high standard flats	160 – 200 ℓ/c/d
(iii)	Suburban; tenement dwellings; low cost housing	90 – 100 ℓ/c/d
(iv)	Urban areas served by standpipes	50 – 700 ℓ/c/d
(v)	Rural: standpipes	30 – 50 ℓ/c/d
(vi)	Rural: distance to source > 1 km	10 – 30 ℓ/c/d

Pricing could reduce the demand, or pressure control, but this is more an economic supply and demand situation than supply to poorer communities which may be limited purely from the affordability point of view (IRC, 1981).

These figures do not include system losses, which can be up to 25%, or industrial or public use. Figures quoted by Twort et al (1994) for industrialized cities of 600-700 ℓ/c/d include industrial use and care should be taken to separate the figures, as industrial use can vary enormously, depending on the type of industry. The figures are averaged over a year and vary depending on the time of year and day of week, time of day and weather.

To conduct a survey of water consumption requires careful calibration of meters and patience to observe over a representative period of time. In-house leaks, e.g. cisterns, taps, washers, should be checked. Different types of zones, e.g. class of residence, residential versus commercial and public use, needs to be demarcated. An approximate breakdown of the use, in litres per capita, in the U.K. is:

Drinking	1 l/c/d
Toilet	45
Bathing	23
Laundry	17
Kitchen	50
Car washing	4
Garden	40
TOTAL	180 l/c/d

Where there is no demographic data available, population estimates for a future point in time can be made by applying a population growth factor, F, to the current population size:

$$F = (1 + i)^n \qquad\qquad (2.1)$$

where F = population growth factor
         i = annual growth rate as a fraction
         n = design period in years.

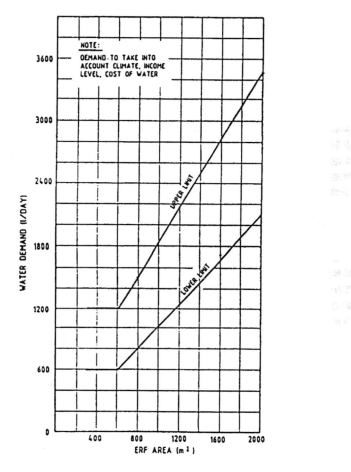

Figure 2.2. Annual average daily water demand for dwelling houses

Change in per capita consumption as well as population total should be considered.

The water demands given in Table 2.2 and Figure 2.2 apply to urban centres with flush sanitation. Values for residential stands of area larger than 2000 m² are not given as these rarely occur in new townships. Values adopted by the designer should be based on local conditions (Dept. Natl. Housing, 1994).

Where upper and lower limits are given, it is envisaged that the upper limit would generally apply to the high-income level township, and the lower limit to the lower income level township.

Designers should note that in adopting water demand figures for a specific design, cognisance must be taken of local factors such as income level, climate and water charges when interpolating between the upper and lower limits provided for these guidelines.

2.1.3 Peak factors

Consumption rate varies during a day, and over a week. For example, with large communities in which a significant number of homes have house connections, water is consumed up to 17 hours per day. To cater for this, peak factors for the distribution system given below should be taken into account in the design.

Population	Peak factor
10 000	1.5
< 5 000	2
< 200	2.5

An additional factor of up to 1.5 may be applicable for seasonal variation, particularly for arid countries.

The peak factor will be reduced by provision of storage (in which case only the storage volume and distribution system beyond the reservoir are designed to cater for peak demands). The provision of intermediate storage will also result in a reduction in the peak flows in the elements prior to the intermediate storage. Peaks can be averaged out over large townships and these figures allow for this. For smaller communities the peak factor should be increased, as in Fig. 2.3.

Many factors can affect consumption however, for example, rainfall or temperature. Metering can also reduce consumption, but whether this is long term, and what degree of meter stop occurs, makes this difficult to estimate. During water shortages (e.g. droughts) consumption can be reduced 10-20% by appeal, but further reduction requires pressure control or punitive tariffs.

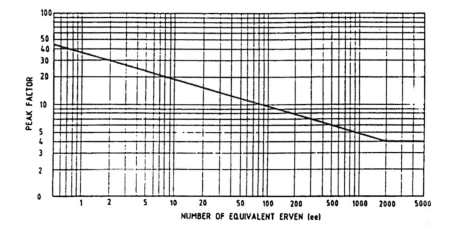

Figure 2.3. Factor for obtaining the peak flow in mains

Table 2.2. Municipal water requirements – annual averages

Type	Per coverage ℓ/d	Per person ℓ/d
Houses	600 – 3 000 per stand (Fig. 2.3)	200 – 300
Multiple units	600 – 1 000 per unit	160 – 200
High rise	450 – 700 per dwelling	90 – 100
Low cost	10 000 – 50 000 per hectare	50 – 70
Offices, shops	400 per 100 m^2	60 – 100 per employee
Hotels		300 – 500 per bed
Hospitals	500 per 100 m-2-	200 – 300 per bed
Schools (day)		25 – 75 per pupil
Parks	100 per 100 m^2	
Factories – light		200 – 500 per worker
– heavy	10 – 50 m^3/ton product	

2.1.4 Pressure requirements

Water supply authorities generally do not guarantee pressures but there are norms required for safety and practical purposes.

The minimum pressure at a household connection should be about 1 bar (10m), possibly decreasing in low cost housing to about 5m, and also reducible in case of emergency, e.g. fire hydrant use nearby. High-rise buildings may thus have to boost pressure to a roof tank. The pressures will be lowest at peak flow time and at the furthest point from the supply reservoir if it is on flat ground. Otherwise it is necessary to check at high points in the supply line and a network flow-head analysis may be necessary for peak flow conditions.

Maximum pressures should be confined to less than 90m (9 bar), but where low cost systems are installed, this upper limit may have to be reduced. Maximum pressures will occur during low flow periods, e.g. at night, and over the low-lying areas. To reduce these maximum pressures, zoning and pressure reducing valves could be employed.

2.2 RESERVOIR STORAGE REQUIREMENTS

Reservoir storage is required for breakdowns in the supply system, and to balance a regular supply rate with a variable draw off. The procedure for calculating such storages is explained later. This chapter provides rule of thumb guides.

Where water is obtained from a bulk water supply authority, the storage capacity provided should comply with the requirements of the authority subject to the requirements of Table 2.3 being met. The figures should be increased to allow for fire flow requirements where applicable.

Table 2.3. Reservoir storage capacity

Method of supply to the reservoir	Suggested minimum storage capacity (hours of annual average daily demand)
Gravity or pumped from 1 water source	48
Gravity or pumped from 2 water sources	24

2.2.1 Elevated storage and pumps

The nominal capacity for elevated storage is given in Table 2.4.

Table 2.4. Elevated storage capacity

Pumping plant service the elevated storage facility	Capacity of elevated storage (hours of instantaneous peak demand)
One electrically driven duty pump, and one identical electrically driven standby pump	2
One electrically driven duty pump	4

The nominal capacity of the duty pump should be equivalent to the sum of the instantaneous peak demand and the fire demand or the instantaneous peak demand plus an allowance of 20 per cent whichever is the greater. The standby power source should operate automatically in the event of an electricity supply failure. As an alternative to providing elevated storage and pumps, a scheme comprising booster pumps delivering directly into the reticulation can be considered.

Large and costly storage facilities would be required if provision were to be made to ensure a continuous supply, no matter what breakdown occurred. However, in practice people can cope quite adequately with a water shortage for a number of hours. Furthermore, the effect of a breakdown will be lessened if there is sufficient system storage (i.e. storage in the intermediate reservoirs, distribution network and terminal storage, if these elements are included in the system).

The expected delay caused by a breakdown must also be considered when determining storage requirements. Table 2.5 gives guideline values for storage provision to cater for breakdowns. Provision must, of course, also be made for balancing requirements as determined by the design.

2.2.2 Balancing volume

The balancing volume is required to cater for peak flows whilst receiving a constant (or variable) inlet flow. Generally, the equalising volume of a storage reservoir is from 1/6 to 1/3 of the total daily demand.

Table 2.5. Reservoir storage capacity

Supply to reservoir	System storage (hours of average daily demand)	Recommended storage capacity (hours of average daily demand)
Gravity	< 5	20
	5 – 10	16
	> 10	12
Pumped	< 5	30
	5 – 10	25
	> 10	20

2.2.3 Other storage reservoirs

Storage reservoirs may also be required for the following primary purposes:
- Water collection from various sources;
- Provide contact time for a certain water treatment operation (e.g. chlorination); or
- Provide water to a pump station (booster reservoir).
- Emergencies, e.g. bursts in supply line
- Fire-fighting
- Break pressure tanks
- Overhead storage for high zones
- Zone storage at different levels

The provision of intermediate storage can have a number of objectives, the most important of which are:

- A reduction in the sizes of the pipes of distribution mains, by reducing the peak flow demand of these mains;
- A reduction in pipeline pressures;
- The reduction of the impact of supply breakdowns;
- The division of the supply network into smaller sub-sections which can be more easily managed by community organisations;
- The reduction in the size of the main storage reservoir, in terms of both balancing storage and emergency storage.

The provision of intermediate storage will usually only be economically feasible in areas where the topography is steep enough to obviate the need for elevated storage, or where the undulating topography dictates the need for different pressure systems for different sections of the community.

2.3 PIPE FLUID MECHANICS

Water is conveyed over large and small distances by pipeline for use in irrigation, hydropower, commercial or domestic use. In a typical water supply system, water is abstracted from a source, purified and pumped to a storage reservoir. From there it is gravitated to consumers connected to a reticulation system.

Early pipelines flowed part full; the pipeline profile therefore being confined to the hydraulic grade line. The reason for adopting a circular cross section was probably more practical than technical. The fact that the circular annulus has a greater hydraulic radius (area/wetted perimeter) than any other shape and its resultant effect on the friction losses would not have been understood. The structural advantage of circular pressurized conduits was probably also overlooked. A circular pressure pipe acts in tension optimising the material properties. Pipelines are nowadays operated under pressure as they convey more flow than a part full pipe.

Water supply engineers have seen a number of advances in pipeline technology over the last century. The most significant advances have perhaps been in fluid mechanics. Apart from a spate of empirical formulae early in the century, the science of boundary layers and roughness has advanced rapidly. The Darcy-Weisbach formula is gaining acceptance as the most reliable method for estimating friction losses.

The analysis of unsteady flow in pipelines progressed with the advent of computers. The theory of water hammer was understood and developed at the beginning of the 20th century, but advances have been most rapid in recent decades. The accessibility of computers to engineers has also facilitated analysis of complex pipe networks, and the computer-oriented techniques of system analysis have been adapted to the optimisation of pipe layouts and sizes.

2.3.1 The fundamental equations of fluid flow

The three basic equations in fluid mechanics are the continuity equation, the energy equation and the momentum equation. For steady, incompressible, one-dimensional flow, the continuity equation is simply obtained by equating the flow rate at a section to the flow rate at another section along the stream tube. By 'steady flow' is meant that there is no variation in velocity at any point with time. 'One dimensional' flow implies that the flow is along a stream tube and there is no lateral flow across the boundaries of stream tubes. It also implies that the flow is irrotational. The energy equation is usually reduced to the friction head loss equation.

The basic energy equation derived by equating the work done on an element of fluid by gravitational and pressure forces to the change in energy. Mechanical and heat energy transfer are excluded from the equation. In most systems, there is energy loss due to friction and turbulence and a term is included in the equation to account for this. The resulting equation for steady flow of incompressible fluids is terms the Bernoulli equation and is written:

$$\frac{V_1^2}{2g} + \frac{P_1}{\gamma} + Z_1 = \frac{V_2^2}{2g} + \frac{P_2}{\gamma} + Z_2 + h_1 \qquad (2.2)$$

where:

V	= mean velocity at a section	
$V^2/2g$	= velocity head (units of length)	
g	= gravitational acceleration	
P	= pressure	
P/γ	= pressure head (units of length)	
γ	= unit weight of fluid	
Z	= elevation above an arbitrary datum	
h_1	= head loss due to friction or turbulence between sections 1 and 2	

The sum of the velocity head plus pressure head plus elevation is termed the total head. Strictly, the velocity head should be multiplied by a coefficient to account for the variation in velocity across the section of the conduit. The average value of the coefficient for turbulent flow is 1.06 and for laminar flow it is 2.0.

For the Bernoulli equation to apply the flow should be steady, i.e. there should be no change in velocity at any point with time. The flow is assumed to be one-dimensional and irrotational. The fluid should be incompressible, although the equation may be applied to gases with reservations.

2.4 FLOW HEAD LOSS RELATIONSHIPS

2.4.1 Empirical flow formulae

The throughput or capacity of a pipe of fixed dimensions depends on the total head difference between the ends. This head is consumed by friction and other (minor) losses.

The first friction head loss/flow relationships were derived from field observations. These empirical relationships are still popular in waterworks practice although more rational formulae have been developed. The head loss/flow formulae established thus are termed conventional formulae and are usually in an exponential form of the type:

$$V = K R^x S^y \ or \ S = K' Q^n / D^m \qquad (2.3)$$

Where: V is the mean velocity of flow
K and K' are coefficients
R is the hydraulic radius (cross-sectional area of flow divided by the wetted perimeter, and for a circular pipe flowing full, equals one quarter of the diameter)
S is the head gradient (in m head loss per m length of pipe).
Superscript x, y, n and m are exponents

Some of the equations more frequently applied are listed in Table 2.6 (Stephenson, 1989). Except for the Darcy formula, the equations are not universal and the form of the equation depends on the units. It should be considered that the formulae were derived for normal waterworks practice and take no account of variations in gravity, temperature or type of liquid, they are for turbulent flow in pipes over 50mm diameter. The friction coefficients vary with pipe diameter, type of finish and age of pipe.

Table 2.6. Pipe friction equations

	Basic equation	SI units
Hazen-Williams	$S = K_1(V/C_W)^{1.85} / D^{1.167}$	$K_1 = 6.84$
Manning	$S = K_2(n/V)^2 / D^{1.33}$	$K_2 = 6.32$
Chezy	$S = K_3(V/C_2)^2$	$K_3 = 13.13$
Darcy	$S = \lambda V^2 / 2gD$	Dimensionless

The conventional formulae are comparatively simple to use, as they do not involve fluid viscosity. They may be solved directly as they do not require an initial estimate of Reynolds number to determine the friction factor. The rational equations cannot be solved directly for flow. Solution of the formulae for velocity, diameter or friction head gradient is simple with the aid of a slide rule, calculator, computer, nomograph or graphs plotted on log-log paper. The equations are of particular use for analysing flows in pipe networks where the flow/head loss equations have to be iteratively solved many times.

The most popular flow formula in waterworks was the Hazen-Williams formula. If the formula is to be used frequently, solution with the aid of a chart is the most efficient way. Many waterworks organizations use graphs of head loss gradient plotted against flow for various pipe diameters, and various C values. As the value of C decreases with age, type of pipe and properties of water, field tests are desirable for an accurate assessment of C.

2.4.2   Rational flow formulae

Although the conventional flow formulae are likely to remain in use for may years, more rational formulae are gradually gaining acceptance amongst engineers. The new formulae have a sound scientific basis backed by numerous measurements and they are universally applicable. Any consistent units of measurement may be used and liquids of various viscosities and temperatures conform to the proposed formulae.

The rational flow formulae for flow in pipes are similar to those for flow past bodies or over flat plats (Schlichting, 1960). The original research was on small-bore pipes with artificial roughness. Lack of data on roughness for large pipes has been one deterrent to the use of the relationships in waterworks practice.

The velocity in a full pipe varies from zero on the boundary to a maximum in the centre. Shear forces on the walls oppose the flow and a boundary layer is established with each annulus of fluid imparting a shear force onto an inner neighbouring concentric annulus. The resistance to relative motion of the fluid is termed kinematic viscosity, and in turbulent flow it is imparted by turbulent mixing with transfer of particles of different momentum between one layer and the next. A boundary layer is established at the entrance to a conduit and this layer gradually expands until it reaches the centre.

The Reynolds number $Re = VD/v$ is a dimensionless number incorporating the fluid kinematic viscosity v, which is absent in the conventional flow formulae. Flow in a pipe is laminar for low Re (less than 2 000) and becomes turbulent for higher Re (normally the case in waterworks practice). The basic head loss equation of Darcy-Weisbach is derived by setting the boundary shear force over a length of pipe equal to the loss in pressure multiplied by the area:

$$\tau \pi DL = \gamma h_f \pi D^2 \qquad (2.4)$$

$$\therefore h_f = \frac{4\tau/\gamma}{V^2/2g} \frac{L}{D} \frac{V^2}{2g} = \lambda \frac{L}{D} \frac{V^2}{2g} \qquad (2.5)$$

Where: $\lambda = (4\tau'\gamma) / (V^2/2g)$ (referred to as the Darcy friction factor)
         τ is the shear stress
         D is the pipe diameter and
         h_f is the friction head loss over a length L

λ is a function of Re and the relative roughness e/D. For laminar flow, Poiseuille found that $\lambda = 64/Re$, i.e. λ is independent of the relative roughness. Laminar flow will not occur in normal engineering practice. The transition zone between laminar and turbulent flow is complex and undefined, but is also of little interest in practice.

Following research by Reynolds, Von Karman and others into turbulence, boundary layer theory was developed to yield a flow-head loss relationship for turbulent flow in pipes. Colebrook (1939) and White fitted an equation to the data, yielding the so-called Darcy-Weisbach friction factor λ (or f in the USA).

$$\frac{1}{\sqrt{\lambda}} = 2\log\left(\frac{k}{3.7D} + \frac{2.5}{Re\sqrt{\lambda}}\right) \tag{2.6}$$

In view of the complicated relationship between the Darcy friction factor, λ, Reynolds number Re and the relative roughness e/D, explicit head loss charts have been prepared (Ackers, 1969), k is a measure of the boundary roughness.

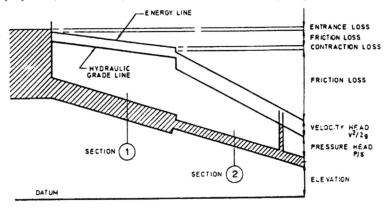

Figure 2.4. Energy heads along a pipeline

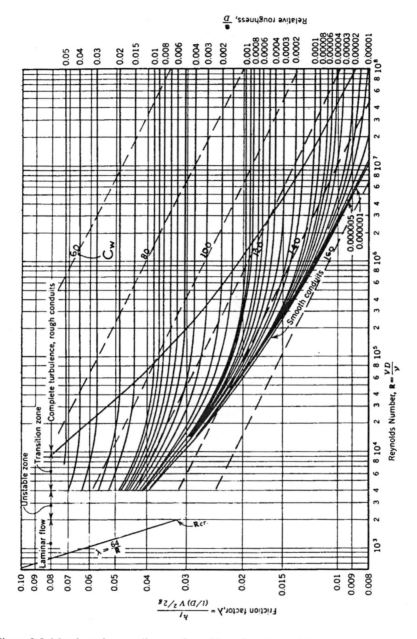

Figure 2.5. Moody resistance diagram for uniform flow in conduits

It will be observed from Figure 2.5 that lines for constant Hazen-Williams coefficient coincide with the Colebrook-White lines only in the transition zone. In the completely turbulent zone with a particular pipe, it varies depending on the flow rate. The Hazen-Williams equation should therefore be used with caution for high Reynolds numbers and rough pipes. It will also be noted that values of C_w above about 155 are impossible to attain in waterworks practice (IWE, 1969). The corresponding C_w line falls below the smooth line on the Moody diagram.

Table 2.7 gives a summary of accepted friction factors for different pipe materials and conditions.

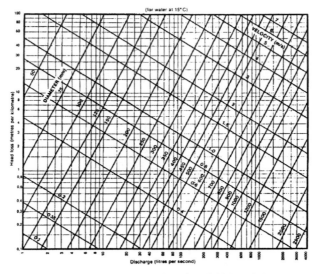

Figure 2.6. Friction loss chart for pipes flowing full, k = 0.3 mm

Table 2.7. Summary of accepted friction factors

	Nikuradse roughness k (mm)	Hazen-Williams C_w
Smooth	0	140
Steel, PVC	0.0115	130
Bitumen lined	0.03	125
Spun concrete	0.25	120
Rusted	1	90
Badly tuberculated	5	60
Hand-packed stone	50	
Rocky	150	
Boulder strewn	500	

2.5 WATER HAMMER AND FLOW CONTROL

The strength of a pipe is often dictated by the pressures resulting from rapid change in flow. Fast closing of a valve may result in high pressures due to the deceleration of the water column. Stopping a pump could lead to vacuum pressures followed by a return surge causing high overpressures. Although Joukowsky understood the basic water hammer theory in the 19th century, the ensuing numerical or graphical calculations were laborious. With the advent of digital computers, numerical solutions became much more feasible. The methods of finite differences and characteristics (Streeter and Wylie, 1967) have gained acceptance. Thus the differential equations of continuity and momentum can be solved in step form for even the most complex pipe systems. Line friction, water separation, varying diameters and pump inertia can be included. Many solutions are possible within a few minutes, making trial and error design simple and economical.

The ultimate objective of the design engineer is to select a system operation policy such that water hammer pressures are tolerable, at the same time installing a practical and economic system. In many pipelines, water hammer protection devices are installed to limit overpressures. Systems in use include:

- Pump inertia
- Non-return valves
- Surge tanks
- One-way discharge tanks
- Compressed air vessels
- Automatic release valves

Which protection to install will depend on the pipeline profile and head. In many situations, a combination of protection devices is used, and in all cases a computer analysis is advisable. The possibility of failure of the protection system must always be born in mind. The safety factor on pipe wall stresses will be influenced by the reliability of the system and the risk the owner is prepared to take.

Line friction can significantly reduce water hammer overpressures. The maximum water hammer head according to the Joukowsky equation is:

$$\Delta h = (c/g)\Delta V \qquad\qquad (2.7)$$

where h is head, g is gravity and V is water velocity. The wave celerity c can be as high as 1200 m/s for rigid steel pipes, and as low as 300 m/s for flexible plastic pipe.

In the case of pumping lines, the water hammer waves travel up and down
before they are reflected as positive increases in pressure. If the line friction is
greater than 70 per cent of cV/g, the water hammer head may never exceed the
original pumping head at the pump end. Figure 2.7 presents results that will be
of use in making initial assessments of maximum water hammer heads along
pumping lines following pump trips.

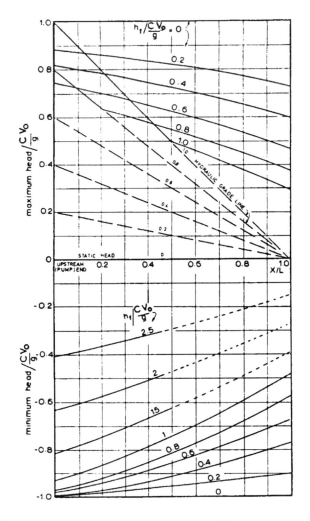

Figure 2.7. Maximum and minimum head enveloped following instantaneous pump
stopping in pipelines with friction

2.5.1 Valves and other fittings

Valves can be used for a variety of functions. They can be beneficially used, but also, under incorrect applications, are detrimental. The following need to be considered:

(1) Flow control
(2) Pressure control/level control
(3) Cavitation
(4) Transients
(5) Safe filling of pipe
(6) Air release
(7) Vacuum relief
(8) Isolating sections of pipe
(9) Preventing reverse flow through pipes
(10) Scour
(11) Preventing loss of water

Different valves have different shaped gates and therefore different apertures, closure rates for constant valve operation. The method of sealing when closed is also relevant to the applicability of the valve for different purposes.

Selecting the proper flow control valves requires information on maximum and minimum flows and the allowable pressure drop across the valve for both the present and projected future demands. It also requires establishing criteria for valve selection. Typical criteria that can be used in selecting a flow control valve (Tullis, 1995) are:

(1) The valve should not produce excessive pressure drop when fully open.
(2) At maximum flow, the operating torque must not exceed the capacity of the operator or valve shaft and connections.
(3) The valve should not be subjected to excessive cavitation, which can result in noise, vibrations, erosion damage and loss of capacity at extreme levels of cavitation.
(4) Pressure transients should not exceed the safe limits of the system.
(5) Some valves should not be operated at smaller openings. Most conventional valves will experience seat damage due to the high-pressure drop and associated high velocity and cavitation near the seat.
(6) Some valves should not be operated near full open. Quarter turn valves like butterfly and ball valves can experience torque reversals near the full open position. This can cause cyclic stresses in the valve shaft and operator that can lead to fatigue.

2.6 PIPELINE OPTIMISATION

Many factors have to be considered in sizing a pipeline. For water pumping mains, the flow velocity at the optimum diameter varies from 0.7 m/s to 2 m/s, depending on flow and working pressure. It is about 1 m/s for low-pressure heads and a flow of 100 ℓ/s increasing to 2 m/s for a flow of 1 000 ℓ/s and pressure heads at about 400 m of water, and may be even higher for higher pressures. The capacity factor and power cost structures also influence the optimum flow velocity or conversely the diameter for any particular flow.

In planning a pipeline system, it should be borne in mind that the scale of operation of a pipeline has considerable effect on the unit costs. By doubling the diameter of the pipe, other factors such as head remaining constant, the capacity increases six-fold. On the other hand, the cost approximately doubles so that the cost per unit delivered decreased to 1/3 of the original. It is this scale effect, which justifies multi-product lines. Whether it is in fact economical to install a large diameter main at the outset depends on the following factors as well as scale:

- Rate of growth in demand (it may be uneconomical to operate at low capacity factors during initial years). (Capacity factor is the ratio of actual average discharge to design capacity.)
- Operating factor (the ratio of average throughput at any time to maximum throughput during the same period), which will depend on the rate of draw-off and can be improved by installing storage at the consumer's end.
- Reduced power costs due to low friction losses while the pipeline is not operating at full capacity.
- Certainty of future demands.
- Varying costs with time (both capital and operating).
- Rates of interest and capital availability.
- Physical difficulties in the construction of a second pipeline if required.

The optimum design period of a pipeline depends on a number of factors, not least being the rate of interest on capital loans and the rate of cost inflation, in addition to the rate of growth, scale and certainty of future demands (Osborne and James, 1973).

In waterworks practice, it has been found economic to size pipelines for demands up to 10 to 30 years hence. For large throughput and high growth rates, technical capabilities may limit the size of the pipeline, so that supplementation may be required within 10 years. Longer planning stages are normally justified for small bores and low pressures.

It may not always be economic to lay a uniform bore pipeline. Where pressures are high, it is economic to reduce the diameter and consequently the wall thickness.

In planning a trunk main with progressive decrease in diameter, there may be a number of possible combinations of diameters. Alternative layouts should be compared before deciding on the most economic. Systems analysis techniques, such as linear programming and dynamic programming, are ideally suited for such studies.

Booster pump stations may be installed along lines instead of pumping to a high-pressure head at the input end and maintaining a high pressure along the entire line. By providing for intermediate booster pumps at the design stage instead of pumping to a high head at the input end, the pressure heads and consequently the pipe wall thickness may be minimized. There may be a saving in overall cost, even though additional pumping stations are required. The booster stations may not be required for some time.

Installing booster pumps at a later stage may often increase the capacity of the pipeline, although it should be realised that this is not always economic. The friction losses along a pipeline increase approximately with the square of the flow, consequently power losses increase considerably for higher flows.

The diameter of a pumping main to convey a known discharge can be selected by an economic comparison of alternative sizes. The pipeline cost increases with increasing diameter, whereas power cost in overcoming friction reduces correspondingly. In the other hand, power costs increase steeply as the pipeline is reduced in diameter. Thus by adding together pipeline and the present value of operating costs, one obtains a curve such as Figure 2.8 from which the least cost system can be selected. There will be a higher cost the greater the design discharge rate.

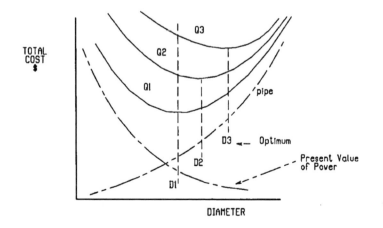

Figure 2.8. Optimisation of diameter of a pumping pipeline

If at some stage later, it is desired to increase the throughput capacity of a pumping system, it is convenient to replot data from Figure 2.8 in the form of Figure 2.9. Thus for different possible throughputs, the cost, now expressed in cents per kilolitre or similar, is plotted as the ordinate with alternative (real) pipe diameter a parameter.

It can be demonstrated that the cost per unit of throughput for any pipeline is a minimum when the pipeline cost (expressed on an annual basis) is twice the annual cost of the power in overcoming friction.

From Figure 2.9, the following will be observed.

(1) At any particular throughput Q_1, there is a certain diameter at which overall costs will be a minimum (in this case D_2).

(2) At this diameter, the cost per ton of throughput could be reduced further if throughput was increased. Costs would be a minimum at some throughout Q_2. Thus a pipeline's optimum throughput is not the same as the throughput for which it is the optimum diameter.

(3) If Q_1 were increased by an amount Q_3 so that total throughput $Q_4 = Q_1 + Q_3$, it may be economic not to install a second pipeline (with optimum diameter D_3) but to increase the flow through the pipe with diameter D_2, i.e. Q_4C_4 is less than $Q_1C_1 = Q_3C_3$.

(4) At a later stage when it is justified to construct a second pipeline, the throughput through the overloaded line could be reduced.

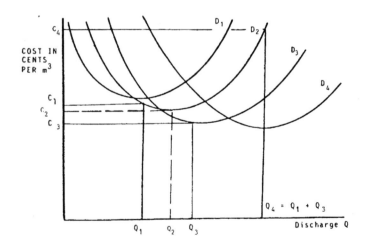

Figure 2.9. Optimisation of throughput for certain diameters

The power cost per unit of additional throughput decreases with increasing pipe diameter so the corresponding likelihood of it being most economic to increase throughput through an existing line increases with size (White, 1969).

2.7 OPTIMUM RESERVOIR SIZES

The sizing of pumping pipelines (optimum diameter), reservoirs (for balancing and emergency storage) and pumping rates is complex as the variables are interdependent. A spreadsheet type solution is required (or worksheet). The problem may be broken down into steps (Stephenson, 1998).

The first problem is to find the optimum diameter as a function of pumping rates, pressure and cost. The optimum pipe diameter will be obtained by adding pipe cost and pumping cost brought to a common time base.

The next problem is to optimise the relationship between balancing storage and pipeline capacity. This is assuming the pipeline delivers into a reservoir at a steady rate from which draw off occurs to suit the consumer. A graphical storage calculation is required for alternative pumping capacities and the optimum balancing reservoir size is obtained as in Figure 2.10.

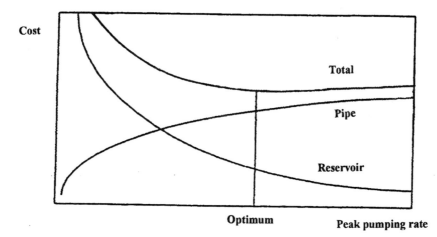

Figure 2.10. Optimum balancing reservoir sizing

To the balancing storage must be added an emergency or reserve storage, to meet shortfalls in times of outage in pumps or pumping lines. This requires a risk analysis of different outages, i.e. duration of outage versus probability. To each outage is attached a cost (loss of water sales, cost of repairs, or economic loss due to water shortage). The probable cost is then compared with the alternative reserve reservoir cost to obtain the minimum, as in Figure 2.11.

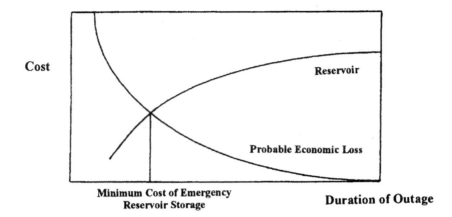

Figure 2.11. Optimum emergency storage

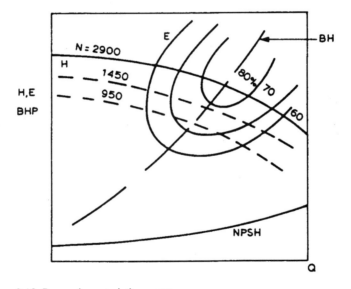

Figure 2.12. Pump characteristic curves

2.8 PUMP CHARACTERISTIC CURVES

The head-discharge characteristic of pumps is one of the curves usually produced by the manufacturer and used to design a viable pumping system. This relationship is useful for pipe size selection and selection of combinations of pumps in parallel (to increase flow) or series (to increase head).

Break horsepower (BHP) or shaft power in kW required by the pump and pump efficiency E are often indicated on the same diagram (see Fig. 2.12). The duty H-Q curve may alter with the drive speed and could also be changed on any pump by fitting different impeller diameters. The BHP and E curves would change then. The resultant head is proportional to N^2 and D^2, whereas the discharge is proportional to N and D.

The required duty of a pump is most easily determined graphically. The pipeline characteristics are plotted on a head-discharge graph. Thus Figure 2.13 illustrates the head (static plus friction) required for different Q's, assuming (a) a high suction sump water level and a new smooth pipe, and (b) a low suction sump level and a more pessimistic friction factor applicable to an older pipe. A line somewhere between could be selected for the duty line. The effect of paralleling two or more pumps can also be observed from such a graph.

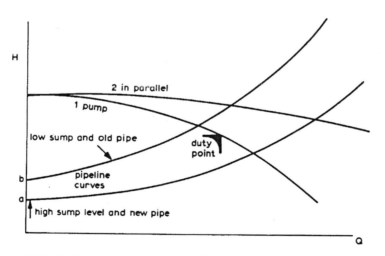

Figure 2.13. Pipeline and pump characteristics

2.9 REFERENCES

Ackers, P. (1969) *Charts for the Hydraulic Design of Channels and Pipes*, 3rd ed., Hydraulic Research Station, Wallingford.

Anis Al-Layla, M. Ahmad, S. and Middlebrooks, E.J. (1985) *Water Supply Engineering Design*. Ann Arbor Science Publishers, Ann Arbor, Michigan.

Colebrook, C.F. (1939) Turbulent flow in pipes, with particular reference to the transition regions between smooth and rough pipe laws. *Jnl. Inst. Civ. Eng*, 11, 133.

Department of National Housing (1994) *Guidelines for the Provision of Engineering Services and amenities in Residential Townships*, CSIR, Pretoria.

Doxiadis, A. (1967) Proc. Water for Peace, Washington, D.C.

IRC (1981) *Small Community Water Supplies.* International Reference Centre for Community Water Supply and Sanitation, The Hague, Netherlands, Technical Paper No. 18, 413 pp.

IWE (1969) *Manual of British Water Engineering Practice*. Instn. of Water Engrs., London.

Meadows, D.H., Meadows, D.L. Randes, J. and Bejrers, W.W. (1972) *The Limits of Growth*, for the Club of Rome, Pan Books, London.

Osborne, J.M. and James, L.D. (1973) Marginal economics applied to pipeline design. *Proc. Am. Soc. Civil Engrs.*, **99**(TE3), 637.

Saparta, D. and Munoz, M. (1995) Water consumption in distribution networks, Short term demand forecast. In *Improving Efficiency and Reliability in Water Distribution Systems*. Ed. Cabrera, E. and Vela, A.F. Kluwer Academic Publications, Dordrecht.

Schlichting, H.I. (1960) *Boundary Layer Theory*, 4th ed., McGraw-Hill, N.Y.

Stephenson, D. (1989) *Pipeline Design for Water Engineers*, 3rd ed., Elsevier, Amsterdam, 241pp.

Stephenson, D. (1998) *Water Supply Mangement*, Kluwer, Dordrecht, 306pp.

Streeter, V.L. and Whylie, E.B. (1967) *Hydraulic Transients*, McGraw-Hill, New York, 329pp.

Tullis, J.P. (1995) Reliability and expected use of dynamic devices in a water distribution system. In *Improving Efficiency and Reliability in Water Distribution Systems*, Ed. Cabrera, E. and Vela, A.F, Kluwer Academic Publications, Dordrecht.

Twort, A.C., Law, F.M., Crowley, F.W. and Ratnayaka, D.D. (1994) *Water Supply*, 4th ed., Edward Arnold, London.

White, J.E. (1969) Economics of large diameter liquid pipelines. *Pipeline News*, N.J.

3

Water demand management and loss control

3.1 CONTROLLING WATER USE

Although there are theoretically water resources sufficient to meet world requirements now and for the foreseeable future, the increasing distances required to pump water, and the storage required to meet droughts, make the cost of exploiting new sources higher and higher. The price of further water is therefore likely to increase. In addition, the expected increasing living standard of many of the population will mean that greater volumes of water will be sought, even though consumption may start at a minimal level. Resources are being depleted or polluted. This affects not only availability for human consumption but also the amount of water in rivers and for nature. A balance will therefore have to be achieved between consumption and new supplies (see e.g. Rademeyer et al., 1997). Sustainability of resources is also threatened and decreased yields may occur. The amount of re-use and re-cycling needs to be considered because this could reduce requirements for raw water. All this cannot be achieved except by considering marginal costs and increasing tariffs. The occasions when water tariffs need to be considered will also affect the instrument used to control usage. During crises (e.g. drought), short-term tariff increases may be applied, whereas in the long-term, the average tariff will depend on historical costs and the cost of new sources.

Water consumption can be limited by physical, sociological or economic means (instruments). Physical means include cutoffs or pressure control by reduced pumping or constrictions in pipes, e.g. orifices or washers. The latter costs money in waste of energy and cost of installations. On the other hand, it may even out the water drawoff variations by making consumers take water uniformly over more hours per day and provide in-house storage to meet peak consumption. The former (curtailing supply over periods of hours), could result in higher peaks when supply is resumed, but this will in turn reduce pressure and therefore peak drawoff. Demand control by pressure reduction could result in different drawoff patterns. Roof tanks could be filled at night. This will save distribution pipe costs but not necessarily reduce total volume of use. It may also be possible to reduce supplies to uneconomical, no longer valued consumers with compensation, in preference to newer consumers. In the long-term, water-saving plumbing devices could be installed or retro-fitted. These include small and double action cisterns, low-volume showers, and automatic tap closers. Invention of water savings devices such as reuse of basin water for toilet flushing, not only saves water in that situation, but they make people aware of water scarcity.

Water savings devices: The use of water savings plumbing systems can do a lot to save water in the household. Their efficiency depends on the design and cost effectiveness will be a function of the value of water saved. In the case of metered connections, these savings directly benefit the consumer. Such devices include:

- Aerators on taps and low flow taps/showers
- Automatic tap closures
- Dual flush toilets
- Low flush toilets (down to 5 ℓ for faeces, 2 ℓ for urine)
- Low water consumption gardens
- Diversion of grey water to gardens
- Diversion of hand basin water to toilet flushing

In industry, the savings and value of reclaiming and recycling is largely one of economics. On the public side, other possibilities arise, such as use of urinals with low intermediate flushing, and replacing night street flushing by sweeping

A distinction should be made between water used and water consumed. Used water returned to sewers is largely a quality and treatment problem. Water lost by evaporation is irretrievable and should be costed at the top marginal cost.

A wastewater tariff based on the concentration of pollutants in effluent could cause industry to dilute its wastewater. But tariffs based on the volume of wastewater could enforce industries to evaporate wastewater. To avoid this, high water tariffs are effective, plus a tariff based on the total pollution load in wastewater.

Sociological methods include appeals, way of living or legal action. Appeals, through the media or on monthly accounts, rarely last long before consumers forget the urgency. Long-term changes in ways of life to reduce water consumption will generally be caused by increasing water costs, together with public relations campaigns. Legal enforcement of water restrictions, in associated with fines, can be effective but costly to apply. It may mean inspectors checking consumers, or relying on spying neighbours. Then fines would have to be imposed by courts unless incorporated in water accounts. Such methods include prohibiting use of water on gardens on specified days, banning filling swimming pools or use of hosepipes for flushing drives. Consumer awareness can encourage local reuse of grey water, e.g. wash-water for gardening, or toilet flushing.

Economic methods include water tariffs, metering or charges on discharges. Theoretically, the best system would be to charge prices which reduce the usage to meet availability. This is, however, an unknown equation since the true value of water may not be known to the supplier or even the consumer. It may also involve tiered tariffs. That is, successively increasing consumption will be charged at higher rates so that the basic requirements of consumers, particularly domestic consumers, are met and more luxurious uses are charged at higher rates. This assumes there will be no trading between consumers (Moore, 1989). It may also encourage consumers to seek alternative sources which, although they may be more costly in total supply, may be cheaper to individual consumers.

Apart from the socio-economic objectives of providing water, there is a long-term value of water. If the world population and standards of living continue to increase, water will become scarcer. It may also occur that climatic change requires more careful use of water owing to reduced availability or greater variability in rainfall.

The traditional approach to supply management is to meet demands with successively more expensive schemes until the demand balances the supply. However, unless marginal pricing is applied, the average supply cost will always be less than the marginal additional cost of water, so that the demand will continue to increase asymptomatically.

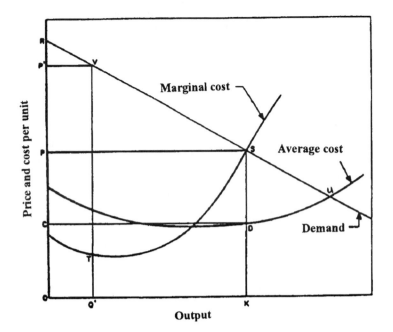

Figure 3.1. Supply and demand with different price structures

3.2 ECONOMIC THEORY OF SUPPLY AND DEMAND

A fundamental concept in economics is the law of supply and demand. Figure 3.1 shows theoretical supply and demand for water. At higher prices, producers would be willing to supply more but consumer demand would decrease; at lower prices, consumers would demand more but producers would cut back on supply. Figure 3.1 shows the theoretical equilibrium condition between the price and the quantity supplied and demanded for average costing and marginal costing (Hirschleifer et al., 1960; Agthe et al, 2003).

With increasing price, the reduction decreases because further reductions may require changes in behaviour that are inconvenient or contrary to personal or social norms. And at even higher prices, there will be no reduction at all if it means cutting into essential uses like cooking and waste disposal. On Figure 3.2, this relationship is shown by an increasingly steep demand curve as prices increase on the left side of the graph. At lower prices, people will buy and use more water, but there is a limit on how much water anyone can use, even if it is free. So again as price falls, demand eventually drops off as well.

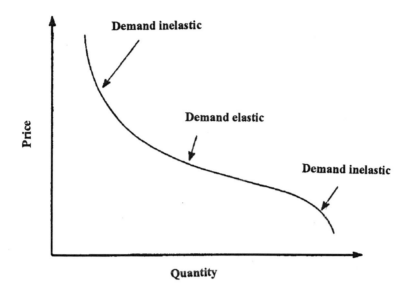

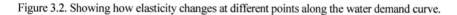

Figure 3.2. Showing how elasticity changes at different points along the water demand curve.

The rate at which demand changes as price changes is called the *price elasticity of demand*. (Similarly, there is a *price elasticity of supply*.) Conceptually, when demand changes a great deal for a given change in price demand is said to be *elastic*. When demand does not change very much compared to the change in price, demand is said to be *inelastic*. Economist calculate the elasticity of demand e as:

$$e = \frac{dQ/Q}{dP/P} \qquad (3.1)$$

where P is price and Q is quantity.

As new water schemes are commissioned, the average cost per unit (long-run average cost or LRAC) is likely to increase due to more expensive projects succeeding cheaper projects. On the other hand, over the life-span of each project the short-run average costs (SRAC) may reduce as consumption increases and more efficient use of facilities occurs (Fig. 3.3).

The short run cost changes are due to construction of new schemes or abnormal one-off costs. For example when a new project such as a dam is constructed it may be able to meet demands for another twenty years but initially the cost (probably annual interest and loan redemption) have to be met.

Without a balancing fund this may result in a price hike for the water. This could in fact be a compound effect, for the increase in cost of water could reduce demand temporally, resulting in a further price increase.

A water company should anticipate these reactions to smooth to price increases. And new schemes could be delayed by applying restrictions for a few years to reduce costs and enable the new project to come in at a higher demand base. If there is more than one source of water the short term cost in Figure 3.3 may not rise to the marginal cost as costs could be averaged over old and new schemes.

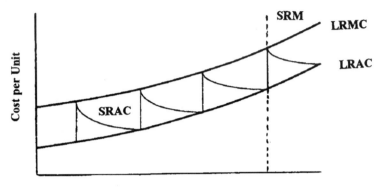

Figure 3.3. Short-run average costs of water supply. SRM = short range marginal, LRMC = long range marginal cost, LRAC = long range average cost.

3.2.1 Effect of metering

Those consumers who pay average cost will tend to use more than those paying higher marginal cost. If the water is metered, it is the (long range) marginal cost to the consumer which influences the consumption (Fig. 3.4) since the supplier can observe who is using water excessively and charge them a higher marginal tariff. If it is unmetered, only public responsibility, which is related to LRAC, controls consumption (Q_1), or the water company can only charge an average tariff which will encourage greater use.

The marginal cost of not metering is area ABQ_1Q_2. This may be compared with the cost of metering.

Actually, the marginal cost varies slightly with metering, so the comparison is a bit more complicated (see Henderson Sellers, 1979).

3.2.2 Management by use of tariffs

If the true value of water to consumers could be assessed, it may be possible to charge a limiting tariff. This method could be applied on a long-term basis or less effectively for short-term (crisis) demand reduction. However, one must be careful of applying crisis criteria persistently. Some consumers may locate their organization based on indicate water tariffs but the use of variable tariffs to manage water during drought, for example, must be explained and incorporated within the overall tariff system.

The level of consumption could be decided at the planning stage, if the cost of assured water is balanced against the cost to the economy of rationing. However, the operational basis will be from a different perspective.

Unfortunately, a uniform tariff cannot be applied in this way to restrict the use of water, for the poorest sectors of the economy may not be able to meet the tariffs which would be imposed on industry in order to force them to restrict water. Therefore, a percentage reduction, or a differential tariff or shadow value may have to be incorporated. The shadow value may not be paid by the poorer sectors but it should be added onto the cost of water to others. The alternative would be to charge a tiered tariff, i.e. the first volume would be at the original tariff and above a lifeline supply rate the tariff would be successively increased as a function of the percent of the lifeline supply rate. In this way, poorer consumers will only pay marginally more for excess consumption, whereas richer or industrial consumers would pay considerably more. The tariffs would have to be based on the economic value to all consumers. Dandy and Connarty (1994) indicate increased tariffs reduce consumption but to a limit.

Hong Kong's experience with tiered tariffs (Chan, 1997) is that the resulting demand management is limited. But they were limited by having to keep charge levels within inflation. Their most successful experiment in saving water was to use sea-water for flushing. Tucson's experiments with block rates also failed due to the politicians' control on maxima (Agthe and Billings, 1997), but their summer rate differential reduced consumption. Australia is also experimenting with demand management (Duncan, 1991).

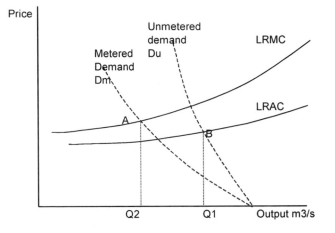

Figure 3.4. Demand curves with and without metering

In fact, water pricing experiences throughout the world (Dinar and Subramanian, 1997) show that external objectives of politicians or administrators can destroy the efficiency of water use control through tariffs. Increasing prices can instead be intended for many purposes, e.g. financing new schemes, becoming financially self-sufficient or cross-subsidization.

In the long run, it may also be that the consumer could find alternatives to being restricted in water usage or paying higher tariffs. He may seek alternative sources such as groundwater. These sources may have a higher operating cost but as they are intermittent it may not be as severe as long-term usage. This results in efficient conjunctive use of alternative resources.

Consumers may also elect to reuse water and if necessary purify the effluent reused. Again this may be a higher operating cost alternative but, owing to the limited duration effect, could be ameliorated.

The effectiveness of economic methods to control use will vary with the consumer. Industry may be most sensitive to price increases, whereas poor people will hardly be able to adjust their consumption even though they may find it difficult to pay. The richer domestic consumer is likely to have most elasticity in demand, but this is likely to constitute a decreasing proportion of the total.

In order to put objectiveness into water tariffs, Bahl and Linn (1992) suggest a five-part tariff based on:

            Variable costs: Consumption
                                     Maintenance

Fixed costs: Connection
                           Development
                           Upgrading

The above basis is, however, not sufficiently detailed to control use or obtain a method of cost allocation. There are other factors which affect water tariffs, e.g.:

- Capital and operating costs
- Opportunity cost
- Time-of-use or peak-load basis (Eskom, 1994)
- Size of property (e.g. Lumgair, 1994)
- Size of connection
- Zoning of district or purpose of use
- Timing of application
- Investment reserve
- Conservation
- Environmental
- Foundation consumers
- Insurance to ensure continuity during shortfalls
- Capacity allocation (Dudley, 1990)
- Tiered
- Cross-subsidization of income groups
- Location

3.3 TIMING

There are three stages during which the tariff for water needs consideration. (Table 3.1 summarizes which methods of demand management are applicable to which occasion.).

Table 3.1. Demand management methods and their use

Method	Crisis management (Drought, non-payment)	Operational time-frame	Long-term (Planning and design)
Technical	Pressure reduction Scheduled supply Valve closure	Flow control Orifices	Metering Loss control Plumbing devices
Social	Appeal Social persuasion Advertisements	Legislation Cross subsidies	Consumer awareness Education
Economic	Fines Punitive measures	Differential tariffs Trade	Supply and demand economics Marginal prices

3.3.1 Long-term (planning and design)

Before a water scheme is constructed, the capital cost of the project is likely to be the most serious economic consideration. Average running costs will be added to discounted capital cost of dams and conduits for alternative schemes in order to select the most economical alternative. If rationing is to be considered at this stage as an alternative to larger resource schemes, the true economic cost to the consumers due to shortfall also needs to be included. (This is not the same as the income to the water supplier which may even increase due to punitive tariffs during shortfall.)

When new water schemes are being considered, the cost of the scheme and consequently the average cost of water to consumers is the prime criterion. Alternative sources and levels of assuredness will be compared (Berthouex, 1971). This section is concerned with the reliability of supply during drought, and typically the more reliable the surface source, the greater the cost will be (see Fig. 3.5).

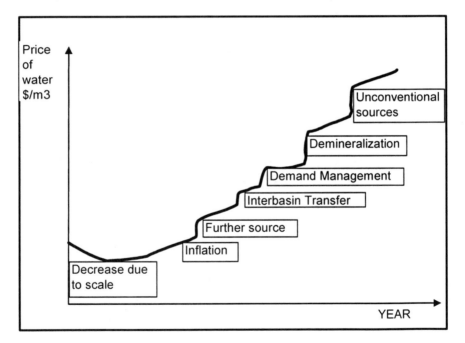

Figure 3.5. Short term and long term trends in water cost.

3.3.2 Operational time-frame

Once the scheme (e.g. dam and waterworks) is built, its cost does not feature in operational optimization. The object of the new optimization exercise is to minimize economic loss due to restrictions. This may mean shuffling the available water around to minimize total economic loss. The result will be an operating policy for a reservoir.

After a water scheme is commissioned, the perspective changes and day-to-day, as well as annual, supply rates change. Each year, the tariff may be revised as the supply rate increases, and hence the tariff could be reduced if it were solely to meet fixed repayment costs. However, funds for future more expensive schemes also have to be raised so it rarely happens that the tariff drops over the years. An operational policy for reservoirs may be designed to enable water to be conserved during drought. The control of usage could be by tariffs. The tariff may be consumer orientated, i.e. lower tariffs for the poor, higher for the rich, or industry. A tiered or sliding tariff structure generally results. (Fig. 3.6 shows the resulting effect on consumption).

Unit cost

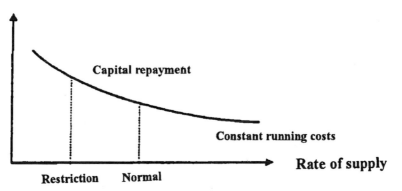

Figure 3.6. Effect of tariff on consumption.

The objective of the water works should be to cover costs. They should not unduly be enriched, by charging marginal costs or basing tariffs on what the market will pay. There may also be a planning component and a stabilizing component in the tariff.

The consumer on the other hand is entitled to minimize his costs. He may seek alternative sources of water, or insure himself against shortfalls. Industry and agriculture would suffer real losses if water was restricted or charged at an uneconomic rate. He may store water, or trade it, or purify and reuse wastewater.

The trading or reselling or buying water could affect the water works efforts in the short term so the supplier needs to think of similar saving measures. Generally an operating cost intensive source will be retained as a standby, as the costs may not be incurred. Such schemes could include recycling or boreholes.

Capacity allocation is not a tariff-based method of controlling water usage provided there is some other way of controlling the volumes used. There are a number of other methods for controlling water use during periods of shortfall or crisis. For example, public appeal has been resorted to with limited success. There are also methods of physically restricting supply of water by control valves, orifices and pressure reduction. The latter have been employed with roof tanks so that consumers can draw at peak rates while inflow is restricted. It may also be that the consumer could find alternatives to restrictions or paying higher tariffs. He may seek alternatives such as groundwater. These sources may have a higher operating cost, but as they are intermittent it may not be a severe penalty. This is efficient conjunctive use of alternative resources.

3.3.3 Crisis management

When there is a shortage at the source, e.g. during a drought, then there could be rationing of water, but at the same time the authorities have to meet fixed costs. The tariff may have to be increased (see Fig. 3.7).

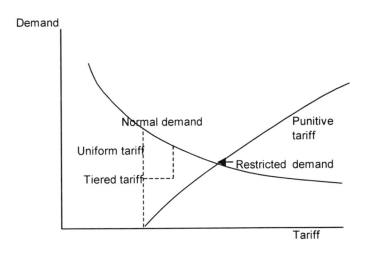

Figure 3.7. Effect of restrictions on cost of water supply

Assuming that an emergency has arisen in the way of drought or some other reason for inability to supply water, then the method of restricting water consumption could be based on an economic system as follows:

3.3.3.1 Penalties or punitive tariffs

Higher tariffs could be charged for total consumption if consumption is above a set figure (Davis, 1995). Alternatively, a marginal penalty could be applied for consumption above a certain figure. This method is not guaranteed to reduce consumption correctly because the supplier has not necessarily estimated the value of water to the consumer.

3.3.3.2 Purchase system

If there were a free market, then consumers could bargain amongst themselves to purchase different allocations of water.

3.3.3.3 Shortfall surcharge

Due to lower sales figures by the water authority, they may have to increase tariffs in some way to meet their costs which cannot all be reduced in proportion to the amount supplied.

The problem of time lag arises with crisis management by means of tariffs. Following the establishment and promulgation of punitive tiered tariffs to meet a certain requirement, it may be months before the tariff is charged, detected and evaluated by a consumer. He will then change his consumption, but possibly not by the amount desired by the biller. So the process may be iterative.

3.3.4 Notes on management by use of tariffs

If the true value of water to consumers could be assessed, there is likely to be a wide range of tariffs and the supplier may unduly benefit from overcharging. One must be careful of applying long-term criteria during crisis. Some consumers may locate their organization based on indicated water tariffs, but the use of variable tariffs to manage water during drought must be explained and incorporated within the overall tariff system. The level of rationing can be decided at planning stage if the cost of assured water is balanced against the cost to the economy of rationing. However, the operational basis during drought will be from a different perspective.

A drought would be identified if the water level in the reservoir is low and the probability of refilling the reservoir during the current operational season is remote. The objective is to minimize the probable economic damage by applying water restrictions. The fact that water restrictions may be implemented by use of tariffs is incidental, but it has the advantage that the tariff can be more easily decided if the level at which the tariff will influence the consumption is known, i.e. the economic value of water to the consumer is known. Unfortunately, this may result in recuperation of excess income or possibly under-recovery of income by the water supply authority and therefore a balancing fund would have to be built up by the water supply authority to ensure he does not make a profit or loss, if it is an autonomous non-profit-making organization. After ranking all consumers, a relationship between minimum damage and level of restriction could be established. Then the objective would be to minimize the probable damage or economic loss due to restrictions. In order to apply the restrictions, the cost of water must be increased to its perceived economic value.

There is some optimum tariff which may be charged for water at which a compromise between consumer and supplier satisfaction is achieved (Riley and Scherer, 1979). This level may never be achieved, because of conflicting objectives, political intervention or the complexity in achieving it. Political objectives could be towards achieving socially acceptable levels of supply (Triebel, 1994).

3.4 THE COST OF WATER

To control use of water by means of tariffs requires estimating the marginal value of water as well as the marginal cost. The components which make up the supply cost of water include (see also Table 3.2):
- Capital costs
- Operating costs
- Quality control, purification, pressure maintenance, supply rate including back-up for droughts
- Funding of indirect projects such as redistribution of wealth or national improvement in health and economy
- Deterrents for conserving resources such as a premium to reduce usage of water
- Components to pay for environmental protection or reclamation
- Community funding including training
- Reserves for future expansion and to ensure continuity of supply or jobs
- To cross-fund, e.g. other department's shortfalls, or redistribution of charges.

The historical basis on which tariffs are calculated is generally the cost of supplying the water (Stephenson, 1995). However, there is the possibility of charging for water before it has been controlled or tapped by man. This is a form of funding as the real cost is zero, seeing it is a renewable source. If the resource is mined, such as the use of groundwater at a rate greater than the natural replenishment rate, then there may be a long-term cost to the environment.

The historical cost has been the one most commonly used for establishing water tariffs (Palmer Development Group, 1994). The income from water tariffs is used to meet the costs of repaying loans, operation, maintenance, fuel, management and often a fund for future expansion. Based on average cost, the water authority will charge a tariff which could be the total expenditure divided by the total sales of water.

A deviation from this method of costing is the marginal cost basis. Based on the fact that additional augmentation costs more than the original source of water, new users may have to pay more. Alternatively, all users may have to meet the additional cost. An alternative marginal effect may be the reduced cost due to bulk supply since the cost per unit delivered from a source decreases the larger the pipeline or the supply system.

If the total income from tariffs is only to meet average costs, then it is purely a financial calculation. However, there are invariably economic components which make the historical or average cost basis rather academic. For example, the non-technical components described above may be added onto the total cost.

The cost of water is not static even though historical costs may be constant, until the loans are repaid. Invariably, there is no reduction in average tariffs when costs are paid off, since expansion increases expenditure faster than the reduction in loan repayments over years.

Costs increase because supplies have to be augmented and these augmentation schemes are invariably from more and more costly sources. There is also inflation of prices causing the unit cost to increase. Policy factors may also cause increasing cost to some of the consumers.

For example, subsidization or redistribution of resources may mean more acceptable costs to some, but others will have to pay more to meet total costs. There may also be cost increases of a temporary nature due to limited sales, for example during drought, which means that the unit price must be increased to meet certain fixed costs.

Historical water costs vary enormously throughout the world, and it is difficult to compare them internationally. They depend on the cost of installations at the time, inflation since then, the standards of supply and the ability of the consumer and government or authority to meet costs.

The cost of municipal water in Europe and North America is of the order of $5/m^3. In South Africa, it is less than R3/m^3 (US$ 0.50c), and in some regions in Africa, it is sometimes free. An affordability of 1 to 2% of income is a yardstick in developed countries, but in some developing communities they may pay up to 10% of their income.

The methods developed for justifying water resource projects, particularly in the United States, in the mid-20th century, were based on comparing benefits and costs of projects before ranking them, or deciding on the scale or priority of development. Whether these techniques can be applied to water supply is doubtful. In particular, the evaluation of benefits which cannot be cashed in could distort the market. It could result in over-expenditure or power-building in government centres which fund water supply projects. At the most for water supply, it should be used for ranking projects but the social impact needs to be evaluated for inclusion in the decision-making process.

When trying to assess the value of water to a user with regard to curtailing supply, the true long-term value may not be the applicable figure. The user will only consider his operating benefits minus costs, since capital expenses cannot be avoided. He will also consider primarily cash benefits, since intangible benefits, e.g. education, are long-term. So, it is important to distinguish between long-term and short-term benefits as well as tariffs.

The principles of economics, however, should be used for comparing projects and optimizing supplies. Thus the possibility of alternative sources or inter-basin transfers may have to be compared in some fashion.

Table 3.2. Factors affecting water prices

Supply costs	Charges	Controls
DIRECT	Sale of natural resource for	Differential tariffs
Dams	income	Subsidization
Pumping	Prevention of over-	- communities
Pipelines	exploitation	- localities
Reservoirs	Cost of alternative sources	- relocation
Purification	Cost of depletion	- use type
Administration	Fines	- higher marginal cost
Upgrading	Pollution – cost of	Drought rationing
Land	purification	
INDIRECT	Environmental restitution	
Financing	Control of usage	
Risk minimization	Economic benefits	
Standby equipment	- health	
Monitoring	- time	
Future more expensive	- education	
sources	- commercial	
Commissions	Taxes	
Mismanagement	Affordability	
Inefficiency	Permits	
HIDDEN (NOT	Willingness	
CHARGED)	Bearability	
Labour disruption during		
construction		
Rerouting communications		
Loss of land surface		
Loss of future potential		
Alternative uses of water		
Environmental impact		
Wastewater disposal		
Siltation		

Table 3.3. The benefits of water supply

Benefits
Income
Health
Improved quality of life
Time for
-    education
-    leisure
-    economically productive
Commercial and industrial development
Agricultural
Power generation
Environmental

The benefits and costs of water supply are not easy to evaluate (see Gibson, 1987). Tables 3.2 and 3.3 list some of these. The costs can vary not only for the direct installation costs but also the social impact costs. These could be as obscure as changing social customs due to different methods of water collection. There are also changing population demographics which are difficult to evaluate, and the interruption of the economy by providing temporary construction employment. The river patterns may change if the water is dammed. This may affect agriculture. The environment is affected whether it is due to burying pipelines or construction of structures. More particularly, it is affected by the change in hydrology if the demand is for surface water. Groundwater is also obviously affected and the effects are not as readily seen in the short-term, but in the long-term, it could have severe environmental implications.

There are also opportunities lost as the water cannot be used for other purposes and also future planning will change owing to the lesser availability of water. Costs of planning also need to be considered in the total system cost, and if all direct, indirect and hidden costs were included, it is likely that the level of water supply would be reduced in many countries.

On the other hand, the benefits of providing water are many. Not only are they those listed below but also they have a multiplying effect in parallel with many services. That is, money is injected into the economy, the level of development increases, standards of living increase, expectations increase and therefore the entire economy is provided with an injection. Of course, there is also the effect of increasing price leading to lower consumption (Postel, 1985).

The human rights issue means that if water is to be provided to all, then it must be marketed at affordable rates, which vary considerably. It therefore appears that some form of differential tariff system would be required whereby the richer subsidized the poor. This could be disguised in various ways. For example, incremental water consumption would be charged at a higher and higher tariff. This assumes that the full cost is to be recovered by the water supplier. Subsidization by the government could further complicate the issue. This in fact may be necessary if the policy is set by the government.

An alternative to the cost recovery pricing system would be the production cost pricing system. This would imply that prices were pushed up to reflect the value of water to the consumer, whereas the price may not be pushed to the limit of affordability, it would reflect some value to the consumer (see Mirrelees et al., 1994).

The third alternative is the water scarcity pricing system whereby the price of water is increased to reflect its value (see Berk, 1981). This may be on a permanent basis or temporarily during drought. Unless a thorough understanding of the affordability of water is obtained then the price to limit consumption during scarcity may be a matter of trial and error.

The problems of setting affordable tariffs, particularly to poorer communities, will draw in the following considerations:

- Adequate quality of service, that is pressure and flow
- The possibility of upgrading the system as living standards or affordability improve
- Labour-orientated construction to inject money into the community
- Flexibility to ensure that various levels of demand are satisfied to their standard
- Charging for services to recover what is possible, but also to instill a sense of value
- This may involve prepayment systems or flat rate systems to simplify collection of rates
- Speed of delivery which is a function of financial resources and technical resources

The problem of non-payment for water complicates the issues – the cost must be borne by others until pressure is sufficient to right the problems causing non-payment.

Methods of subsidizing water costs vary. If the subsidizer does not want to become involved in the politics, he may subsidize the water supplier and this could be by means of direct payments or reducing taxes or cost of raw water. The alternative of payment to the consumer is complicated not only by administration or the need to appear equitable and just, but also in the method of payment. It would appear more logical to subsidize indirectly, that is by reducing taxes or providing other services to reduce expenditure. Donors often subsidize the capital cost of the system, particularly rural water supply schemes.

It is also not easy to decide how to discriminate between recipients subject to different levels of subsidization. In many cases, abuse of the systems needs consideration (misappropriation or resale).

The value of water to a consumer is influenced by risk (Cortruvo, 1989). If there are frequent interruptions (due to breakdowns) or lengthy rationing (drought) or pressure drops or pollution, or high tariff increases, the value is diminished. Unfortunately, supply authorities generally give no indication of these or the associated probability of occurrence. Some are catered for, e.g. emergency storage, and others may be completely unknown, e.g. future price increases.

3.4.1 Future trends

The future is likely to see increasing water costs. This will automatically reduce consumption. The theoretical correct way to control consumption would be to charge marginal costs on the top consumption, but the administration and lifeline requirements make this difficult.

Conflicting objectives make economic methods impractical for accurate day-to-day control, but economics can be used in the longer term.

Physical ways of limiting consumption (pressure reduction, cutoff) are only applicable in periods of crisis, and long-term education of consumers is seen as a necessity.

3.5 VALUE OF WATER

Whether water is a 'free good' or an 'economic good' is a paradox which has been the subject of much debate (see Janusz, 1998). The International Conference on Water and Environment in Dublin, 1992, listed the following guiding principles:
- Water has an economic value in all its competing uses.
- There is a basic right of all human beings to have access to water and sanitation, at an affordable price.
- Managing water as an economic good is an important way of achieving efficient and equitable use and of encouraging conservation and protection of water resources.
- The value of water can be measured in terms of:
    (1) its utility, which results in an economic benefit;
    (2) its exchange value;
    (3) its scarcity. The scarcer the resource, the greater its value.

The price is not the same as the value of water. Public organizations use one or more of three prices (Aswathanarayama, 2001):

(1) The market value based on supply and demand.
(2) The administered price based on cost recovery or political decisions.
(3) The accounting price, which could be the shadow value of the water, or marginal value.

The shadow value is a function of the value in alternatives uses. For example it could be the replacement value (not possible for domestic use), or the value for alternative consumers.

3.6 LOSS CONTROL

Losses of water can be up to 50% in older reticulation pipe systems. And in irrigation systems it can be equally alarming bearing in mind the large quantities used. Many cities quote losses of 20 to 30% but the figures are seldom less than 15%. The losses may not only be the fault of the supply system, there may be consumer plumbing leaks or requirement of water which cannot be charged for. This is not strictly water lost, but could result in a loss of revenue.

The term unaccounted for water is also used, but some losses can be measured but not avoided e.g. flushing out pipelines or reservoirs, or water used for street washing. If a proper interdepartmental charging system could be developed it may reduce some of these losses. And it is not only water which is lost. It could be revenue, or energy or customer relations which is lost, all of which represent financial loss to the water company.

Whereas in the past losses used to be taken for granted, with improvement in business sense the cost of losses, as well as the waste of resources is realized and appropriate steps are taken to minimize losses. The following methods are used to reduce losses of water (Stephenson, 1998):

- Passive maintenance (repairs when notified of leaks)
- Active maintenance (vigilant inspection programme)
- Water audits
- Monitoring
- Zoning to limit maximum pressures
- Pressure reduction by valves
- Metering
- Targets, e.g. 5 ℓ/hs/hr or 500 ℓ/km/hr
- Education of consumers

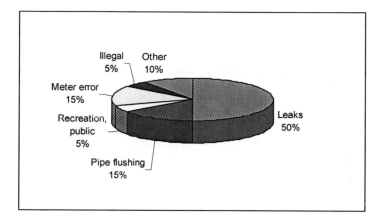

Figure 3.8. Components of water loss in reticulation systems.

Table 3.4. Types of losses in water supply

Loss of:	Water	Revenue	Other
To Supplier	Leaks, Bursts	Non metered, Meter slip	Pilfering
	Flushing streets	Illegal connections	Office inefficiency
	Reservoirs	Public use, Fire fighting	Energy, Friction
	Backwash filters	Shortage of water	Data
To Consumer	Demand management	Plumbing leaks	Confidence in supplier
	Stolen	Meter misreading	Pressure
	Leaking fittings	Corruption	Damage to property
	Wastage	Ineptitude	Water quality

Apart from the inertia of supply organizations, there are physical factors affecting the loss rate. The following factors affect losses:

- Pipe age
- Pipe material
- Corrosion protection, internal and external
- Wall thickness, pressure class, standard
- Jointing system
- Changes in pressure
- Soil type, moisture
- External damage

- Number of consumers, connections
- Maintenance

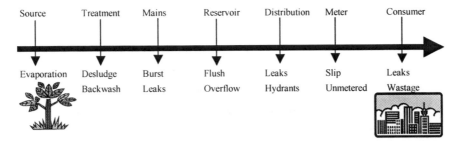

Figure 3.9. Location of losses in a water supply system.

Many water supply companies now take active steps to find losses and reduce them. The following leak detection methods are applied:
- Night flow measurement
- Step pressure/flow measurement
- Remote sensing using infra red waves
- Visual inspection
- Gas/tracer injection and detection
- Seismic refraction
- Resistivity surveys
- Noise frequency detection
- Noise correlation

In all cases cost effective loss control should be aimed at. That implies careful management to ensure the correct level of maintenance. Apart from revenue loss the supplier should conserve resources as they are a national asset and of latent value in their natural state. In fact national monitoring of resources and if necessary imposition of control measures from the resource and pollution point of view are desirable.

Loss reduction improves the economics of supply as the costs are reduced. This assumes cost effective loss control. But there are also hidden benefits. Less resources are consumed. Older systems can have their life prolonged. Structural damage at leaks is minimized. And if losses are due to theft, better consumer response is achieved. Rehabilitation of pipelines and reservoirs is a specialized task which has benefits beyond the saving in water. Relaying of pipes by plastic sleeves or in-situ applied mortar reduces friction and energy losses. The capacity of pipes is increased. So the pipes may supply more consumers and disruption of roadways by new pipes is avoided.

Rehabilitation of water works forms a component of loss control. Pipeline relining, repair or replacement needs to be considered in the alternatives. The subjects of asset management and optimum level of maintenance are covered in Chapter 11.

3.7 WATER HARVESTING

The interception of water in its various locations on a small scale is often referred to as water harvesting. This may be by means of tanks below the gutters or roof. Then it is called rainwater harvesting. It could be done on a larger or commercial scale with plastic sheets to collect the water.

In arid areas and on the coast, e.g. Namibia, dew harvesting has been attempted. There are winds carrying moist air and large temperature drops at night. During the night, moisture condenses on vertical mesh screens and is led into containers.

Runoff can be harvested by catching runoff in furrows. This is done on agricultural plots and the water is permitted to infiltrate or is diverted to dams.

Forestry harvests water by catching precipitation on leaves or on the mulched ground. Trees use a lot of water and forests reduce runoff. The question arises as to whether the water may legitimately be intercepted to reduce runoff downstream. In some instances the forester may be charged for abstracting from the catchment runoff. The original runoff has to be measured previously or modelled. And the possibility of use downstream should be established.

Similarly there may be attempts by authorities to charge for use of boreholes. Aquifers contribute to total runoff and by abstracting, the water table drops and lateral discharge diminishes. Greater infiltration may occur so there is less surface runoff too. However, there is no record of considered compensation for recharge of groundwater by discharging water or waste water onto the ground. There are also possible charges for discharging polluted water into sewers so this makes up for otherwise disposing of wastes, but quality control is a problem which will not easily be solved by tariffs.

3.8 REFERENCES

Agthe, D.E. and Billings, R.B. (1997) Equity and conservation pricing policy for a government run water utility. *J. Water SRT-Aqua*, **46**(5), 252-60.
Agthe, D.E., Billings, R.B. and Buras, N. (Eds). (2003) *Managing Urban Water Supply*. Kluwer, Dordrecht.

Aswathanarayama, V. (2001) *Water resources management and the environment.* Balkema. Lisse.

Bahl, R.W and Linn, J.F. (1992) *Urban Public Finance in Developing Countries.* Oxford Univ. Press.

Berk, R.A. (1981) *Water Shortage.* Alot Books, Cambridge, Mass.

Berthouex, P.M. (1971) Accommodating uncertain forecasts. *J. Am. Water Works Assoc.,* **66**(1), 14.

Chan, W.S. (1997) Demand management. *Water Supply,* **15**(1), 35-39, Blackwell Science.

Cortruvo, J.A. (1989) Drinking water standards and risk assessment. *J. Inst. Water and Environ. Management,* **3**(Feb), 6-12.

Dandy, G.C. and Connarty, M.C. (1994) Interactions between water pricing, demand managing and sequencing water projects. *Proc. Conf. Water Down Under,* Adelaide, Austr. Inst. Eng., 219-224/

Davis A. (1995) Rand Water's response to the drought. *IMIESA,* **20**(9).

Dinar, A. and Subramanian, A. (Eds) (1997) *Water Pricing Experiences.* World Bank Tech. Paper 386, 164 pp.

Dudley, N. (1990) Alternative institutional arrangements for water supply probabilities and transfers. *Proc. Seminar Transferability of Water Entitlement.* Univ. New England, Amidal.

Duncan, H.P. (1991) Water demand management in Melbourne. *Water* (June) 26-28.

ESKOM (1994) *Time-of-Use Tariffs.* Megawatt Park, South Africa.

Gibson, D.C. (1987) The economic value of water. *Resources for the Future,* Washington.

Henderson-Sellers, B. (1979) *Reservoirs.* MacMillan, London.

Hirschleifer, J., de Haven, J.C. and Milliman, J.W. (1960) *Water Supply Economics Technology and Policy.* Univ. Chicago Press.

Janusz, K. (1998) Economic value of water. In H. Zebidi (Ed) *Water, a looming crisis?* IHP-V. Tech., Dec, Paris, Unesco, 407-416.

Lumgair, G. (1994) *Water Supply Tariff Formulae for Local Authorities.* MSc(Eng) report, Univ. Witwatersrand, Johannesburg.

Mirrelees, R.I., Forster, S.F. and Williams, C.J. (1994) *The Application of Economics to Water Management in South Africa.* Water Research Commission, Report No. 415/1/94, Pretoria.

Moore, J.W. (1989) *Balancing the Needs of Water Use.* Springer-Verlag, New York.

Palmer Development Group (1994) *Water and Simulation in Urban Areas. Financial and Institutional Review. Report 1: Overview.* Water Research Commission, Pretoria.

Postel, S. (1985) *Conserving Water: The Untapped Alternative.* Worldwatch. 67, Washington.

Rademeyer, I.J., Van Rooyen, P.G. and McKenzie, R.S. (1997) Water demand and demand management, *S.A. Civil Eng.* Sept, 13-14.

Riley, J.G. and Scherer, C.R. (1979) Optimal water pricing and storage with cyclical supply and demand. *Water Res. Research,* **15**(2), 253-259.

Stephenson, D. (1995) Factors affecting cost of water supply to Gauteng. *Water SA,* **21**(4), 275-280.

Stephenson, D. (1998) *Water Supply Management.* Kluwer Academic Press, Dordrecht.

Triebel, C. (1994) Tariffs for water supply and sanitation. *Proc. Natl. Water Supply and Sanitation Policy Conf.,* Kempton Park, SA.

4

Sewerage

Sewage is generally removed to wastewater treatment works by underground pipes or sewers flowing under gravity. Pipes are in many ways a very convenient means of removing drain water. They are buried so that they are unobtrusive, they are structurally strong, and the hydraulic properties of circular pipes are favourable in comparison with other types of closed conduits.

Although the design of pressure pipes is beyond the scope of this section, basic principles of hydraulics of circular pipes are presented together with some design rules.

4.1 FLOW IN CIRCULAR DRAINS

4.1.1 Manning equation

The Manning equation is widely used for head losses in open channel flow calculations and for part full pipes. The equation is:

$$V = \frac{K}{N} R^{2/3} S^{1/2} \tag{4.1}$$

where K is 1.0 in SI (metre) units and 1.486 in ft units.

The Manning roughness coefficient N is assumed to depend only on the boundary roughness, *k*. Hence it avoids the iterative solution needed for the Colebrook White equation (see Chapter 2 for pipe flow equations), wherein the friction coefficient is a function of flow.

The Manning equation is therefore only applicable to turbulent flow with a rough boundary. It is however easier to use than the Darcy equation and has thus retained popularity despite its limitations. Typical values of *N* are given in Table 4.1.

Table 4.1. Manning's *N*

Smooth glass	0.010
Concrete, galvanized or lined steel	0.011
Cast iron	0.012
Slimy or greasy sewers	0.013
Rivetted steel, vitrified	0.015
Rough concrete	0.018

4.1.2 Non-circular cross sections

A circular pipe is normally the most economic if it is to be designed to resist internal pressures. A circular shape has the shortest circumference per unit of cross sectional area, consequently it requires least wall material, as well as being easy to manufacture.

Elliptical or horseshoe shapes are often adopted for sewers or drains. They have different strength and hydraulic characteristics to circular pipes. Vertical elliptical pipes (major axis vertical) have smaller wetted perimeters when running partly full with low flows. Consequently, the velocity is higher than for a circular pipe, which assists in flushing. The vertical load on a vertical elliptical pipe is less than on a circular pipe with the same cross sectional area, and the strength is greater because the curvature is sharper at the top. Horizontal elliptical pipes (major axis horizontal) are sometimes used where vertical loads are low or clearance is limited. Running partly full, they will discharge relatively high flows at small depths of flow which may be an advantage if head is limited.

Arch shapes with flat bottoms have similar hydraulic characteristics to horizontal elliptical shapes for low flow under partly full conditions. The arch shape is usually the most practical shape in tunnelling.

Provided the cross-sectional shape does not differ much from circular, the Darcy equation is applicable. *4R* is substituted for *D* in the equation and in the Reynolds number. *R* is the hydraulic radius *A/P* where *A* is the cross sectional area of flow and *P* the wetted perimeter. For a circular pipe flowing full, *R = D/4*.

4.1.3 Uniform flow in part-full circular pipes

Most friction formulae for full pipe flow have been used for part-full flow (Mussalli, 1978). However, the cross sectional area and wetted perimeter have to be calculated. For a circular pipe of diameter D running at depth y, the cross-sectional area of flow is:

$$A = \frac{D^2}{4} \cos^{-1}\left(1 - \frac{2y}{D}\right) - \left(\frac{D}{2} - y\right)\sqrt{yD - y^2}$$ (4.2)

The wetted perimeter is:

$$P = D \cos^{-1}\left(1 - \frac{2y}{D}\right)$$ (4.3)

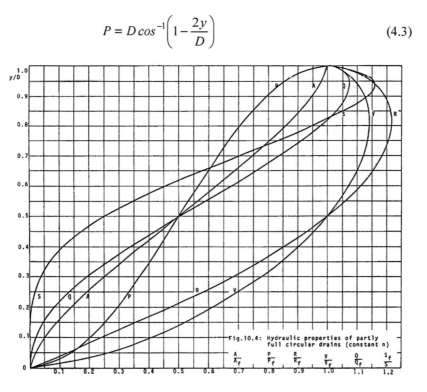

Fig.10.4: Hydraulic properties of partly full circular drains (constant n)

$$\frac{A}{A_f} \quad \frac{P}{P_f} \quad \frac{R}{R_f} \quad \frac{v}{v_f} \quad \frac{Q}{Q_f} \quad \frac{S_f}{S}$$

Figure 4.1. Hydraulic properties of partly full circular drains (constant *n*)

Using these equations, charts may be prepared yielding a dimensionless relationship between flow depth and cross sectional area and hydraulic radius as a proportion of the full depth value, i.e. A/A_f and R/R_f versus y/D, as given in Figure 4.1.

Also indicated on the chart are lines denoting the velocity ratio V/V_f and the discharge ratio Q/Q_f versus y/D for uniform flow. For non-uniform flow, the ratio of friction gradient S_f/S versus y/D is indicated assuming $Q/Q_f = 1$. These lines are dependent on the assumed friction loss equation. If Manning's equations is used with roughness N independent of depth, then the resulting relationships are as indicated. In fact, if Strickler's approximation for N is used, where $N = 0.13k^{1/6}/g^{1/2}$, then N is independent of flow depth and dependent only on boundary roughness k. If the Darcy-Weisbach equation is adopted, then assuming a constant friction factor λ, i.e. independent of depth of flow, similar relationships could be plotted. The friction factor λ is known to vary with Reynolds number though, especially for shallow depths and correspondingly low Reynolds numbers. In such cases the relationships between V/V_f, Q/Q_f and y/D are not unique unless a varying λ is used, i.e. λ is a function of two variables, k/R and Reynolds number, so a different line will apply for each case.

Camp (1946) performed tests to determine the variation of N and λ with depth. His charts are presented by ASCE (1969), but it should be borne in mind those relationships are not completely in accordance with the Colebrook-White equation for the reasons indicated above.

Using a head loss chart for full pipe (see Chap 2) and Figure 4.1 and given any three of the five variables Q, D, S, V and y, the other two may be determined. The flow conditions for full-bore flow $(y/D = 1)$ are yielded simultaneously. Designate Q_f = flow at full bore, and V_f = velocity at full bore. Now assume the flow, pipe diameter and slope (Q, D and S) are known, and y/D and v are to be determined. Calculate Q_f and V_f from an equation such as 4.1, and using the ratio Q/Q_f, read y/D from Fig. 4.1. Also read V/V_f from, Fig. 4.1 and calculate V knowing V_f.

As another example, given $Q = 50\ell/s$, $S = 0.0005$ and $y/D = 0.25$, find the necessary diameter D and corresponding velocity: From Fig. 4.1, $Q/Q_f = 0.135$, so $Q_f = 370\ell/s$, and from Chapter 2, $D = 525mm$ and $V_f = 1.7m/s$. Now from Fig. 4.1, $V/V_f = 0.7$, hence $V = 1.2m/s$.

An interesting fact is illustrated in Fig. 4.1. The flow for a partly full pipe is greater than the flow through a fully charged pipe if the depth of flow is between 82% and 100% of the diameter. The reason for this is that the wetted perimeter increases rapidly but the area does not, as the pipe fills up over the last portion. The additional capacity should not be relied upon however, as the slightest irregularity may cause the pipe to run full.

4.2 DRAINAGE NETWORK OPTIMISATION

One of the engineer's objectives is to produce a satisfactory design such that the overall cost is a minimum. This is referred to as cost optimisation.

Economic pressures and the advent of electronic computers have both prompted many researchers to search for cost saving design methods. The conventional design approach is based on a set of design standards and criteria, with no alternatives compared. Optimisation is in practice often approximated by investigating a series of designs. The designer selects on the basis of his best professional judgement a system layout and a combination of pipe diameters and grades. He will design and estimate the cost of a number of alternative layout-size-slope combinations. Using conventional design methods, it is not feasible to design and evaluate more than a few alternatives. Computers enable a larger number of alternatives to be evaluated, but an optimal solution is not guaranteed.

An ideal optimisation model would produce the minimum cost design by simultaneously varying both network layout and pipe design. This has been attempted by a number of researchers. The fact remains, however, that no method is in existence for obtaining such an overall solution. There are, however, design methods which produce the minimum cost design for a system of a given network layout. I.e. if flows are known, the system becomes 'linear' and can be optimised directly.

4.2.1 The variables

A drainage network for any area can be anything from a single pipe to an interconnected tree-like network (Fig. 4.2) depending on the size and configuration of the catchment, and the relative economic advantages and cost of the drainage system.

The designer of a drainage network must decide the following:
- Layout plan of the township or catchment
- Location of drains within the permitted zones
- Spacing of inlets
- Location of manholes at bends and changes in grade and diameter
- Capacity of inlets
- Size of surface gulleys
- Diameters of subsurface drains
- Gradient of drains
- Design details of inlets, manholes, drops, branches, etc.

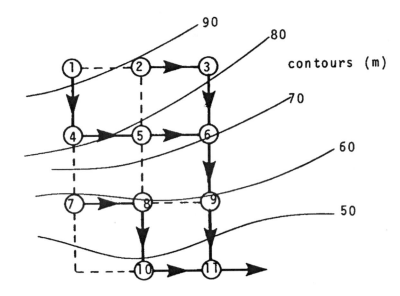

Figure 4.2. Drainage network with alternative layouts

Many of the design parameters can be selected independently, thus eliminating the number of alternatives. Drain layouts are frequently fixed by the road layout. Drains ae located on the downhill side of roads or on both sides. Surface gulley sizes depend on the design flows and drain inlet spacing.

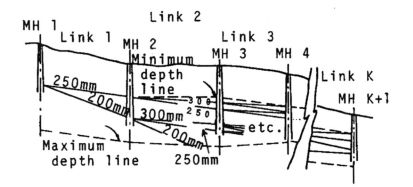

Figure 4.3. Drain profile showing solution procedure for dynamic programming optimization

Assuming each inlet accepts the total gulley flow at that point, the problem is to select gulley capacity and subsurface drain capacity for least total cost. In the case of systems with fixed layouts and consequently known design flows, the problem is to determine the optimum combination of diameter and grade for each link pipe (Fig. 4.3).

Costs of drainage components depend primarily on the flow in the conduit. This may be a variable if the network layout and inlet spacing are to be determined, or may be fixed if the layout and inlet spacing are pre-selected. Drainage layouts are more often than not of the tree-like layout, with the flow in each successive branch being cumulative. There is often no problem in selecting between alternative routes as the most economic layout is usually with drains flowing downhill. This is unlike water supply networks where the flows in each pipe and the layout have many possibilities (one of which is optimal) (Stephenson, 1989; Yen and Sevuk, 1975). Cost of drains depends on:

- Diameter
- Depth, which influences excavation and pipe wall thickness
- Manhole spacing
- Locality, such as in built-up or open ground

Dajani and Hasit (1974) and Merritt and Bogan (1973) present some typical cost data, but these are highly dependent on locality and cost escalation, so the engineer is urged to compile cost data to suit the particularly project. There will be a number of practical constraints on the system to be optimized. These include:

- Permissible pipe depth depending on ground conditions
- Available pipe strengths
- Minimum grade or velocity to avoid deposition
- Maximum grade to avoid erosion or noxious gas release
- Minimum diameter for access purposes
- Maximum surcharge
- Pipe diameters must be those commercially available
- Gradient and diameter must be consistent with flow rate and friction factor
- Invert levels of successive pipes at intersections (manholes) must be equal or fall
- Head losses at inlets and branches must be allowed for

Whereas there are many mathematical methods for the selection of a least-cost system, some of the methods cannot accommodate all the constraints rigorously. In particular, linear programming and geometric programming require continuous variables. Mixed integer programming can be used to select discrete values, as can dynamic programming.

Implicit in the following is that the runoff rate into each drain or per unit area of catchment, is known. This means the design flow intensity has been estimated beforehand. Flow rates are used as input to the analyses. The computations subsequently produce pipe sizes and grades, and correspondingly, flow velocities. In order to determine the concentration time of the system, the flow should strictly be routed through the system. This may indicate a new concentration time which should equal design storm duration, upon which it would be necessary to revise precipitation rates in the light of the known intensity-duration relationships. In fact, the situation is even more complicated as the concentration time and consequently design storm vary down the system.

It is also assumed the pipes will run full at design flow in the following sections. This could be varied though to suit the design standards.

4.2.2 Dynamic programming for optimising compound pipes

One of the simplest optimisation techniques, and indeed one which can normally be used without recourse to computers, is dynamic programming, e.g. Meredith (1971); Walsh and Brown (1973); Kally (1980). It is in fact only a systematic way of selecting an optimum program from a series of events and does not involve any mathematics. The technique may be used to select the most economic diameters of a compound pipe which may vary in diameter along the length depending on grades and flows. For instance, consider a drain fed by a number of inlets. The diameter of the main is increased as input takes place along the line. The problem is to select the most economic diameter for each section of pipe.

A simple example demonstrates the use of the technique; consider the line in Fig. 4.4. Two inlets feed stormwater to a drain, and the hydraulic gradient, assuming pipes flow full, should not be above ground level. The elevations of each point and the lengths of each section of pipe are indicated. The cost of pipe is 10 cents per millimetre per metre length of pipe. The analysis will be started at the upstream end of the pipe (point A). The most economic arrangement will be with minimum cover (zero say), at point A. The depth at point B may be anything between 13m and 31m above the datum, but to simplify the analysis, we will only consider three possible heads with 5m increments between them at points B and C.

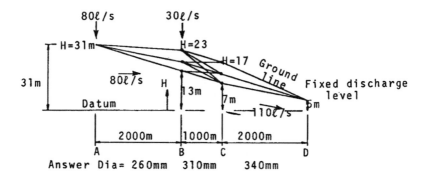

Figure 4.4 Profile of drainpipe optimised by dynamic programming

The diameter of the pipe between A and B, corresponding to each of the three allowed head may be determined from a head loss chart and is indicated in Table 4.1 (I) along with the corresponding cost.

The number of possible hydraulic grade lines between B and C is 3 x 3 = 9, but one of these is an adverse gradient so may be disregarded. In Table 4.1 (II), a set of figures is presented for each possible hydraulic grade line between B and C. Thus if H_B = 23 and H_C = 17m, then the hydraulic gradient from B to C is 0.006 and the diameter required for a flow for 110 ℓ/s is 310 mm. The cost of this pipeline would be 0.1 x 310 x 1 000 = $31 000. Now to this cost must be added the cost of the pipe between A and C, marked with an asterisk. It is this cost and the corresponding diameters only which need be recalled when proceeding to the next section of pipe. In this example, the next section between C and D is the last and there is only one possible head at D, namely the discharge level.

In Table 4.1 (III), the hydraulic gradients and corresponding diameters and costs for Section $C–D$ are indicated. To the costs of pipe for this section are added the costs of the optimum pipe arrangement up to C. This is done for each possible level at C, and the least total cost selected from Table 4.1 (III). Thus the minimum possible total cost is $151 000 and the most economic diameters are 260, 310 and 340 mm for Sections A-B, B-C and C-D respectively. It may be desirable to keep pipes to standard diameters, in which case the nearest larger standard diameter could be selected for each section as the calculations proceed. Or each length could be made up of two sections; one with the next larger standard diameter and one with the next smaller standard diameter, but with the same total head loss as the theoretical result. Alternatively, one could select head intervals to result in real diameters.

Table 4.1. Dynamic programming optimisation of a compound drain

I

HEAD AT B	HYDR. GRAD.	DIA. mm	COST
H_B	h_{A-B}	D_{A-B}	COST $
13	.009	250	50000
18	.0065	260	52000
23	.004	300	60000

II

$H_C =$	7			12			17		
H_B	h_{B-C}	D_{B-C}	COST $	h_{B-C}	D_{B-C}	COST $	h_{B-C}	D_{B-C}	COST $
13	.006	310	31000 50000 81000	.001	430	43000 50000 93000	-	-	-
18	.011	270	27000 52000 79000*	.006	310	31000 52000 83000*	.001	430	43000 52000 95000
23	.016	250	25000 60000 85000	.011	270	27000 60000 85000	.006	310	31000 60000 91000*

III

H_C	h_{C-D}	D_{C-D}	COST $
7	.001	430	86000 79000 165000
12	.0035	340	68000 83000 151000*
17	.006	310	62000 91000 153000

It will be seen that the technique of dynamic programming reduces the number of possibilities to be considered by selecting the least-cost arrangement at each step. Of course, many more sections of pipe could be considered and the accuracy would be increased by considering more possible heads at each section. A computer may prove useful if many possibilities are to be considered, and there are standard dynamic programming programs available.

4.3 DESIGN OF SEWERS

The design of gravity and rising sewage mains poses problems not encountered elsewhere. Special types of pipes of pipes are required to resist corrosion, and particular attention has to be paid to the laying of pipes. Design criteria to prevent deposition, aeration or erosion are largely empirical. This section serves to describe some of the differences between sewers and other pipelines. Full coverage of sewerage is given elsewhere (Escritt, 1972; White, 1978; Bartlett, 1970).

Sewers are designed to carry sewage and wastewater from residential and industrial developments. Stormwater is frequently conveyed in a separate system although a certain amount of infiltration and stormwater ingress may occur in any system.

The volume of domestic sewage normally corresponds to the water consumption and the per capita water use may vary from 100 litres per day in undeveloped countries to 500 litres per day in developed cities, with an average value of 200 litres per day. Peak daily flows from towns may be two or three times the daily average, but this is attenuated in large sewerage systems with travel times over a few hours. On the other hand, the sewers at the head of the system may take peaks ten or twenty times the daily average. Ground water infiltration through cracks, joints and manholes may amount to up to 1/3 of the average dry weather flow.

It is common practice to design sewers to run half full at peak dry weather flow, the remainder of the capacity being reserved for stormwater inflow. The allowance of 50% of the sewer capacity for illegal stormwater inflow is conservative in well planned and constructed systems, and sometimes up to 60% of the sewer depth in used for dry weather flow to reduce the cost.

Sewers are graded to maintain the velocity at peak dry weather flow at least 0.6 m/s and preferably 0.75 m/s to prevent deposition of solids. If 0.75 is used, the velocity during early years when the sewer may be running under-capacity will be reasonably high too. Sewer flushing (Watson, 1937) could be done with fire hydrants, or peak flows. Grading is normally designed to keep sewers shallow and excavation to a minimum.

Rule of thumb grades are used for the upper reaches of the sewerage system and the engineer should refer to the relevant by-laws. House drains which may be 100 mm dia. are restricted to minimum grades of about 1/100, depending on the number of connections feeding into it. Larger sewers should be designed using the flow formulae such as that of Manning.

Maximum grades are also imposed on sewers to prevent liquids running away from the solids in the sewage. Velocities should be kept less than about 2.5 m/s which limits the maximum grade. On steep ground, it may therefore be necessary to construct drops or ramps in the sewer before manholes. Provision should be made in this case for rodding the sewer, by providing a rodding eye at the level of the sewer before the drop.

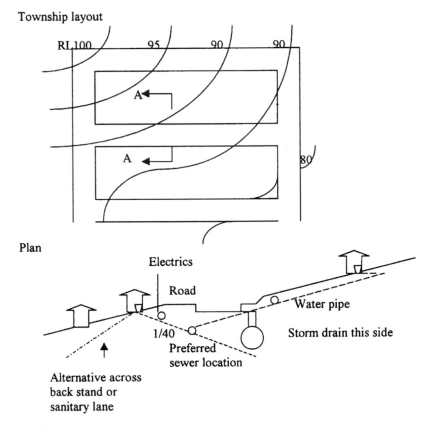

Figure 4.5. Township layout plan and section.

The notion that high velocities in sewers led to erosion was disproved by CIRIA (1968). The rate of erosion is largely independent of fluid velocity, although it is a function of the grit load. Erosion at bends and discontinuities is sometimes a problem though. Tests indicate that pitch-fibre is least resistant to erosion, followed by asbestos cement, concrete, clayware, PVC and cast iron, in that order.

In planning townships, the location of sewers and storm drainage become critical if cross fall is steep (see Fig. 4.5). The depth of sewer at any connection is dictated by the minimum depth and can fall from the house, which could have a toilet at its remote corner.

For household plumbing, there should be traps (u tubes) to seal gases, and at the top end, a vent pipe or vacuum breaker to prevent sucking in seal or trap water. U-tubes also act as solids traps for grit, but open air inlets require screens and larger traps, particularly in poorer areas, where there are no other disposal means for pot scrapings, etc.

4.3.1 Sewer flows

The peak inflow rates into sewers are generally higher than the water supply rate to the building. This is because of rapid releases, e.g. due to cistern, baths, discharging. Hence, the upper reaches of sewers are considerably bigger than water pipes (e.g. 100 mm vs. 12 mm for a house). But there are other reasons for bigger sewer pipes, namely, ease of cleaning by rodding, lower gradients due to free surface flow, and necessity for higher minimum velocities to maintain solids in suspension.

Design flows from houses are typically 1 liter per minute to 1,5 litre per minute (depending on township standards) peak dry weather flow. There is a reduction factor for large towns due to out of phase contributions. Other design flows are:

Schools	45 l/d/p over 6 hr
Hotels, hospitals	200 l/d/p over 8 hr
Industry	10 l/gross ha/min, with special consideration for 'wet' industries

In addition, there is external water ingress and leakage to cope with (e.g. 0.1 l/min/house) and stormwater inflow (up to 1% of the precipitation). Chapter 5 discusses these problems, but the standard is to allow 30-50% of sewer capacity for these.

4.3.2 Construction

Sewers are normally designed structurally to resist external loads only, although for large diameter sewers the self-weight and weight of water may be considered.

Salt glazed earthenware, high density polyethylene, UPVC, pitchfibre, asbestos cement and cast iron pipes are used for diameters up to 300 mm or even 450 mm, while for larger diameters, concrete or asbestos cement sewers are stronger and more economical. The ability to resist corrosion and algal growth and encrustation are also important considerations in the selection of the type of pipe. Steel is seldom used for sewers as it corrodes and is expensive. Mortar lined steel pipe may, however, be used where the pipe has to span between piers across a valley.

Although most sewers nowadays are made circular in cross section, many large sewers used to be constructed with horseshoe, U or egg-shaped cross sections. The merits of various shapes were discussed above, but generally the advantage of circular shapes, especially cost and ease of laying, are accepted.

Large sewers used to be constructed in cast iron or precast concrete segments, or with tiles or vitrified bricks. Modern construction tends to use jointed precast concrete pipes or placed in-situ reinforced concrete, both of which can most easily and economically be constructed circular.

Although the general methods of pipe laying outlined by Stephenson (1979) are applicable, special precautions are necessary in laying sewers. It is important that the pipe is laid to the true grade, and that jointing is done as neatly as possible. Special attention will have to be paid to soil loads and excavation methods, as sewers are frequently laid very deep. Unlike pressure lines, sewers have to be laid at a prefixed gradient, often against the natural ground gradient, resulting in deep excavations.

Sewer pipes are frequently laid on a concrete or granular bed to minimise settlement. The trench is excavated in the invert level of the pipe, and pegs are driven in on the centre-line of the pipe, at a spacing not exceeding one pipe length and with the peg level at the invert level of the pipe. To obtain the correct levels, sight rails are placed across the trench at manholes and at intermediate points. The sight rails may be offset if the excavation is done mechanically. Boning rods are used to measure the height from the line of the sight rails to the sewer invert level.

After pegging the bottom of the trench, the bottom is hand trimmed to just below the bottom of the pipe barrel if it is to be laid on the ground, with joint holes at the sockets. If the pipe is to be laid on a concrete bed or bench, the trimming is taken down an additional 75 mm. The concrete bed is cast leaving holes for pipe sockets and the pipe laid on a mortar bed on the concrete screed. Alternatively, the pipe sockets may be laid on the concrete screed, and the pipe barrel supported with wedges until joints have been made. The pipes are apt to move during jointing if the latter method is used though.

The pipes are laid from downhill end, with sockets facing uphill. After the spigot end has been forced into the socket and caulked up, a concrete haunching to the pipes may be poured on the screed, up to $\frac{1}{4}$, $\frac{1}{2}$ or even the full height of the pipe. Any mortar or jointing material which may have penetrated the pipe during jointing should be raked out.

The sewer should be tested for watertightness after construction. The ends of a section are bunged closed and water pumped in through a cock on one of the bungs. Concrete sewers are allowed to stand thus for at least 48 hours to let the walls absorb water. The amount of leakage should then be measured by noting the additional amount of water required to be pumped in to maintain the level in a stand-pipe constant over say 24 hours. The leakage should normally be less than 350 litres per hour per metre diameter per kilometre of pipe. Testing with compressed air and a manometer is also used. It is usually more economic and practical than testing with water. If there is leakage, it can be detected by adding dyes into the water, or using pungent gas.

In routing a sewer, it should as a rule follow the contour, falling slightly to maintain the grade. Where valleys are encountered, the sewer should be laid around them, or across them on piers. In towns, the sewers should follow the roads (see Fig. 4.5) or be laid in servitudes. In some cases, the sewer may gravitate to the bottom of the valley where a pumping station would be required to lift the sewage up the next rise. Pumpstations for sewage are problematic.

Where the pipe is laid across a bridge or piers, mortar lined steel, or cast iron pipe is preferred. Spans should be short to minimize sagging. Access tees may have to be provided at intervals. Inverted siphons through dips are not favoured as they are susceptible to settling of solids, especially during low flows. To avoid low velocities, a flushing tank and priming siphon may be used (Fig. 4.6), although the arrangement wastes a certain amount of head. During low inflow, the inverted siphon will not flow until the flushing tank is empty, and the process is repeated. The capacity of the flushing tank should be at least equal to the capacity of the inverted siphon to ensure the pipe is completely flushed each cycle. The diameter of the siphon pipes will depend on the head available. A similar system is sometimes used at wastewater treatment works filters to maintain circulation of rotating sprinklers.

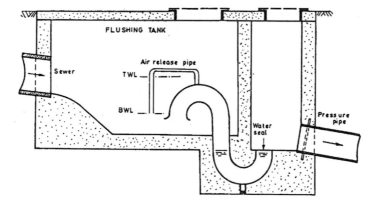

Figure 4.6. Flushing siphon for valley crossing

4.3.3 Access and ventilation

Sewers should be well ventilated to remove corrosive, poisonous or explosive gases and to prevent air locks. The vent pipes provided at the head of domestic connections are normally sufficient for reticulation systems. Outfall sewers may need ventilating manhole covers, or preferably ventilation columns which disperse the gases into the air and avoid nuisance smells.

Manholes are constructed at a maximum spacing of 90 m for 150 mm diameter sewers, increasing to 300 m for 1 m diameter and larger sewers. They provide ventilation points, and permit access for clearing blockages by rodding or flushing. The sewer is laid dead straight between manholes, although sewers larger than about 750 mm diameter are nowadays sometimes laid with large-radius bends. The alignment can be checked after construction with mirrors, or nowadays with laser instruments.

Manholes are normally constructed at changes in direction or grade and at junctions. An exception to the last is house connections which may connect in with an oblique junction with an access eye.

Although manholes used to be made of brick, nowadays precast concrete annular rings are used for simple manholes, while special manholes are cast in concrete or even PVC. The floor is made of mass concrete with a 25 mm granolithic screed. A pre-formed half-round channel may be set in the floor or else it is formed in the concrete. The floor should be benched towards the channel at the level of the soffit of the pipe. Bends and junctions are swept (radius to centre line twice the sewer diameter) and space for standing should be provided to one or both sides of the channel. Soffits of pipes of unequal diameter are set on the same line. The floor channel should be sloped at the same grade as the sewer and an additional drop of 25 mm should be allowed at bends over 45°.

Manholes should be at least 1 000 mm dia for sewers under 300 mm dia, and 1 200 mm for larger sewers, and at bends, junctions and drops of smaller sewers. Sewers over 750 mm dia often have manholes with even larger dimensions, cast in-situ. Deep manholes may have a reduced diameter shaft (about 700 mm inside) from 2 m above the top of the benching (see Fig. 4.7). Cast iron step-irons are set into the wall on the side of the manhole cover. In the case of large sewers, care should be taken in positioning the manhole cover and safety chains should be put across the downstream sewer.

Spacing of manholes should be such that the sewers can be inspected and cleaned by rodding. On larger sewers, gentle curves are possible. Distances of 100 m for small sewers to 200 m for large sewers are common.

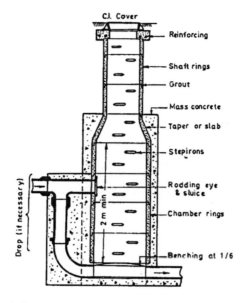

Figure 4.7. Sewer manhole.

4.3.4 Computer Design and Grading

Sewerage networks or mains are well suited for automatic design by computer
and drawing of sections by a linked plotter. A sewerage network is designed
largely on empirical bases which can be readily programmed as constraints. The
fact that a sewerage network is open ended and not pressurized like a water
reticulation network, considerably simplifies optimization by dynamic or linear
programming.

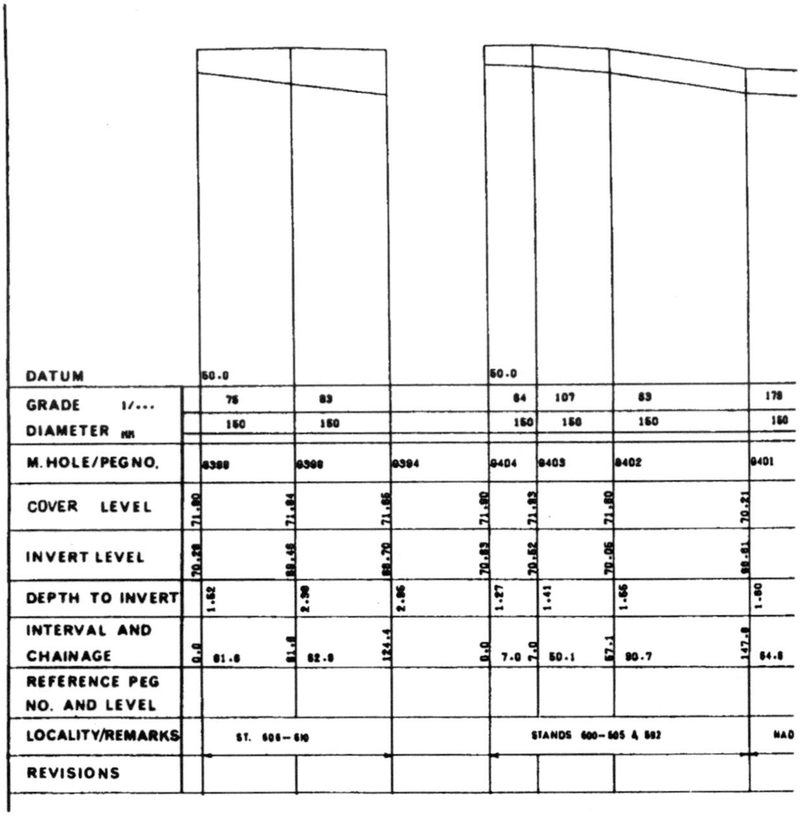

DATUM	60.0			60.0			
GRADE 1/...	75	83		84	107	83	178
DIAMETER мм	160	160		160	160	160	160
M.HOLE/PEG NO.	8380	8388	8394	8404	8403	8402	8401
COVER LEVEL	71.80	71.84	71.86	71.80	71.83	71.80	70.81
INVERT LEVEL	70.28	69.48	69.70	70.83	70.82	70.06	68.81
DEPTH TO INVERT	1.52	2.38	2.06	1.27	1.41	1.66	1.80
INTERVAL AND CHAINAGE	0.0 81.8	81.8 82.8	124.4	0.0 7.0	7.0 50.1	57.1 90.7	147.8 84.8
REFERENCE PEG NO. AND LEVEL							
LOCALITY/REMARKS	ST. 606–610			STANDS 600–605 & 602			NAO
REVISIONS							

Figure 4.8. Computer plot of graded sewer

Although it is possible to start from data from a contour plan or GIS map and develop the plan layout within the computer, it is preferable to lay the network out on a plan, peg and survey it in the field and feed the data from pegging sheets directly to the computer. A suitable program will then select the most economic depths and pipe diameters. The data may be displayed in summary form for visual inspection and adjustment if desired, taking off quantities. The results are then submitted to a separate plotting routine for drawing longitudinal sections. Such a plot is illustrated in Figure 4.8.

4.4 REFERENCES

ASCE and WPCF (American Society of Civil Engineering and Water Pollution Control Federation) (1969) *Design and Construction of Sanitary and Storm Sewers*, NY, 332pp.

Bartlett, R.E. (1970) *Sewerage*, Elsevier, London.

Camp, T.R. (1946) Design of sewers to facilitate flow. *Sewerage Works Jnl*, **18**(3).

CIRIA (1968) *Erosion of Sewers and Drains*, London.

CP 2005 (1968) *Sewerage Design*. BSI, London/

Dajani, J.S. and Hasi, Y. (1974) Capital cost minimization of drainage networks. *Proc. ASCE*, **100**(EE2), 10448, p325-337, April

Escritt, L.B. (1972) *Sewerage and Sewage Disposal*, Macdonald and Evans, London.

Harris, G.S. (1970) Real time routing of flood hydrographs in storm sewers. *Proc. ASCE*, **96**(HY6), 7327, p1247-1260, June.

Kally, E. (1980) Automatic computerized computation of an optimal sewerage network by dynamic programming. *Water Services*, **84**(1018), p723-725, Dec.

Meredith, D.D. (1971) *Dynamic programming with case study on the planning and design of urban water facilities*. Treatise on Urban Water Systems. Colorado State Univ., Fort Collins, p599-652.

Merritt, L.B. and Bogan, R.H. (1973) Computer based optimal design of sewer systems. *Proc. ASCE*, **99**(EE1), 9578, p35-53, Feb.

Mussalli, Y.G. (1978) Size determination of partly full conduits. *Proc. ASCE*, **104**(HY7), 13862, p959-974, July.

Stephenson, D. (1989) *Pipeline Design for Water Engineers*. Elsevier, Amsterdam, 3rd ed.

Walsh, S. and Brown, L.C. (1973) Least cost method for sewer design. *Proc. ASCE*, **99**(EE3), 9796, p333-345, June.

Watson, J.D. (1937) Sewer flushing. *Water Works and Sewerage*. Aug.

White, J.B. (1978) *Wastewater Engineering*, Arnold, London.

Yen, B.C. and Sevuk, A.S. (1975) Design of storm sewer networks. *Proc Am. Soc. Civil Eng.*, **101**(EE4), p 535-553, Aug.

5

Sewer leakage and rehabilitation

5.1 STORMWATER AND GROUNDWATER INGRESS

Stormwater and groundwater infiltration accounts for large peak flows (up to 3 times the AADWF) experienced in sewers. The source of this infiltration can be attributed predominantly to household stormwater being directed into the sewer system through gulleys, and to a lesser extent, due to missing or damaged manhole covers (see Ainger et al, 1998; Metcalf and Eddy, 1979; White, 1987). Ground water infiltration produces a steady base flow in the sewers, which increases treatment costs. Both surface and subsurface ingress reduce the operating capacity in downstream sewers.

The main causes of stormwater inflows into the sewer systems have been identified as follows:

- Manholes without covers
- Stormwater systems connected directly to sewer reticulation system
- Sewer gulleys below ground level or rainwater downpipes connected to gulleys
- Broken/cracked sewer pipelines
- Infiltration through benching in manholes
- Infiltration through pipe/channel connections at manholes

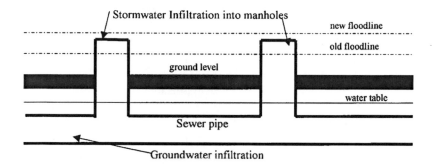

Figure 5.1. Effects of urbanization on existing infrastructure.

5.2 PROBLEMS IN WASTE WATER COLLECTION SYSTEMS

Sewer pipeline stoppages and collapses are a health hazard and structurally costly problem. Roots that penetrate joints and grow inside pipes cause over 50% of stoppages, while a combination of roots, corrosion, soil movement and inadequate construction are the cause of most of the structural failures.

Besides stoppages and collapses, infiltration and inflow (I/I) can reduce capacity of a sanitary sewer system and negatively affect operation of the entire sewerage system. I/I are generally defined as stormwater that enters the collection system through direct connections (e.g. roof leaders, cellar and yard area drains, foundation drains, commercial and industrial the so-called "clean water" discharges, drains from springs and swampy areas, etc.). Infiltration is defined as water that enters from the ground. In the late 1980's, the term "rainfall-induced infiltration" was first used to describe infiltration with flow characteristics resembling inflow (i.e. a rapid increase in flow which coincides with a rain event). Rainfall-induced infiltration results when stormwater runoff causes a rapid groundwater recharge around sewers, including manholes and building connections, which then enters the system through defective pipes, pipe joints, or manhole walls (see Li and McCorquadale, 1999; Reynolds, 1995; Sinske and Zietsman, 2002)

I/I can greatly increase flows and cause unnecessary burdens on the treatment plant. In some systems, sewer lines become overloaded and uncontrolled sanitary sewer overflows (SSOs) occur, since most sanitary sewers do not have overflow structures designed into the system. SSOs often occur through manholes and defective lines in residential neighbourhoods causing backups into homes and streets. I/I also causes surcharging of

wastewater treatment plants and pumping stations. In comparison to combined
sewer overflows, SSOs generally contain higher percentages of raw sanitary
sewage (which can contain high levels of infectious (pathogenic)
microorganisms, suspended solids, toxic pollutants, floatables, nutrients,
oxygen-demanding organic compounds and faecal matter, oil and grease, and
other pollutants) and a lower-level percentage of stormwater. SSOs represent a
human health risk because of pathogenic organisms.

Old sewer systems using mortar or mastic jointing materials, can
substantially contribute to exfiltration as well as to I/I. Exfiltration can occur
when the elevation of the sewer liquid level is above the groundwater table.
This positive elevation head can cause raw sewage to exfiltrate through open
joints to ground water and other areas of the environment. Stoppages and
collapses can cause flows to exit the sewerline and concentrate in the trenches
dug for the sewerlines or other utility lines. The flow is then conveyed by
gravity along the trench to surface water or the flow can infiltrate into the
ground and threaten drinking water supplies. Loose joints and deteriorated
pipes are the main sources of egress. House lateral connections to street
sewers are other prime points of leakage. Exfiltration from both combined and
sanitary sewers can be a substantial source of pollution in terms of impact on
groundwater and surface water quality.

5.3 DETERMINING EXTRANEOUS FLOWS

5.3.1 Infiltration into sewers

Sewers usually follow the watercourses close to (and occasionally below) the
beds of streams. As a result, old and collector sewers may receive
comparatively large quantities of groundwater, whereas sewers built later at
high elevations will receive relatively small quantities of groundwater. With
an increase in the percentage of area in a community that is paved or built
over, comes (1) an increase in the percentage of stormwater conducted rapidly
to the storm sewers and watercourses, and (2) a decrease in the percentage of
the stormwater that can percolate into the earth and thus tending to infiltrate
the sanitary sewers

The rate and quantity of infiltration depends on the length of the sewers,
the area served, the soil and topographic conditions, and the population
density. Although the elevation of the water tables varies with the quantity of
rain percolating into the ground, the leakage through defective joints, porous
concrete and cracks has been large enough, in some cases, to lower the
groundwater table to the level of the sewer.

The amount of groundwater flowing from a given area may vary from a negligible amount for a highly impervious area or an area with a dense subsoil to 25 or 30 percent of the rainfall for a semipervious area with a sandy subsoil permitting rapid passage of water. The percolation of water through the ground from rivers or other bodies of water sometimes has a considerable effect on the groundwater table, which rises and falls continually.

The amount of flow that can enter a sewer from groundwater or infiltration may range from 0,01 to 1,0 m³/s mm-km or more.

Expressed another way, infiltration may range from 0,2 to 20 m³/ha/d. During heavy rains, when there may be leakage through manhole covers, or inflow, as well as infiltration, the rate may exceed 500 m³/ha/d. Infiltration/inflow is a variable part of the wastewater depending on the quality of the material and workmanship in constructing the sewers and building connections, the character of the maintenance and the elevation of the groundwater compared with that of the sewers.

5.3.2 Inflow into sewers

Direct inflow can cause an almost immediate increase in flow rates in sanitary sewers. The effects of inflow on peak flow rates that must be handled by a wastewater treatment plant would be up to 5 times higher than average dry weather flow (Stephenson, 1982 and 1985, APWA, 1970; Civil Eng. Ref. Book, 1995).

Inflow rates are usually determined by using a network of continuous flow meters that operate before and during a significant storm. The inflow rate can be determined from the flow hydrographs by subtracting the normal dry weather domestic and industrial flow and the infiltration (including steady flow) from measured flow rate.

5.3.3 Determination of ingress events

When the infiltration and inflow rates are determined for each sub-area, it will usually be found that some small areas of a collection system may contribute significantly to the infiltration/inflow. It can be that about 75 percent of the inflow comes from 20 to 30 percent of the system, whereas 75 percent of the infiltration comes from 40 percent of the area.

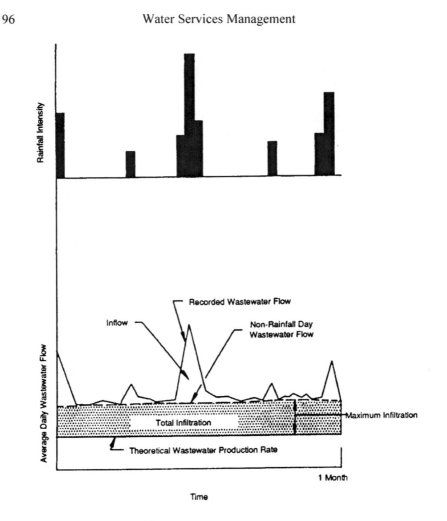

Figure 5.2. Representation of ingress events

Total inflow is the sum of the direct inflow at any point in the system plus any flow discharged from the system upstream through overflows, pumping station bypasses and the like.

Delayed inflow is stormwater that may require several days or more to drain through the sewer system. This category can include the discharge of sump pumps from cellar drainage as well as the slowed entry of surface water through manholes in ponded areas.

5.4 IMPACT OF STORMWATER AND GROUNDWATER INGRESS

5.4.1 Wastewater Treatment Works (WWTW)

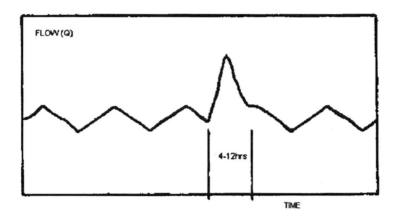

Figure 5.3a: Flow pattern for short duration of infiltration

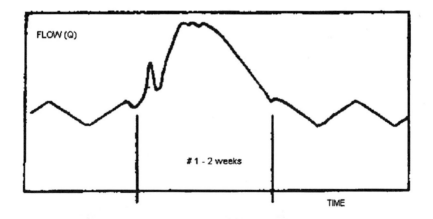

Figure 5.3b: The flow pattern due to excessive infiltration.

Treatment works are often designed for short duration of infiltration, about 15% of the Dry Weather Flow (DWF), which is normally caused where gullies are channeled into the sewers on properties. The flow pattern with this type of infiltration is shown in Figure 5.3a and b and the works are able to cope without any negative effects on effluent quality.

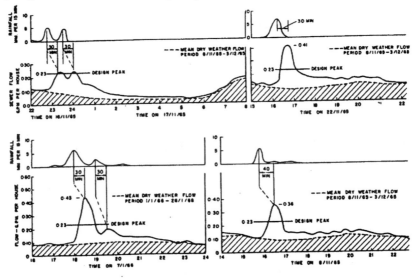

Figure 5.4. Influence of rainfall on sewer flow – Bloemfontein (Crabtree, 1970)

The effect on the flow patterns due to change in infiltration at the treatment works as explained previously, is shown in Fig. 5.4. Due to higher stormwater and groundwater infiltration, the peaks of infiltration are higher and of a much longer duration. The effect on treatment works is dramatic such that the effluent does not comply with the effluent standards applicable (Skipworth et al, 2000; Ashley, 2000).

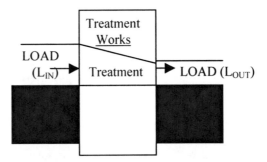

Figure 5.5. Effective treatment under normal flow conditions

Under normal flow conditions, the treatment works receives a flow with a pollution load. After treatment the effluent quality will be reduced to comply with the effluent standards and no pollution of the receiving water stream will take place as illustrated in Fig. 5.5 above.

When infiltration occurs, an additional volume of water flows through the treatment works reducing the retention time. This will reduce the effectiveness of the biological process leading to a higher load leaving the works as illustrated below in Fig. 5.6.

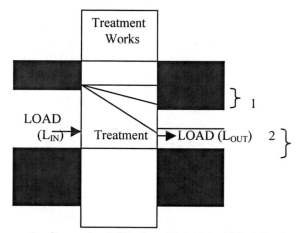

Figure 5.6. Overloading as a result of excessive infiltration. (1) Additional load leaving system due to poor treatment not complying with effluent standards. (2) Normal effluent after treatment complying with effluent standards.

This additional load leaving the works is normally worse than the effluent standards and needs to be addressed. In terms of legislation, it is a duty to ensure that no environmental pollution or damages takes place and the load in excess of the effluent standards must be removed. The pollutants in Table 5.1 are measured in terms of Effluent Standards and must be addressed by either chemical treatment or increased plant capacity.

Table 5.1. Different pollutants and their resulting effluent standards

Pollutant	Effluent standard
SS (Suspended Solids)	Retention time, i.e. treatment capacity for settling
COD (Chemical Oxygen Demand)	Retention time, i.e. treatment capacity for biological action
N (Nitrates)	Retention time, i.e. treatment capacity for biological action and adsorption
P (Phosphates)	Chemical treatment, ferric chloride
E. Coli	Chemical treatment, chlorine

The hydraulic components of wastewater treatment works need to be enlarged if inflow is to be allowed for. The cost of pipework and hydraulic components can be 25% of the total cost and a 50% increase in capacity could increase those costs 10%. In addition, settling basins, grit channels and screens need enlarging. The effect on aeration, filters, digesters and disinfection is less as the load is not increased much.

5.5 REDUCING STORMWATER INFLOW

Methods aimed at attenuating or reducing inflows into a sewerage system are listed below. Many are widely recognised as source control techniques or so-called best management practices (BMPs) (see also Duchesne et al, 2001; Ermolin, 2001; Geustyn, 2002). Examples of these techniques include:
- attenuation of roof runoff
- disconnection of paved areas and/or roof areas
- reductions in the numbers of gullies and stormwater inlets
- control of inflow rates at gullies
- identification and correction of malconnections

5.5.1 Local structure improvements

Surcharging in lengths of sewer can sometimes be caused by local restrictions in sewers or manholes. Examples include lateral connections that project into a main sewer and thereby cause blockages, or poor benching in manholes that produces extra head losses or backing up of flow in side branches. Once these types of restriction have been located, they can usually be remedied at a relatively low cost.

5.5.2 Effective utilisation of existing storage in sewerage systems

Options for making better use of existing storage in sewerage systems tend to be specific to the particular circumstances but can either be passive or active. An example of a passive type is the addition of flow control devices in the upstream part of a system to make use of unused storage in manholes and thereby reduce peak flows farther downstream. Lowering of velocities and deposition needs to be considered.

Active solutions interact between flow conditions and the operation of equipment such as pumps, gates and off-line storage tanks. The interaction may be achieved through the application by operations staff of written rules based on past experience, perhaps supported by analysis of the behaviour of the system using a hydraulic model. Alternatively, the operating rules may be implemented automatically by means of electronic links between the flow control equipment and sensors located at key points in the system. In the next generation of solutions, real-time control may become an option, with a computer model forecasting flow conditions in the sewerage system and evaluating alternative strategies for operation of the control equipment; this type of option will normally tend to be applicable only on a catchment-wide basis.

5.5.3 Pro-active maintenance

This is maintenance work carried out in a planned way at key points in a sewerage system to ensure that the hydraulic capacity is not reduced by blockages or by the build-up of sediment deposits or excessive sliming in the pipes. Monitoring of the results of the work must be carried out to determine its effectiveness and, if necessary, to adjust the frequency of cleaning.

5.5.4 Overflows or redirection of flows within a system

External overflows, bifurcations or diversions can be located at points of hydraulic overloading to remove excess flows from sewerage systems. External overflows are generally designated as combined sewer overflows (CSOs), and entail discharge of excess flow to a surface water sewer, land drain or watercourse. Diversions, together with bifurcations, are used to divert excess flows either into another part of the same system with spare capacity or into another adjacent system (see Hoangnam, 1982; Horvarth, 1994; Larson, 1979).

5.5.5 Holding ponds

Storm peak overflow facilities can be provided before the Head of Works plus elsewhere on the plant (i.e. from balancing tanks). These overflows should flow to two containment dams and be recycled back to the inlet of the works when the storm flows have subsided. These dams can also be used for unavoidable overflows from tanks and sumps and polluted runoff (storm or wash-down water) from the works. The dams must be designed so that they can be periodically drained, dried out and cleaned (desludged) during the dry season. The outlets of the dams should be fitted with scum baffles.

   They require more attention than clear stormwater ponds as infiltration and odours require control. Skimming and sludge abstraction methods need consideration.

5.5.6 Enhancement of sewer network

Together with the provision of storage and construction works to increase sewer capacity, this is the most commonly used method of solving flooding problems. Existing systems are replaced or enhanced to remove the hydraulic restriction(s) that cause the sewer flooding problems. In some cases this may be achieved by replacing an existing length of sewer by one of high flow capacity (i.e. having a larger diameter or smaller hydraulic resistance). Alternatively, a length of by-pass sewer may be constructed to carry some of the flow from the existing sewer over the section where it has insufficient capacity. There is a significant risk that works to improve conditions at one location may transfer the problem farther downstream unless equivalent improvements are made all the way through the system.

Flows upstream of a critical part of the system are restricted to the capacity of the pipes by controlling and storing excess flows until the system can cope. Peak flows may be attenuated by providing purpose-built storage, usually in the form of on-line or off-line detention tanks, or by temporarily holding back surface water run-off (e.g. in detention ponds or in open areas such as car parks). Flow control devices are used to control the onward flow to the downstream part of the system and/or to divert flows into storage.

5.6 ANCILLARY SYSTEMS

5.6.1 Pumping systems

Pumping systems for dealing with sewer-flooding problems can be considered in three categories of size and complexity:

(1) Packaged systems, consisting of pumps and storage chambers that can take gravity flow from a group of properties and, if necessary, pump it under pressure into a surcharged public sewer (either directly or via a gravity pipe).

(2) Intermediate-size pumps installed in inspection chambers that can discharge flow from a single property or basement into a surcharged sewer (either directly or via a gravity pipe).

(3) Small macerating pumps with small-bore pipework that can discharge flow under pressure from individual units such as baths, sinks, WCs and showers. Higher-rated units are able to pump from below ground level into surcharged sewers, either directly or via a gravity pipe.

Pumps used for categories (1) and (2) are normally either small submersibles with macerators or grinder pumps.

5.6.2 Vacuum systems

These systems transport sewage by inducing and maintaining a vacuum in the collecting pipes by means of central vacuum pumps and a reservoir. Conventional gravity drains connect one or more properties to a sewage collection chamber. When the sewage reaches a preset level, a pneumatic "interface" valve opens and the contents of the chamber are sucked into the vacuum line. When the chamber is almost empty, the valve closes.

Vacuum systems should normally deal only with domestic wastewater because satisfactory performance depends on their being sized accurately in relation to the maximum design rate of flow. These systems should not be expected to cater for overland flow, infiltration or roof run-off, and additional flows should not be added without considering the limitations of the original design.

5.6.3 Protective structures

A wall or bund can be constructed around a single property, or a group of properties, to offer protection from sewer flooding, but this is only used in specific circumstances. However, bunds have been placed around manholes in foul sewerage systems as a temporary measure to minimise the extent of flooding at low points in gardens and open spaces.

5.6.4 Purchase of land/properties

This involves the sewerage undertaker purchasing a property either to remove it from a list of occupied properties affected by flooding or to change its usage so as to mitigate the effects of flooding. In some cases where flooding is confined to an occupied basement, only a change of usage for that part of the building might need to be negotiated. In all cases, the proposed course of action should not be imposed on the owners of the properties but must be carried out by mutual agreement; it is only likely to be adopted in exceptional circumstances.

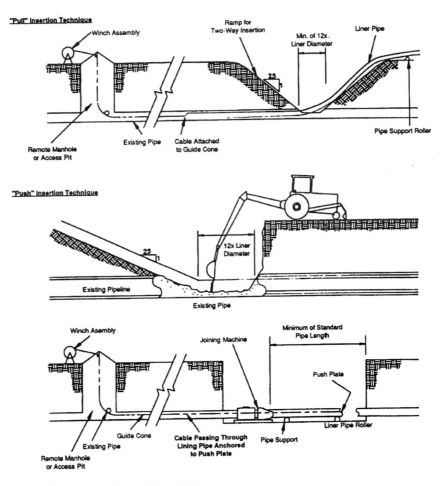

Figure 5.7. Insertion methods (EPA, 1991)

5.7 MANAGEMENT OF INGRESS

There are a number of methods that can be used to prevent stormwater and groundwater infiltration into sewers. These methods can generally be classified as being either remedial or preventative in nature. Remedial methods are those methods that tend to accommodate infiltration and inflow, e.g. by building storage tanks at waste water treatment plants (WWTP), etc. Preventative methods on the other hand tackle the problem at the source and aim to eliminate infiltration instead of accommodating it. This section investigates the economics of the various methods of controlling inflow of groundwater and stormwater into sewers. It also provides a basis upon which to decide whether to solve or ignore inflow problems.

5.7.1 Preventative measures

There are various methods that can be used to mitigate groundwater infiltration and stormwater ingress into sewers instead of remedying it. These methods are summarised below:

Sealing of sewers. The sealing of sewers involves sealing cracked sewer pipes and manholes to protect them from groundwater infiltration. It also involves realigning sewers that have become misaligned and sealing other defective joints.

Replacing missing or broken manhole covers. Missing or defective manhole covers provide an access for stormwater to enter into sewer systems. Missing manhole covers on pavements are of concern because most of the surface runoff during heavy storms finds access into the sewer system through manholes that are not covered.

Raising manholes above flood lines. Manhole levels are designed to be above flood lines so as to stop stormwater infiltration. The increased flood lines due to urbanization are in some instances higher than the manholes, thus increasing the amount of stormwater ingress through manholes. Raising manhole levels with increasing flood lines significantly eliminates the amount of stormwater ingress into sewers.

Training buildings and plumbers. Builders and plumbers can be trained to familiarize them with combined and separate sewer systems and to prevent households from diverting flow from household stormwater drains into sewers.

Regulatory measures. The last method would be to enforce laws that prevent the diversion of stormwater gullies into sewers. Offenders must be liable to a fine and/or imprisonment.

Table 5.2. Summary of costing preventative measures

Preventative measure	Description of problem	Estimated cost
Sealing of sewers	Remedial measures by rehabilitation methods	See Table 5.3
Replacing missing or broken manhole components	Covers only	$ 50
	Covers and frames	$100
	Covers and cover slabs	$150
	Reconstruct manhole	$500
Raising manholes above flooding		
Training of maintenance staff		
Regulatory measures	Policing, etc.	

5.7.2 Remedial measures

There is a growing recognition of the problems of groundwater and stormwater inflow into sewers. The methods that most authorities use to control infiltration are currently remedial and not preventative in nature. Remedial solutions are undesirable because they are merely temporary arrangements and do not solve the problem. Some of these methods involve:

Building holding tanks at WWTPs. This method is the most economical and simplest of the methods to be discussed. Building holding tanks at WWTPs has the effect of levelling high volumes during WWFs. Holding tanks function like balancing reservoirs used in water reticulation systems. They are there to ensure a constant supply rate at all times. This method involves building additional tanks and piping.

Increasing the capacity of WWTP. With this method, the capacity of the wastewater treatment plant is increased rather than trying to control the rate of inflow. The United States of America already has a number of wastewater treatment technologies, which can be integrated to create a WWTP of any desired capacity and treatment rate. However, the cost involved is high.

Increasing the capacity of the sewer system. Once the capacity of the sewer is exceeded due to the inflow of groundwater and stormwater, sewers are re-laid or duplicated to restore acceptable flow conditions.

5.8 ALTERNATIVE METHODS FOR REMEDIATION

5.8.1 Rehabilitation methods

Pipeline rehabilitation methods use the existing pipe either to form part of the new pipeline or to support a new lining (WPCF, 1983; WRC, 1983). Rehabilitation is preceded by cleaning the pipe to remove scale, tuberculation, corrosion and other foreign matter. Linings, to be effective, must make intimate contact with the pipe surface. Proper surface preparation significantly affects the strength and bonding of lining. These methods can be divided into two categories: nonstructural and structural. (See also Grigg, 2003 and Saegrov and Schilling, 2004.)

5.8.1.1 Nonstructural lining

Nonstructural lining involves placing a thin coating of corrosion-resistant material on the inner surface of the pipe. The coating is applied to prevent leaks and increase the service life. However, coating does not increase the structural integrity of the pipe.

Cement mortar lining. Cement mortar linings are unique because they are porous. Corrosion protection is achieved by the development of a highly alkaline environment within the pores, which is a result of the production of calcium hydroxide during cement hydration. Cement mortar is applied using a variety of equipment, depending on pipe size and overall project length. Access to the pipeline is accomplished by excavation and removal of a length of pipe.

Epoxy lining. Epoxy resin lining of water mains is an alternative to cement mortar lining. It has not been widely used in the United States. However, it has been practiced in several other countries, including the United Kingdom and Japan. Epoxy lining has an estimated life in excess of 75 years. Glass fibres can add to the durability.

5.8.1.2 Structural lining

Structural lining involves placing a watertight structure in immediate contact with the inner surface of a cleaned pipe. A variety of technologies are available, including sliplining, cured-in-place pipe, fold and form pipe, and closed-fit pipe lining. This is the only rehabilitation technique that improves the structural integrity of a pipe.

Sliplining. Sliplining is the oldest rehabilitation method. In this process a new pipeline of a diameter smaller than the pipe being repaired is inserted into the defective pipe and the annulus grouted. It has the merit of simplicity and is relatively inexpensive, but there is a reduction in flow capacity (35 to 60%), depending upon pipe size. Excavation is required for insertion and receiving pits. All service connections, valves, bends and appurtenances must be individually excavated and connected to the new main.

Cured-in-place pipe. Cured-in-place pipe (CIPP) involves placing a fabric tube, impregnated with a thermosetting resin that hardens into a structurally sound jointless pipe when exposed to hot circulating water or steam into a cleaned host pipe, using the inversion process described below. Access to the pipeline is accomplished by excavation and removal of a length of pipe. There is no reduction in flow capacity. However, the flow must be completely stopped or by-passed during installation and curing. All service connections, valves, bends and appurtenances must be individually excavated and connected to the new main.

Fold and form pipe. Fold and form pipe (FFP) utilizes thermoplastic materials polyvinylchloride (PVC) or polyethylene (PE) that are heated and deformed at the factory from a circular to a U-shape to produce a net cross section than can be easily fed into the pipe to be rehabilitated. The FFP is fed from a spool into the existing pipe, where hot water or steam is applied until the liner gets heated enough to regain its original circular shape and create a snug fit within the host pipe (Spero, 1999). All service connections, valves, bends and appurtenances must be individually excavated and connected to the new main.

Close-fit pipe. Close-fit pipe lining involves pulling a continuous lining pipe that has been deformed temporarily so that its profile is smaller than the inner diameter of the host pipe. This lining method is often referred to as the modified sliplining approach. Close-fit pipe lining makes use of the properties of PE or PVC to allow temporary reduction in diameter and change in shape prior to insertion in the defective pipe. As with sliplining, excavation is required for insertion and receiving pits. All service connections, valves, bends and appurtenances must be individually excavated and connected to the new main. Close-fit pipe had a design life of greater than 50 years.

5.8.2 Replacement methods

Replacement of pipelines can be accomplished by using either trenchless or open-trench techniques.

Trenchless replacement. Trenchless replacement involves inserting a new pipe along or near the existing pipe without requiring extensive excavation of soil. Trenchless replacement can be done with minimal disruption to surface traffic, business and other activities, in contrast to open trenching. There is a significant reduction of the social costs associated with construction. The best-known trenchless replacement techniques are pipe bursting, microtunneling and horizontal directional drilling.

Pipe bursting. Pipe bursting is a method for replacing pipe by bursting from within while simultaneously pulling in a new pipe. The method involves the use of a static, pneumatic or hydraulic pipe-bursting tool drawn through the inside of the pipe by a winched cable, with the new pipe attached behind the tool. The bursting tool breaks the old pipe by applying radical force against the pipe and then pushes pipe fragments into the surrounding soil. The liner pipe can be the same size or as much as two pipe sizes larger than the existing pipe. Excavation is required for insertion and receiving pits.

Microtunneling. Microtunneling involves the use of a remotely-controlled, laser-guided, pipe-jacking system that forces a new pipe horizontally through the ground. This trenchless method is used for construction pipelines to close (± 1 inch or ± 2.54 cm) tolerances for line and grade. This method can be cost-effective compared to open-cut construction when pipelines are to be installed in congested urban or environmentally sensitive areas, at depths greater than 1,5 ft (0,6 m) in unstable ground, or below the water table. Microtunneling can be used in a variety of soil conditions from soft clay to rock, or even when there are boulders to deal with, and can be used at depths of up to 100 ft (30,48 m) below the water table without dewatering.

Horizontal directional drilling. Horizontal directional drilling (HDD) consists of a rig that makes a pilot bore by pushing a curing or drilling head that is steered and guided from the surface. Drilling fluid is pumped through the drill/push rods and displaces the cut soil. When the pilot bore is completed, pulling back a reamer enlarges the hole. Progressively larger back-reamers are used until the hole is large enough to pull in the pipe. HDD is suitable for installing pipes under waterways, major highways and other obstacles.

Open-trench replacement. Open-trench replacement is the most commonly used method for replacement of water mains; it involves placing new pipe in a trench cut along or near the path of the existing pipe. Open-trench replacement is cost intensive and is plagued with the expected problem of working within developed areas where pipes may be beneath streets, sidewalks, customer landscapes, utility poles and so on. There are two basic types of open trench replacement: (i) conventional, and (ii) narrow. The conventional open-trench method uses the same approach as that used to place new pipe. The narrow-trench replacement method is similar to conventional open-trench method, but the trench width is kept to the absolute minimum possible. It is primarily used for installing polyethylene pipes.

To replace a sewer is an expensive exercise, not only because the initial cost of construction is completely wasted, but also because over time, the area through which the sewer was laid becomes built up and disruption of traffic and even access are expensive.

The next best method of replacing pipes is trenchless technology, such as by jacking pipes through the soil and by using vibration or front excavation methods.

In some cases, it is necessary to replace the pipe, particularly where the holes are large and the joints cannot be resealed. It may also be that the flow rate has increased so much that the whole system is inadequate. It may be possible to re-route the sewer but this may involve new pumping stations as the downward slope of the ground may not be followed. Pumping stations pose their own particular problems, more particularly they require an open sump to balance the flow. The problem in between pumping, i.e. when the pumps stop, is that that solids tend to settle and may gravitate to the bottom end and even cause blockages which confront attempts to restart the pumps. Pumping mains operate at a higher velocity than gravity mains and therefore there is turbulence and gas formation so that venting of the top end is important. Air valves and other valves are susceptible to corrosion and a minimum number of fittings is therefore recommended for pumping systems.

Sewer pipe laying is considerably more expensive than water pipe laying because the pipe has to be laid to a constant gradient and therefore has relatively deep excavation in some areas. This may result in the necessity to shore the sides of the trench, particularly if the danger of collapse appears in built up areas and it may endanger contracting staff. Sewer pipes are larger than the water pipes for any area because they operate in a part full condition and the water velocities are lower. It is not easy to put in diversions while new construction happens, and for this reason rehabilitation of sewers is particularly expensive compared with water pipes.

Water pipes are often constructed in the form of loops or closed networks so that alternative routes exist.

Table 5.3. Summary of rehabilitation/replacement methods

Method	Pipe size range (dia/mm)	Common materials	Generic cost ($/m dia/m length)
Cement mortar lining	100-1 500	Cement-sand	100-300
Epoxy lining[a]	100-300	Epoxy resin	900-1 500
Sliplining	100-4 000	HDPE, PVC, fibre-glass reinforced polyester	400-600
Cured-in-place pipe	150-2 000	Polyester resins	600-1 400
Fold and form pipe	200-500	HDPE. PVC	600
Close-fit pipe	50-1 200	PE. PVC	400-600
Pipe bursting	100-1 000	HDPE. PVC. ductile iron	700-900
Microtunneling	300-4 000	HDPE, PVC, concrete, steel, fibreglass	1 700-2 400
Horizontal directional drilling	50-1 500	HDPE, PVC, steel, copper, ductile and cast iron	1 000-2 500

Note: HDPE = high density polyethylene; PVC = polyvinyl chloride, PE = polyethylene
[a] Cost is in $/mm

The use of slip liners or other pipes inserted into older pipes may increase the capacity of the system because the friction is less. However, for smaller pipes, the reduction in diameter is significant and the capacity of the system may be reduced because of this.

The most popular materials used for slip lining are polyolefins, fibreglass reinforced polyesters (FRP), reinforced thermosetting resins (RTR), poly vinyl chloride (PVC) and even ductile iron for large pipes. Polyethylene is the most common polyolefin used and it is one of the most flexible, enabling it to be collapsed and drawn through older pipes. The injection of steam under pressure is one of the methods to re-round the collapsed pipe and press it against the older original pipe. Where the pipe cannot be expanded to fit the original pipe exactly, grouting may be used. This is often used for PVC liners.

The cleaning of the original sewer prior to relining is important. This goes beyond pure water cleaning and water jetting. Abrasive materials should be used to rub the surface but the use of pigs (large round objects forced through the pipe under pressure) is one method of scraping the pipe before doing the final cleaning. Sand blasting or chipping may be required when the pipe is to be relined with mortar. Mortar can be applied internally using the centrifugal or short blast process and spinning trowels can provide a smooth surface. Even slightly non-rounded pipes can be lined this way because the trowels will be mounted on flexible arms. The mortar is fed from the surface through a pipe.

There are three types of special concrete used for lining pipes, i.e. cement mortar, shotcrete and cast concrete. Rapid hardening agents are useful for shotcrete and trowelled mortar. Some separation is expected and even some cracking after the concrete has cured owing to shrinkage. However, there is an autogenous healing in most concretes owing to the surplus cement, and cracks seal although there may remain gaps between the lining and the old pipe. Spiral reinforcing can be applied in thick mortars for large pipes but care is needed to ensure there is an adequate cover to the steel. The aggregate size should be small, the smaller the diameter of the pipe, to ensure a smooth finish, and a compromise is required between liquidity to ensure transport into the pipe from the surface, and rapid setting which is achieved by adding reagents.

Grouting of old pipes and linings has the following advantages:
- Grouting increases the structural strength by providing continuity
- It can get into inaccessible places
- It sticks different materials such as lining to original sewer surface.
- It increases service life
- It is long lasting and durable
- It is chemically resistant and temperature resistant
- It increases the structural strength by providing a continuous wall instead of two separate walls as in the case of slip lining.

5.9 BENEFIT-COST ANALYSIS

5.9.1 Cost-effectiveness analysis

When peak I/I rates are less than 0,15 m^3/d (mm km) including service connections, the infiltration/inflow is not usually considered excessive and a cost-effectiveness analysis is not usually required.

In a typical I/I analysis, the following information is required to determine the extent of the infiltration/inflow and the evaluation survey program that should be conducted if the infiltration/ inflow is excessive:
- Peak inflow rates by sub-drainage area
- Average and peak infiltration rates by sub-drainage area
- Estimates of flows bypassed from system including locations
- Projected peak flows tributary to major transport components
- Projected average and peak flow tributary to treatment facilities
- Capacities of all major existing transport components and treatment facilities
- Estimates of I/I reduction levels and costs by sub-drainage area.

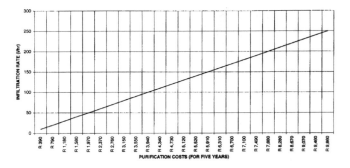

Figure 5.8. Benefit-cost analysis for infiltration repairs

Water infiltration in sewer pipelines is common and should be included in the peak design flow. A norm of 15% of the dry weather flow for water infiltration and stormwater ingress is a generally acceptable standard. Infiltration exceeding the 15% norm will result in pipe capacity problems and an unnecessary increase in sewer discharge volumes and treatment costs (e.g. Cashman et al., 2004).

A reduction in the infiltration rate, on the other hand, will not only save on sewerage treatment costs, but may defer capital expenditure for the upsizing of sewer pipelines and water care works (Brown and Caldwell, 1984; Farmer, 1975; Phalafala, 2003; Schpoll, 1982).

The decision to solve or ignore an infiltration problem should therefore be based on a cost-benefit analysis and the graph in Figure 5.8 has been prepared for this purpose. This graph is based on the rate of infiltration and the additional cost required to treat the volume of water over a period of one year. The rationale is therefore that the cost of the remedial work must be recovered through savings in the treatment costs within a period of no more than five years. The example illustrates benefit-cost analysis based on local conditions and unit costs.

Typical losses which can be applied in absence of relevant values for a specific case are listed in Table 5.4 below:

Table 5.4. Typical values for unaccounted-for water (UAW)

Parameter	Typical value (%)	Reference
Unaccounted for water (UAW)	10 to 35	Of the total water input
Error in water purchased	1.6 to 3.0	(+) or (-) of UAW
Error in water sold	10 to 12	(+) or (-) of UAW
Theft	Average 10	As for UAW
Unmetered (e.g. fire-fighting)	15 to 25	As for UAW
Bursts and wastage	Average 20	As for UAW
Leakage and overflows	Average 30	As for UAW

Source: Barta (2000)

5.10   SEWER MAINTENANCE

Sewers experience infiltration of groundwater through cracks and leaking joints. They receive inflow from gullies, open manholes and leaking taps in buildings. Alternatively, sewage can escape through leaks into the ground, or will surcharge through open manhole lids, or back through household toilets or basins or vents.

The analysis may indicate that sewer rehabilitation is desirable or even sewer replacement. Alternatively, some form of management may be possible. This could include stormwater detention, and/or separation. Alternatively, the system may have to be enlarged to cope with the additional flows and this would include the wastewater treatment works.

The problem of egress is primarily health-related as sewage could contaminate soil and ground water and even reach the surface. On the other hand, the problems of infiltration and inflow are primarily economic. It may even be that the design of the system is such that it can cope with a degree of external water, but the consequences should be evaluated.

Problems may arise due to limited hydraulic capacity, structural failure possibly due to corrosion or overloading and quality. The quality problem may affect the environment if there is overflow from the system, for example when the wastewater treatment works cannot cope with the flow rate. Or it may affect humans if there is overflow in streets of buildings.

Problems of leakage can be expected to increase over time. This is with pipe movement, drying and saturating of soil, ingress of roots, spalling or corrosion of sewer pipe walls and deterioration of joints between pipes.

The method of coping with the problem is also likely to change with time. Whereas the world was more tolerant of leakage in the past, tightening of health regulations and greater cost consciousness have forced us to be more vigilant. The methods of coping with the problem may therefore change from storage or diversion to reducing the leakage. Sophisticated inspection methods are now available including closed-circuit television cameras. Chemical grouting of the joints and soil and methods of relining the sewers are advancing with improved chemicals.

5.10.1 Preliminary analysis

In order to evaluate the situation, the following data should be assembled:
* Maps and drawings, and preferably GIS data
* Maintenance records
* Geophysical and weather data
* Wastewater works inflow rates and quality
* Relevant information from the stormwater section and water supply section such as loss control

- Records of development of the catchment, i.e. rate of industrial and residential construction.

An analysis should follow including demarcation of different areas on the map and correlation of flow rates with various possible causes. Historical trends in per capita flows and changes in water quality should be plotted. At the same time, an assessment of the requirements for a more detailed study should be made. These will be identification of monitoring points, investigation of the possibility of different methods of inspection, consideration of how flows can be diverted during inspections and repairs and assessment of staff capabilities. In fact, trained personnel are required for such maintenance as inspection of sewer surfaces and identification of different forms of deterioration require knowledge of the materials. Methods of isolating sewer sections, including rubber plugs, sandbags and temporary diversion pumps should be costed.

5.10.2 Sewer testing

During construction, the testing of sewers should be standard as leaks are rarely detected afterwards (see Stein, 2001) Regular tests during the life would be desirable but are more difficult. The following methods are available.
- *Smoke tests*: used during initial construction as well. Smoke can be detected outside of the sewer using sniffer dogs or sensitive sensors.
- *Flooding and tracing using dye or salt*: Pressurization will improve this as the water has to be detected outside of the sewer.
- *Plumbing inspections*: This requires the cooperation of communities and is therefore a laborious process.
- *Manhole inspections*: This is the easiest way of identifying leakage into manholes.
- *Flow isolation*: This requires blocking off inflowing laterals or outflowing diversions.
- *CCTV* (*Closed Circuit Television Cameras*): The sewer has to be cleaned to enable the camera trolley to travel up it, so it has to be isolated.
- *Visual inspection*: This requires an even cleaner sewer and dangers of gassing and contaminations should be borne in mind and avoided.

Flow monitoring in sewers is not as easy as in water pipes and special apparatus is required which may not normally be installed in the system for normal operation. Water depth is usually the indicator of flow rate and the Manning equation or other channel flow equation is used to convert a water level to a flow rate, knowing the gradient, roughness and nature of the liquid. Although spot depth readings can be made by staff, it is preferable to have non-intrusive systems such as Doppler or electronic methods mounted in the manhole for continuous recordings.

Simple tests used at design stage including rolling a ball down the sewer or inspection by mirrors from ground level must not be forgotten.

Table 5.5. Different types of sewer rehabilitation methods

Structural	Excavation and replacement
	Insertion
	Speciality concrete
Misaligned pipes	Excavation and replacement
Additional capacity needed	Excavation and replacement or duplication
Avoid reduction in capacity	Excavation and replacement
Damaged pipes	Excavation and replacement
High infiltration	Grouting
	Slip lining
Leaking joints	Grouting
	Lining
Circumferential cracks	Grouting
	Lining
Small holes	Grouting
	Lining
Radial cracks	Grouting
Roots	Slip lining
	Re-jointing
Corrosion	Slip lining
	Mortar lining
Broken pipes in busy areas	Cured in place lining
	Slip lining
	Coating
	No-dig pipes
Non-circular pipe	Cured in place inversion lining
Mild deterioration	Cured in place inversion lining
Corrosion	Lining
	Coating
	CIPL
Corrosion by wastes	Slip lining
	Speciality concrete
	Lining
	Coating
	Cured in place inversion lining
Misaligned pipes and bends	Cured in place inversion lining

Notes:

Speciality concretes include sulphate resistant additives such as potassium silicate and calcium akuminate

CIPL = Cured in Place Lining, such as resin impregnated felts

Inversion lining is formed by inserting the resin-impregnated felt tube into the pipe and inverting it against the pipe allowing it to cure

5.10.3 Corrosion of sewers

Lack of oxygen in sewers promotes the formation of sulphides. Sulphur compounds are hydrolized to sulphides such as hydrogen sulphide gas. The action is especially noticeable in warm climes and with high BOD sewage. The hydrogen sulphide is released from the sewage and oxidized on the exposed walls of the pipe manholes to sulphuric acid. Sever corrosion of steel or concrete pipes often occurs just above the water line. The cement in concrete pipes is attacked, leaving siliceous aggregates intact.

The corrosion may be counteracted by the use of calcareous or preferably limestone aggregates in the concrete. The lime neutralizes the attack and evens the corrosion. The life of sewers may be doubled in this way. As an additional precaution, an extra thickness of concrete, terms a sacrificial layer, may be specified in the manufacture of the sewer pipe. An additional 12 to 50 mm on the inside is often specified, and this thickness is not taken into account in structural design. Limestone is often used as concrete aggregate.

Sewage pumping mains are a bad source of hydrogen sulphide. The gas is generated in the rising main and does not escape until the sewage is discharged into a manhole or gravity sewer. Pumping mains should rise continuously as dips may cause gas to collect at the high points as well as deposition of solids at the low points. Depositions turn septic and promote generation of gas. Although there are available air valves for sewage pumping mains, the valves have to be flushed often, and open-ended vent pipes are preferred. A vent pipe is especially necessary at the discharge end, as well as a stilling chamber and water seal. rising mains should be designed to have the shortest possible length and the velocity should be high (0.8 to 2.5 m/s) under all conditions, to prevent deposition and to flush out gases. The volume of the pipe should preferably be less than the volume pumped in a cycle.

Ventilation and uniform hydraulic conditions are required to minimize this type of corrosion, but more particularly it should be identified at design stage and appropriate pipe materials selected. Fibre cement such as asbestos cement is also prone to attack and the walls of these pipes are often thinner than concrete pipes would be so that they may deteriorate relatively rapidly. Fortunately, the corrosion occurs on the soffit or top of the pipe and it is relatively easy to pick up with cameras or visual inspection. Corrosion of metal fittings such as manhole step irons and reinforcing in concrete pipes is also likely due to low pH and therefore corrosion resistant step irons and manhole covers are advisable.

The following pipe materials are corrosion-resistant:

- Vitrified clay, which is however not able to take high tensile stresses and therefore cannot be placed under large loads or built in large diameters, the latter also due to manufacture limitations.

- PVC and polyethylene pipe which can be made in long lengths to reduce number of joints. These are flexible and require careful backfill.
- Glass fibres in resins and polyesters.

5.10.4 Rehabilitation of manholes and sumps

Larger structures can be accessed manually and do not suffer loss of capacity with a relatively thick lining. Concrete is therefore often an option here and even cast against shuttering is possible. Although many of the techniques used for the pipe can be applied, it is also easier to access the outside of manholes, i.e. they can be excavated and exposed, whereas this cannot be done for pipes without endangering their stability.

One of the ways of reducing infiltration into manholes, especially at the top, is to raise the manhole above ground level and flood levels. It is a relatively simple matter to add extra rings to concrete ring type manholes. Alternatively a smaller access chimney can be built on top of the slab where the original manhole lid may have been.

In built up areas, it is necessary to seal manhole lids to avoid odours. However, in open areas ventilation may be permitted. Then the danger of surcharging and outflow from the vent should be considered. High posts above ground level may be a way of venting and preventing overflows and inflow.

5.11   REFERENCES

Ainger, C.M. et al. (1998) *Dry Weather Flow in Sewers*, London, CIRIA.

APWA (American Public Works Association) (1970) *Control of Infiltration and Inflow into Sewer Systems,* 11022 CFF 12/70.

Ashley, R.M. (2000) The management of sediment in combined sewers, *Urban Water,* Elsevier Science Ltd, **2**(4), (February 2001), 263–275.

Barta, B. (2000) Urban water cycle in a South African metropolis. *Proc. Xth World Water Congress,* Melbourne, Australia, March.

Brown, B. and Caldwell, A. (1984) *Utility infrastructure rehabilitation,* National Technical Information Service Publication PB86-N14642, Washington.

Cashman, A., Savic, D., Saul, A., Ashley, R., Walters, G., Blansby, J., Djordjevic, S, Unwin, D., and Kapelan, Z.(2004). Whole life costing in sewers; developing a sustainable approach to sewer network management. *Proc.Novatech,*Lyon, 449-456.

Civil Engineering Reference Book (1995) *Sewage and Sewerage Disposal.* 4th Ed., Butterworth Heinemann Ltd., UK, pp.29/26.

Crabtree, P.R. (1970) *Flow and Infiltration Gauging in Sewers,* NBRI, Pretoria.

Duchesne, S.(2001) Mathematical modeling of sewers under surcharge for real time control of combined overflows, *Urban Water,* Elsevier Science Ltd., **3**(3), 241–252.

EPA (Environmental Protection Agency) (1991) *Sewer System Infrastructure Analysis and Rehabilitation.* EPA/625/6-91/030.

Ermolin, Y.A. (2001) Estimation of raw sewage discharge resulting from sewer network failures, *Urban Water*, Elsevier Science Ltd., **3**(4), May 2000, 271–276.

Farmer, H. (1975) Sewer system evaluation and rehabilitation cost estimates. *Water and Sewage Works,* April.

Geustyn, L. (2002) *Seminar on pro-active management of pipe systems*, Rand Afrikaans Univ., 19 September, Johannesburg.

Grigg, N.S. (2003) *Water, Wastewater and Stormwater Infrastructure Management*, Lewis, McGraw Hill, N.Y.

Grobicki, A.M.W. and Cohen, B. (1999) A flow balance approach to scenarios for water reclamation. *Water SA*, **25**(4), October, 473-482.

Hoangnam, N.V. (1982) Integrated control of combined sewer regulators, *Jnl of the Enviro. Engg. Div, ASCE*, **108**(EE6), December, 1342–1359.

Horvath, I. (1994) *Hydraulics in Water and Wastewater Treatment Technology*, J. Wiley, Chichester.

Larson, K.E. (1979) Calculating overflows from combined sewer systems, *Jnl of the Enviro. Engg. Div, ASCE*, **105**(EE5), October, 829–841.

Li, J. and McCorquadale, A. (1999) Modelling mixed flow in storm sewers, *Jnl of Hydraulic Engineering*, **125**(11), November.

Metcalf and Eddy (1979) *Wastewater Engineering*, 2nd Ed, McGraw-Hill, New York.

Phalafala, M.R. (2003) *Impact of stormwater ingress and groundwater infiltration on municipal sanitation services*. MSc(Eng), Univ. Witwatersrand, Johannesburg.

Reynolds, J.H. (1995) Infiltration: a case study of control, *Municipal Engineer*.

Saegrov, S. and Schilling, W. (2004). Computer aided rehabilitation of sewer and stormwater networks. *Proc. Novatech Conf.*, Lyon, 1641-1646.

Scholl, J.E. (1982) 1980 Needs survey: combined sewer site studies, *Jnl of the Enviro. Engg. Div, ASCE*, **108**(EE2), April, 315–326.

Sinske and Zeitsman (2002) Sewer system analysis with aid of GIS. *Water SA*, **28**(3).

Skipworth, P.J. et al. (2000) The first foul flush in combined sewers: an investigation of the causes, *Urban Water*, Elsevier Science Ltd., **2**(4), 317–325, February.

Stein, D. (2001)*Rehabilitation and Maintenance of Drains and Sewers*, Ernst and Sohn, Germany.

Stephenson, D. (1982) Computer analysis of Johannesburg sewers,*Proc.IMIESA*,4,13-23.

Stephenson, D. and Hine, A.E. (1985) Sewer flow hydrographs, *Mun. Engineer*, ICE.

Vaes, G. and Berlamont, J. (2001) The effect of rainwater tanks on design storms, *Urban Water*, Elsevier Science Ltd., **3**(4), 303–307, July.

WPCF (Water Pollution Control Federation) (1983) Existing sewer systems evaluation and rehabilitation. *ASCE Manual no. 62/WPCF Manual FD-6*.

Water Research Centre (WRC) (1983) *Sewerage rehabilitation manual*, Swindon, UK.

White, J.B. (1987) *Wastewater Engineering*, 3rd ed, London, E. Arnold.

6

Alternative urban drainage systems

6.1 INTRODUCTION

There are certain standards in the western world for drainage of cities which may not be appropriate for rural or poorer areas, or areas with different topography and climate. Best drainage practice would typically comprise a water-borne sewerage system which transports all household and industrial wastewater to a wastewater treatment works which is designed to convert the waste water to an environmentally acceptable effluent. In addition, there may be a separate stormwater drainage system comprising pipes under the ground within the city area, and channels in open places conveying runoff but which is seldom treated.

There are also many cities which operate with a combined sewerage system. That is, waste waters from domestic and industrial sites are only part of the capacity of the drainage system, and there is additional capacity for stormwater to enter the system. This is a separate issue to the undesirable ingress of stormwater and seepage into separate sanitary sewers. Combined sewerage systems pose the problem of how to purify the effluent. There will be large fluctuations in flow rates from wet periods to dry periods and these are frequently accommodated in storage ponds prior to treatment. See also Brombach et al (2004) on pollution due to combined systems.

6.2   DRAINAGE STREAMS

There are a number of alternatives to these systems however, and the distinction between the streams is often not as clear as we would like to think in developed urban areas. On the other hand, some communities discharge the entire stream into water courses or the sea, and others attempt a physical separation, for example by using skimming weirs or other methods of separating cleaner water from more polluted strata.

If the streams are broken down into the individual components, it will be realized that each of the above streams consists of contributions of different quality and quantity and therefore the cost of conveying and treating each component individually could be considered.

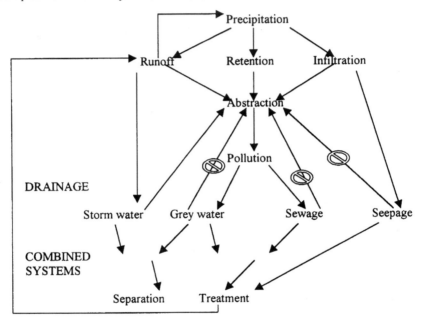

Figure 6.1. Flow paths.

With the realization that it is not going to be economically feasible to provide the entire world's population with water-borne sanitation, comes the realization that many of the processes westerners have become used to are luxuries or are unnecessary. Serious consideration is now being given to subjects such as on-site disposal of sewage, separation of sewage and direct re-use of different streams of wastewater. The conventional systems of bulk wastewater treatment is aimed at an average influent, which is not necessarily what the most economical solution. The same applies to the transport systems, i.e. separate or combined systems, or alternative systems prove viable in some circumstances and may bring recognition to the fact that the present systems in many cities may not be the most economic.

Typical flow contributions from a middle-upper class western home are tabulated in Table 6.1. The actual sewage is a relatively small component and even that can be divided into white water (clean flush water), brown water (faeces) and yellow water (urine). The total sewage stream is termed black water. Flushing water is the greater component. In addition, homes discharge a considerable amount of wash water which is termed grey water.

Table 6.1. Reasons for source control; loading in household wastewater (Matsui et al., 2001)

Yearly Loads kg/(person/year)		Grey water 25000-100000 ℓ/p/year	Urine 500 ℓ/p/year	Faeces 50 ℓ/p/year
N	4 - 5	3%	87%	10%
P	0.75	10%	50%	40%
K	1.8	34%	54%	12%
COD	30	41%	12%	47%
Treatment		Treatment	Treatment	Biogas-Plant Composting
Disposal		Reuse/Water cycle	Fertiliser	Soil conditioner

The wash or rinse water from showers, baths and washing machines is primarily polluted by detergents with nitrates, phosphates, oils and fats in many of these. Kitchen sink waste also contains biodegradable and even non-degradable substances but very few pathogens, so that only biological treatment would be necessary. The fats and greases may also be removed by relatively simple physical means. The removal of the minerals and phosphates in wash water is more of a problem.

The possibility of the separate transport of the different streams or some of them should be considered. In fact, some of the streams may not necessarily be transported at all and this can solve a disposal problem. Research is being undertaken into the separation of yellow water from black water from toilets (Matsui et al, 2001). On-site treatment of brown water is common in China.

Urine contains nitrates and ammonia which can be stabilized with retention over a few months. Black water contains pathogens but composting or biodigestion can neutralize this so that it can be disposed of in or on ground. On the other hand, the high nutrient loading of urine makes it attractive for irrigation provided care is taken to prevent any small concentrations of pathogens to have any affect.

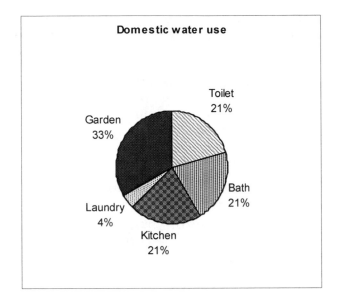

Figure 6.2. Typical household water use proportions.

Table 6.2. The wastewater palette (fractions from households (Henze and Ledin, 2000)

Type	Content
Classic	Toilet, bath, kitchen, wash
Black	Toilet
Grey	Bath, kitchen, wash
Light grey	Bath, wash
Yellow	Urine
Brown	Faeces

Alternatively, one or both of these streams may be transported with flush water separately to sewage works or to holding tanks for separation and transport at an average flow to a wastewater treatment works. Even the flush water can be reduced considerably, bearing in mind that the majority of flush water is used for urine disposal and the amount required for faeces removal is relatively low. Therefore, household urination facilities such as urinals for men could reduce the amount of flush water considerably, thereby reducing transport problems and water consumption.

The on-site disposal of black or preferably brown water such as in pit latrines in poorer areas with enough space is therefore not only economically attractive for these communities but also reduces the problems of wastewater transport and treatment and reduces water consumption. The alternatives for densely built-up areas are vacuum tanks or chemical toilets which may be expensive on the face of it, but considering water savings and transport and wastewater treatment costs, could prove the most economical solution.

6.2.1 Grey water

The largest volume of water then apart from the toilet flush water is the so-called grey water contribution from washing. This warrants consideration of separate low technology treatment. The treatment would be primarily physical in the way of screens, skimmers either surface or sub-surface, settlers, or vortex separation facilities. The removal of the phosphates would need chemical precipitation unless it was combined with biological waste to produce bacteria for assisting in the removal process. Of course, the alternative of changing detergent types and washing methods should be considered, especially if large communities are to be converted to grey water drainage systems.

This alternative is again attractive for poorer communities because the cost of grey water drainage is considerably less than conventional water-borne sewerage. Not only are the flows much lower because of the reduced peak load from toilet flushing, but also detention of peaks is very feasible, enabling discharge pipes to be relatively small bore. Gradients and depths are not as important as for conventional sewerage and laying can be undertaken by relatively unskilled communities. Blockages are less likely, and screens and traps will remove any solids.

Table 6.3. Breakdown of domestic water consumption in Northern Europe, and for a poor community (after Henze and Ledin, 2000)

Consumer	Existing ℓ/p/d	With savings	With savings plus B grade water for flush and wash	Poor community
Toilet	50	25	0	0
Bath	50	25	20	7
Kitchen	50	25	25	8
Laundry	20	5	5	5
Garden	80	20	-	0
Total	250	100	50	20

Therefore the quality and methods of treating grey water need more consideration. These can be informal type processes as they are primarily physical processes, and in some areas land disposal or infiltration and ground filtration are possible.

6.2.2 Stormwater

The remaining stream is the stormwater stream. This can be a relatively high flow rate in tropical and typically developing areas, in fact much bigger than the sewerage flow rates. Whereas in temperate climates, storm flow tends to be more regular and therefore can be accommodated with slight increases in the sanitary sewers, this is not possible or desirable where the stormwater flows are high. Some storms could even cause surcharge and overflow of sewers posing health risks and aesthetic disapproval. In these circumstances, separate stormwater systems are therefore desirable but these can be of varying technological solutions.

The necessity to convey all storms in underground storm drains is not there, as during intense storms streets are usually abandoned and the street could serve as a major storm drainage transport system.

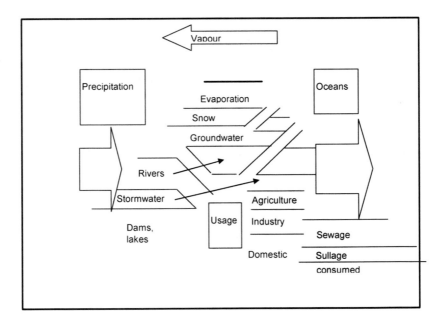

Figure 6.3. Flow streams in the hydrological cycle.

Minor storms could be conveyed by underground pipes to enable streets to be used and to minimize the pollution load down gutters and streets.

The quality of storm water varies considerably between different communities. Whereas in modern cities, street washing and litter prevention are common, in lesser developed communities the streets or pavements are often the only or simplest place for the disposal of garbage. Litter which is randomly discarded, items such as bags and cans, and household garbage therefore contaminate the storm runoff and can cause problems ranging from health and aesthetic problems to hydraulic blockages and threats to property and life due to flooding.

In such circumstances, open channels may be the simplest form of transport of stormwater but they should be accompanied by screens and other forms of pollution control. The danger of sewage in the drain is of concern.

Table 6.4. Types and contents of domestic wastewater

	Grey water			Blackwater (toilets)			Gardens	Storm
	Shower bath	Clothes washing	Kitchen sink	White Flush	Yellow Urine	Brown Faeces	Evapo-ration	runoff
Water ℓ/p/d	50	60	16	50	3	1	20	200
TDS					√			
BOD			√			√		
N					√			
P		√	√					
SS								√
Fat/oil	√		√					
Food			√					
Paper				√	√	√		√
Plastic								√
Fibre		√						
Metal								√
Hormones					√	√		
Pathogens						√		
Transport	Small bore	Small bore	Small bore	Sewer	Container	Container	Vapour	Channel
Treat	Filter	Skim	Vortex	Works	Retention	Compost		Screen
Disposal	Soak-away	Soak-away	Soak-away	Rivers	Land	Land		Rivers

There are therefore a number of combinations for township drainage ranging from best practice to appropriate technology. Which system to adopt depends on the stage of development of the township, i.e. whether the drainage systems can be designed and installed from scratch or whether they have to be installed after occupation of the land.

6.2.3 Sewerage

Financial resources will be a major influence on the selected design, but acceptability, customs and topography are also influences. Sanitary customs can influence the form of sewerage. Communities in arid environments are used to land disposal, whether buried or surface, and the space is often sufficient to permit this. In such cases, the use of paper may not be common and alternative cleansing forms are used. Eastern washing systems can be used for flushing as well. The squat pit-type latrine with hose washing can be the only form of flushing and this is often less water intensive than western seat toilets which require up to 20 litres for flushing. The use of hand wash water from basins to toilet is also feasible.

Low flush toilets are now being considered even in western countries and 9 litres is a common maximum for full flush but toilets with 5 litres are being used in some countries, such as Sweden and Japan. Dual flush systems are also possible whereby a lower flush, example 2 litres, is used for urine flushing and a higher volume for faeces removal. In such circumstance where low flush is used for the faeces removal, mechanical cleansing of the bowl may be necessary and such facilities should be an integral part of the toilet system subject to domestic acceptability.

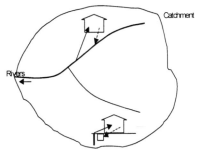
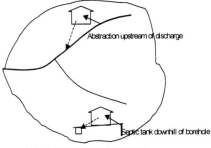

a. Bad catchment planning resulting in polluted water supplies   b. Good catchment planning for clean water

Figure 6.4. Catchment management

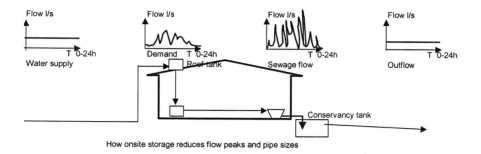

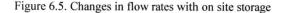

Figure 6.5. Changes in flow rates with on site storage

It should also be noted that paper is a large volume in the sewage system in western systems and alternative cleaning and/or drying systems, e.g. hot air, may be considered. Some of these may however be energy intensive, which could cost more than the savings in the sewerage system.

The topography will also influence the drainage system cost and therefore acceptability. Steep slopes may be conducive for smaller drains, but erosion and living on slopes can be problematic, particularly in high rainfall areas.

Poorer communities are frequently located around the periphery of cities in developing countries and the drainage system from the centre of the city may be a sophisticated water-borne sanitation system. It is easier to link more dense poorer communities into this system if they are at the outer periphery and lower level of the city. However, the incremental cost should be compared to separate drainage and disposal systems for these communities.

Often, poorer communities are in low-lying or water-logged areas and close to natural water conduits. Frequently, the drainage in these conduits even in the street gutters contains wash water or even sewage from upper lying buildings and therefore clear separation of water sources from the drainage system is important. The idea of transporting all water-borne waste into channels and outer in peri-urban areas was discussed by Parikh (1998).

6.3 CASE STUDY IN ALEXANDRA TOWNSHIP

A large proportion of runoff from low cost peri urban townships in South Africa is grey water. Table 6.5 gives a typical analysis of water samples from the roads of Alexandra township North East of Johannesburg. It appears that the water is primarily wash water with Nitrate and Phosphate detergents. There is evidence of sewage in the water also, and other suspended solids such as grit and organic matter.

The simplest way of removing the latter would be by settling, but some disinfection also appears necessary. Research into compact and unattended separation processes is under way. The similar density of organics and water makes screening appear more attractive than gravitational methods. However, ground disposal offers potential as the danger to residents and groundwater is not as great as for sewage.

Table 6.5. Constituents of Grey water from Alexandra, Typical

Parameter	Concentration
Conductivity mS/m	100
Ammonia as N mg/ℓ	30
Nitrate N mg/ℓ	2
Phosphate P mg/ℓ	5
Suspended solids mg/ℓ	500
Total organic Carbon C mg/ℓ	100

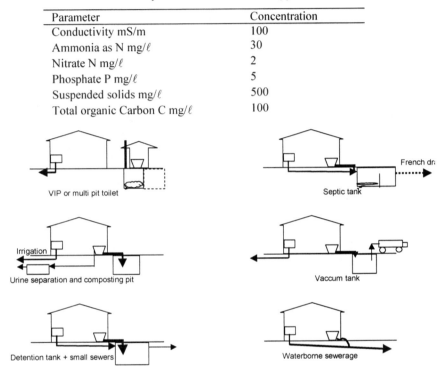

Figure 6.6. Alternative sewerage systems, in order of cost

6.4 SYSTEM MASS BALANCES

An appreciation of losses, uses, savings possible and sources of water can be obtained by doing a mass balance on selected areas. Draw a boundary cutting as few water pipes and drains as possible. Measure inflows from water supply sources, precipitation and cross watershed flows, and outflows via sewers, storm drains, seepage, evaporation and runoff, e.g. at taps, toilets, leakages.

This balance can be done on an annual basis to start with, then over specific time periods. By separating dry weather flows, rainfall ingress is eliminated. By looking at night flows, most usage usage is eliminated. By comparing drought periods with post rain periods, groundwater ingress can be assessed. And quality monitoring can also be used for pollutant mass balance. Then specific ions could be tracked or even injected and used as tracers.

Data can be arranged in a spreadsheet, e.g. sources as rows, destinations as columns, links and losses can be added, and a total balance achieved before a more detailed audit.

Table 6.5 Subcatchment mass balance in ℓ/s:

Catchment:	1 km²					
Rain:	5 mm/h = 1 500 ℓ/s					
Evaporation:	0.1 mm/h					
Source	Total	A Sewer 1	B Sewer 2	C Storm drain 1	D Evaporation	E System storage
Pipe 1	250					
Pipe 2	170					
Rain	1 500					
TOTAL	1 920	120	160	1 100	30 Calculated	610 Difference

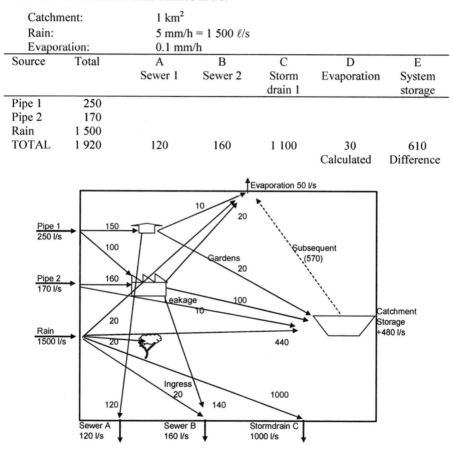

Figure 6.7. Urban water balance example

6.5 DOMESTIC MASS BALANCE

Even greater savings are possible on a domestic scale. Re-use of grey water and in combination with rainwater can halve the water requirements of a sophisticated home. The grey water could be screened and pumped up to a storage tank for use for flushing and gardening.

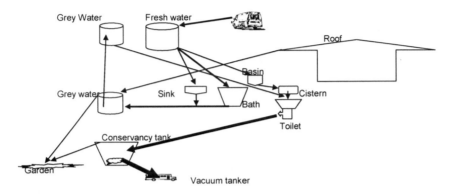

Figure 6.8. Domestic water recycling

6.6 REFERENCES

Brombach, T., Weiss, G. and Fuchs, S. (2004). Combined or separate sewer systems? *Proc. Novatech Conf.*, Lyon, 599-606.

Henze, M.P., Harremoës, J., La Cour Jansen, J. and Arvin, I. (2000) *Wastewater Treatment, Biological and Chemical Processes*. 3rd Ed., Springer Verlag, Berlin.

Henze, M. and Ledin, A. (2000) Types, characteristics and quantities of classic, combined and domestic wastewaters. Chap. 4 in *Decentralized Sanitation and Re-use*. Eds. Lens, P. and Lettinga, G., IWA Pubs., London.

Matsui, S., Henze, M., Ho, G. and Otterpohl, R. (2001) Emerging paradigms in water supply and sanitation. Chap. 5 in *Frontiers in Urban Water Management*, Eds. Maksimovic, C. and Tejada-Guibert, J.A., IWA-UNESCO, London.

Otterpohl, R. (2001) Design of highly efficient source control sanitation and practical experiences. In *Decentralized Sanitation and Re-use*. Eds. Lens, P. and Lettinga, G., IWA Publs., London.

Parikh, H. (1998) Shen networking, an alternative way to reach the urban poor. *Proc. Int. Conf. Urban Storm Drainage*, IWS-IAHR, Sydney.

Stephenson, D. (2001) Problems of developing countries. Chap 6 *Frontiers in Urban Water Management*, Eds. Maksimovic, C. and Tejada-Guibert, J.A., IWA-UNESCO, London.

Winblad, U. (Ed). (1998) *Ecological Sanitation*. SIDA, Stockholm.

7

Wastewater treatment

7.1 SEWAGE QUALITY

Domestic sewage contains organic matter requiring oxygen to stabilize it, as well as other suspended matter and dissolved chemicals. Of most concern, are pathogenic bacteria and viruses which can be transmitted to other humans.

The BOD (Biochemical Oxygen Demand) of city sewage is a measure of its putrification and the load is approximately 50 g per day per head of population, at a concentration of about 250 mg/ℓ of sewage. BOD_5 refers to the 5-day oxygen used in the laboratory. Other quicker indicators of oxygen requirement are chemical oxygen demand (COD) and total organic carbon (TOC).

Domestic sewage is characterized by high nitrate loading. Ammonia in urine is oxidized to nitrite, then nitrate, which is a chemically stable, undesirable pollutant. The ways of removing nitrate are chemical and biological.

In addition to facilities for BOD removal, sewage works generally have channels for grit removal, screens for solids removal, and final stabilization ponds or disinfection facilities. Many countries now require nutrient removal at sewage works, and chemical precipitation of phosphates is possible.

©2005 IWA Publishing. *Water Services Management* by David Stephenson. ISBN: 1843390809. Published by IWA Publishing, London, UK.

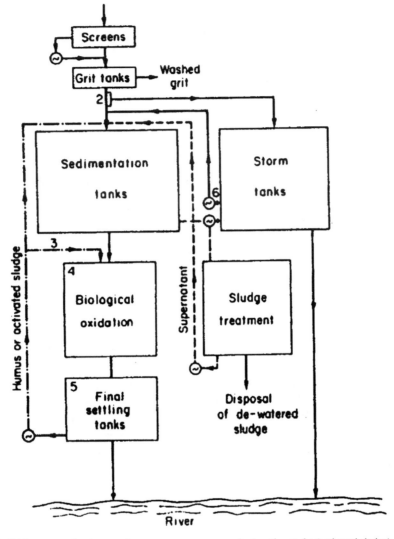

1. Disintegrator shreds screenings
2. Storm weir (flow >3 x DWF overflowed to
   storm tanks)
3. Returned activated-sludge pipe (not needed
   in filter plant)

4. Aeration tanks (activated sludge) or
   percolating filters
5. 'Humus' tanks on filter plant
6. Empties storm tanks after each storm

Figure 7.1. Flow sheet for conventional sewage treatment

However, biochemical reduction of nitrates and phosphorus is now achieved in many treatment processes. This requires sludge recirculation and differentiated zones in the aeration tanks. (Design guides are given by Benefield *et al,*, 1984; Metcalf and Eddy, 1991.)

The process to select will depend on the societies' balance between economy and standards. Poorer communities may opt for lagoons or trickling filters. These require big areas, have odour problems at times, and effluent quality may be limited. Richer towns with more limited land space will prefer activated sludge and or extended aeration systems. Following are some of the processes used (CSIR, 1991; Franceys *et al*, 1992; Mara, 1996).

7.2 TREATMENT PROCESSES

Oxidation ditches offer a simple extended aeration activated sludge process. Screened sewage is circulated around a loop by rotating rotors which also act as aerators. Typical detention time is 24 hours with sludge age of 10–30 days. Loading is 0.15–0.3 kg BOD/m³. Oxygen is input with mechanical aerators giving 1–2 kg O_2/kWh. A large space is required on account of the configuration and careful control of sludge return rate is needed. Acceptable effluent quality is possible but some disinfection may be necessary.

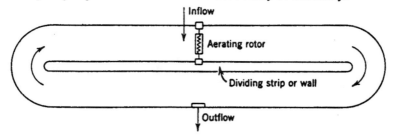

Figure 7.2. Oxidation ditch

Trickling filters or biological filters are simple and energy efficient. Filter volumes should provide 2.5 m³ per m³/d inflow. Media depths of 2m or more are used and application rates of up to 30 m/d. After screening and settling (2–12 hr retention) the sewage is sprayed onto broken stone media but a biological film on the media does most of the purification. The stone media is aerated by updraft. Double filtration and recirculation can reduce the volumes of filters required or can improve efficiency.

Sludge is pumped from the primary settling tank to digesters (retention time up to 30 days). About 60% of the BOD is removed with the sludge. There is often a secondary clarifier after the filter, and possibly an aerated sludge contact tank with sludge return from the secondary clarifier.

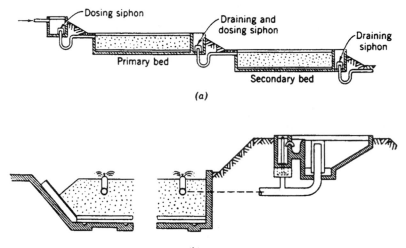

(a)

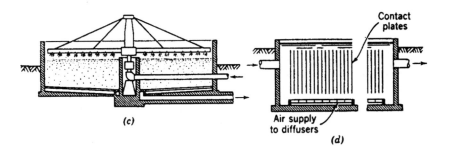

Figure 7.3. Contact beds and trickling filters. (a) Double-contact beds with dosing and draining siphons. (b) Trickling filter with fixed nozzles and automatic dosing tank. (c) Trickling filter with rotary distributor. (d) Contact aerator (Fair *et al.*, 1968).

The clarifier and digester make the process expensive compared with other methods, the space required is large, and the odour and aesthetics often unattractive.

The volume of a trickling filter is about 10 times the volume of an activated sludge-extended aeration unit, i.e. about 0.5 m^2 of filter per capita is needed versus as low as 0.025 m^2/person for activated sludge extended aeration.

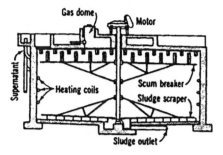

Figure 7.4. Digestion tank with fixed cover, scum breaker and scum scraper (Harding Co.)

Extended aeration – activated sludge is a popular sewage treatment process with good quality effluent. The process is typically preceded by screens, and followed by aeration, clarification and disinfection. Sludge from the clarifer is part returned to the aeration tank and part to a digester. Hydraulic retention time is 6 to 24 h and sludge retention 30 days.

About 2% of the raw sewage appears as sludge. Thus about 100 ℓ of digester is needed per capita. Temperature and mixing are important for proper digestion. The neutral sludge is pumped to drying beds and thence can be used on land.

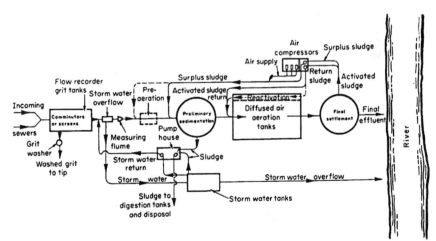

Figure 7.5. Flow diagram of typical diffused-air plant

Various forms of recirculation and compartmentalization are possible in the aeration tank. Removal efficiency of BOD and SS can be 90%. Nitrification and removal of many biodegradable toxic compounds occurs because of the long aeration time. Based on 1.5 kg O_2 per kg BOD_5 and 5 kg O_2/kg $NH_4.N$, 200 m^3 of air per kg BOD_5 is required. About 0.5 kWh is needed to transfer one kg of oxygen to the sewage.

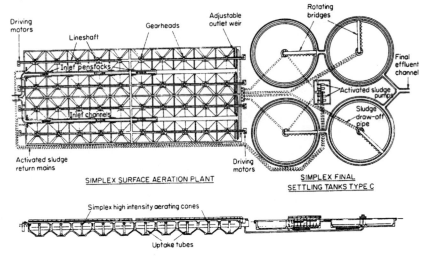

Figure 7.6. Layout of aeration tanks and final sedimentation tanks of a Simplex surface aeration plant

Stabilization ponds (e.g. Meiring et al, 1968; WISA, 1988) are simple to construction and operate although space requirements are large and effluent may not be of the highest standard at all times of the year. Effluent BOD can be less than 30 mg/ℓ. Algae near the surface of the ponds in sunlight takes up carbon dioxide and releases oxygen. Ammonia release also occurs near the surface.

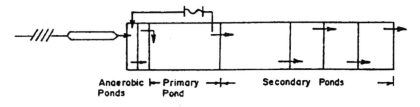

Figure 7.7. Anaerobic-aerobic pond system

Retention times can be 30-180d and surface area of ± 20 m²/capita is needed. Loading can be 30-50 kg BOD/ha/d. Depths are 1-2m. Multiple ponds in series are recommended. Ponds may be mechanically aerated at 0.1- 0.3 kW/m³, reducing retention times down to 3–20 days.

Wetlands (Batchelor, 1994 and 1995) of reeds or other plants are known to purify wastewater but design criteria are still scanty. In fact, there is a problem of environmental pollution if wastewater is intentionally discharged into natural wetlands, so constructed wetlands are more acceptable. There are two types, one being a free water surface type (FWS) and the other a subsurface flow type (SF) using primarily the root zone for purification.

Design criteria for FWS's are:

      Retention time: 7-365 days

      Area: 4 ha/(ℓ/s) or 100 m²/person contributing 200 ℓ/d.

Typical design criteria for SF's are:

      Area: 5 m²/person contributing 200 ℓ/d

      Depth: 0.3-1m

Expected Performance:
- BOD removal: 90%
- N removal: 70%
- P removal: 85%

Wetlands act in the following ways:
- Sediment settling
- Adsorption by sediments, microbes and plants
- Surface aeration, and
- Disinfection by sun

Sediments absorb phosphorus, hydrocarbons and trace metals. Plants absorb nitrogen and nutrients, but decaying vegetation can recontribute these chemicals. Pre-screening and settling are advisable before discharging to wetlands. The reaction of wetlands to shock storm loads and high concentrations of pollutants needs study. In selecting the correct treatment process, economic and environmental factors will be considered. The life cycle and operation costs are important, perhaps more so than water works (see WHO, 1994; Winblad, 1998).

Table 7.1. Type of sewerage – selection chart

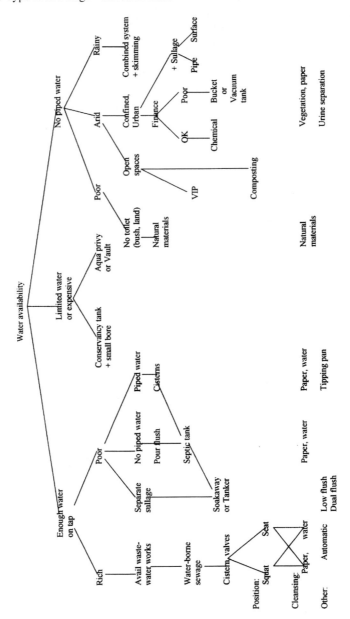

7.3 APPROPRIATE SANITATION

Conventional sewerage, for so long considered by engineers as the only sanitation technology for urban communities, has two major disadvantages (Mara, 1996; Perkins, 1989; Pickford, 1995; Serageldin, 1995):

- High cost: the World Bank found sewerage investment costs in eight capital cities in developing countries to be US$600–4 000 per household (1980 prices), with total annual costs per household of US$150–650, and
- The need for in-house (or at least on-plot) water

It is thus simply not an option for low-income urban communities, unless they are subsidized (Alaerts et al, 1991; Cairncross, 1990; DWAF, 1994; Kalbermatten, 1980b; Nichols, 1991; Palmer. 1992; Shrinivasen, 1990). Fortunately, several well-tried and robust alternative sanitation technologies exist that are not only cheaper than conventional sewerage, but are also able to deliver the same health benefits and offer the opportunity for community participation (self-help) to reduce costs. These are:

On-site systems: VIP latrines
                           Pour-flush toilets
                           Septic tanks

Off-site systems: Settled sewerage
                           Simplified sewerage.

With VIP latrines and pour-flush toilets, arrangements have to be made for sullage disposal. Sullage is the non-toilet wastewater generated by a household (i.e. the water from sinks and showers). In high density urban areas, simplified sewerage can be cheaper than on-site systems.

Safe sanitation facilities are needed:

- to improve the health and quality of life of the whole population;
- to integrate the development of a community in the provision of sanitation;
- to protect the environment; and
- to place the responsibility for household sanitation provision with the family or household.

The safe disposal of human excreta alone does not necessarily mean the creation of a healthy environment. Sanitation should go hand in hand with an effective health care programme. The five main criteria to be considered when providing a sanitation system for a community are:

- reliability;

- acceptability;
- appropriateness;
- affordability;
- sustainability.

With these criteria in mind, designers should note the following principles:
- Basic services are a human right.
- Government should provide the infrastructure.
- Individuals should pay for their sanitation service.
- Development should be demand-driven and community-based.
- Regional allocation of development resources, including limited resources, should be equitable.
- Attention must be given to the economic value of water.
- Sanitation systems must be financially viable for service provider; payment by the user must ensure this.
- Sanitation provision must not take place in isolation from other services.
- The environment must be considered and protected in all development activities.
- Sanitation provision should include health and hygiene education.

7.4 CATEGORIES OF SANITATION SYSTEMS

There are two places to treat human waste. It can either be treated on site before disposal, or removed from the site and treated elsewhere. In either case, the waste may be mixed with water or it may not. On this basis, the following four groups may be distinguished:

Group 1:   Not water added – requiring conveyance.
Group 2:   No water added – no conveyance.
Group 3:   Water added – requiring conveyance.
Group 4:   Water added – no conveyance.

Table 7.2 lists sanitation systems associated with each of the above groups.

Table 7.2. Categories of sanitation systems

	REQUIRING CONVEYANCE (treatment at central works)	NO CONVEYANCE REQUIRED (treatment, or partial treatment, on site)
No water added	GROUP 1 Chemical toilet	GROUP 2 Ventilated improved pit toilet Ventilated improved doible-pit toilet Ventilated vault toilet Continuous composting toilet

Water added	GROUP 3
	Water-borne sanitation
	Flushing toilet with conservancy tank
	Settled sewage system
		GROUP 4
		Flushing toilet with septic tank and subsurface soil absorption field
		Low-flow on-site sanitation systems
		    Aqua-privy toilet
		    Pour-flush toilet
		    Low-flush system
		    Low-flow septic tank

The advantages and disadvantages of each system will depend on its particular application. What may be a disadvantage in one situation may in fact be an advantage in another.

7.4.1 Group 1: No water added – Requiring conveyance for treatment at central works

7.4.1.1 Chemical toilets

A chemical toilet stores excreta in a holding tank which contains a chemical mixture to prevent odours caused by bacterial action. The contents of the holding tank must be emptied periodically and conveyed to a sewage works for treatment and disposal. Some units have a flushing mechanism using some of the liquid in the holding tank to rinse the bowl after use. The chemical mixture usually contains a powerful scent as well as a blue dye. Chemical toilets can range in size from the very small portable units used by campers to the larger units supplied with a hut. The system can provide an instant solution and is particularly useful for sports events, construction sites or other temporary applications where the users are accustomed to the level of service provided by a water-borne sanitation system. The system can also be used where emergency sanitation for refugees is required, in which case it can give planners time to decide on the best permanent solution.

Factors to consider before choosing this option are the following:
- Chemical toilets have relatively high capital and maintenance costs.

- It is necessary to add correct quantity of chemicals to holding tank.
- Periodic emptying of the holding tank is essential, requiring vacuum tanks.
- The system only disposes of human excreta and not other wastes.
- Units can be installed quickly and easily.
- They can be moved from site to site.
- They require no water connection and very little water for operation.
- They are hygienic and free from flies and odour, if maintained correctly.
- Chemicals could have a negative effect on treatment works.
- No power is needed except lighting.

7.4.2 Group 2: No water added – no conveyance (treatment or partial treatment on site before disposal)

7.4.2.1 Ventilated improved pit (VIP) toilet

The VIP toilet is a pit toilet (Bester and Austin, 1997) with an external ventilation pipe. It is both hygienic and inexpensive, provided that it is properly designed, used and maintained. The vent pipe should be black and at least 100 mm, but preferably 150 mm, in diameter. The upper end of the vent should be covered with a gauze screen stretched over the end of the pipe. Wind blowing over the open end of the pipe causes negative pressure, which draws air through the toilet pedestal and the pit, and up the vent pipe; this keeps the toilet free of odours. This venting action is assisted by the thermal effects of the sun shining on the black vent pipe. Since odours are generally absent, flies are not attracted to the pit. The superstructure should be kept relatively dark inside so that any flies that get into the pit via the pedestal will be attracted to the light entering the pit through the vent pipe. When they try to leave the pit via the vent pipe, they are trapped by the gauze screen and will eventually die and fall back into the pit.

It is possible to construct the entire toilet from local materials although it is more usual to use commercial products for the vent pipe and the pedestal. Several toilet superstructures are also commercially available. When the pit is full, the superstructure, pedestal, vent pipe and slab are normally moved to a freshly dug pit and the old pit covered with soil. The VIP toilet can be made more permanent by lining the pit with open-jointed brickwork or other porous lining. The pit can then be emptied when required, using a suitable vacuum tanker. Water (sullage) poured into the pit may increase the fill-up rate and should be avoided. The VIP is therefore not recommended where a water supply is available on the site.

Details information on VIP design is available in a publication by Bester and Austin (1997). Factors to consider before choosing this option:

- If plot is small, there may be insufficient space for relocation.
- Unlined pit walls may collapse.
- Excreta are visible to the user.
- System may be unable to drain all liquid if large quantities of wastewater are poured in.
- The cost is low; it is cheapest form of sanitation maintaining health standards.
- Toilet can be constructed by the community, even if unskilled.
- Locally available materials can be used.
- Components can be manufactured commercially and erected rapidly.
- The system is hygienic provided it is used and maintained correctly.
- The system can be used in high-density areas if a pit-emptying service exists.
- The system can be upgraded at a later stage to increase user convenience.
- The system cannot ordinarily be installed inside a house.
- The quantity of water supplied to the site should be limited.
- If a pit-emptying service exists, access to the pit should be provided.

7.4.2.2 Ventilated improved double-pit (VIDP) toilet

The VIDP toilet, also known as the twin-pit composting toilet, was developed mainly for use in urban areas where, because of limited space on smaller plots, it may be impossible to relocate the toilet every time the pit becomes full. The VIDP is a low-cost, simple but permanent sanitation solution for high-density areas. Two lined shallow pits, designed to be emptied, are excavated side by side and are straddled by a single permanent superstructure. The pits are used alternately. When the first pit is full, it is closed and the prefabricated pedestal is placed over the second pit. After a period of at least one year, the closed pit can be emptied, either manually (if this is culturally acceptable) or mechanically, and then it becomes available for re-use when the other pit is full. Each pit should be sized to last a family two to three years before filling up. It is important that the dividing wall between the pits be sealed, to prevent liquids seeping from the pit currently in use into the closed pit, thus contaminating it. The VIDP toilet can be built partially above the ground in areas where there is a high water table.

Factors to consider before choosing this option are similar to those for the VIP, but include the following:

- Training is required to ensure that the pit lining is properly constructed.
- The contents of the used pit may be safely used as compost after one year.
- User may not be prepared to empty the pit, even though contents have composted.
- The superstructure can be a permanent installation.
- The system can be regarded as a permanent sanitation solution.
- The system can be used in high-density areas.
- The system can be used in areas with hard ground, where digging a deep pit is impractical.

7.4.2.3 Ventilated vault (VV) toilets

The VV toilet is basically a VIP toilet with a watertight pit which prevents seepage. It can be regarded as a low-cost, permanent sanitation solution, especially in areas with a high groundwater table or a poor capacity for soil infiltration, or where the consequences of possible groundwater pollution are unacceptable. The amount of wastewater disposed of into the vault should be limited to avoid the need for frequent emptying, so it is advisable not to use this type of sanitation technology where a water supply is available on site. Should an individual water connection be provided to each stand at a later date, as part of an upgrading scheme for the residential area, then the vault can be utilized as a solids-retention tank (digester) and liquids can be drained from the site using a settled sewage system, also called a solids-free sewer system (see the section on "Settled sewage systems" under Group 3). Ventilation, odour and insect control operate on the same basis as the VIP toilet.

Factors to consider before choosing this option are similar to those of the VIP, but include the following:

- It is necessary to use a vacuum tanker service for periodic emptying of the vaults.
- Builder training and special materials are required if a waterproof vault is required.
- The system can be used in areas with a high water table if properly constructed.
- The system can be used where pollution of the groundwater is likely if VIP toilet is used.
- The system can be used in high-density areas.
- This system provides good opportunities for upgrading since the vault can be used as a solids-retention tank when upgrading to solids-free sewers.

7.4.2.4 Continuous composting (CC) toilets

Continuous composting toilets make use of air to enable aerobic bacteria to break excreta down to fertilising material. The excreta is contained in a vault below the pedestal and liquid input must be kept to a minimum. Vegetable matter must be added in the right quantity and compost must be removed at the right rate.

Factors to consider before choosing this option are the following:

- This toilet is considered sophisticated and user education may be required.
- A high standard of maintenance is required.
- The excreta are visible to the user.
- The toilet cannot be constructed by unskilled labour.
- The amount of water added to the system must be limited, which may require that separate provision be made to dispose of waste water.
- The user is required to remove the compost manually.
- The toilet is a permanent installation.
- Compost is produced.
- The toilet is usually free of insects.

7.4.3 Group 3: Water added – requiring conveyance (treatment at a central works)

7.4.3.1 Full water-borne sanitation

This is an expensive option and requires ongoing maintenance of the toilet installation, the sewer reticulation and the treatment works. The system requires a water supply connection. Water is used to flush the excreta from the toilet pan and into the sewer, as well as to maintain a water seal in the pan. The excreta is conveyed by the water, in underground pipes, to a bulk treatment works which may be a considerable distance from the source. The treatment works must be able to handle the high volume of liquid required to convey the excreta. The quantity of water required (usually 6–10 litres per flush) can be reduced by using low-flush pans designed to flush efficiently with as little as 3 litres. Research has indicated that the operation of the sewer system is not adversely affected by low-volume flush toilets. Flush volumes of 8–9 litres are normally used. In an area where water is costly or scarce, it may be counter-productive to purify water only to pollute it by conveying excreta to a treatment and disposal facility.

Factors to consider before choosing this option are the following:

- The system is expensive to install, operate and maintain.
- The system can be designed and installed only by trained professionals.

- The treatment works must be operated and maintained properly to prevent pollution.
- The system requires large amounts of water to operate effectively and reliably.
- The system is hygienic and free of flies and odours, provided it is properly maintained.
- A high level of user convenience is obtained.
- The system can be regarded as a permanent sanitation system.
- The toilet can be placed indoors.
- This system can be used in high-density areas.
- An adequate uninterrupted supply of water must be available.

7.4.3.2 Flushing toilet with conservancy tank

This system consists of a standard flushing toilet which drains into a storage or conservancy tank on the property; alternatively, several properties' toilets can drain into one large tank. A vacuum tanker regularly conveys the excrement to a central sewage treatment works for purification before the treated effluent is discharged into a watercourse. The appropriate volume of the conservancy tank should be calculated on the basis of the planned emptying cycle and the estimated quantity of wastes generated. The volumes of tanks are sometimes prescribed by the service provider.

Factors to consider before choosing this option are the following:
- The system is expensive to install, operate and maintain, although the capital cost is lower than the fully reticulated system.
- The system can be designed and installed only by trained professionals.
- A treatment works must be operated and maintained properly to avoid pollution.
- Vacuum tankers must be maintained by the local authority.
- Regular collection is essential.
- The system is hygienic and odour-free, provided it is properly operated and maintained.
- A high level of user convenience is obtained.
- The toilet can be placed indoors.
- The system can be used in high-density areas.
- This system has good potential for upgrading since the conservancy tank been be used as a digester when upgrading to a settled sewage system.
- An adequate, uninterrupted supply of water must be available.

7.4.3.3 Settled sewage systems

In settled sewage systems, as known as solids-free systems, the solid portion of excreta (grit, grease and organic solids) is retained on site in interceptor tanks (septic tanks) while the liquid portion of the waste is drained from the site in a small-diameter sewer. Such sewers do not carry solids, and have very few manholes. Tolerances for excavation and pipe laying may be greater than for conventional sewers, allowing lesser skilled labour to be used. Although the liquid portion of the waste must be treated in a sewage works, the biological design capacity of the works can be greatly reduced because partial treatment of the sewage will take place in the retention tank on the site. The tank will also result in a much lower peak factor in the design of both the reticulation and the treatment works. The retention tanks should be inspected regularly and emptied periodically, to prevent sludge overflowing from the tank and entering the sewers. This system is an easy upgrading route from septic tanks with soakaways, conservancy tanks, and other on-site systems, as they can be connected to a settled sewage system with very little modification. Tipping-tray pedestals and water-saving devices can also be used in settled sewage schemes.

Factors to consider before choosing this option are the following:
- The system requires a fairly large capital outlay if new interceptor tanks have to be constructed (i.e. if there are no existing septic tanks which can be used).
- Care must be taken to ensure that only liquid waste enters the sewers.
- Vacuum tankers must be maintained by the local authority.
- The system is hygienic and free of flies and odours.
- A high level of user convenience is obtained.
- The system can be regarded as a permanent sanitation system.
- The toilet can be placed indoors.
- All household liquid waste can be disposed of via this sanitation system.
- The sewers can be installed at flatter gradients and be designed to flow under pressure.
- Provides a reasonably priced option when on-site sanitation need to be upgraded because of increased water consumption and higher living standards.
- The system can be used in high-density areas.
- An adequate, uninterrupted supply of water must be available.
- Regular inspection of the septic tank/digester is required to prevent an overflow of sludge.

7.4.4   Group 4: Water added – no conveyance (treatment or partial treatment on-site before disposal)

These systems generally dispose of all or part of the effluent on site. Some systems retain only the solid portion of the waste on site and the liquids are conveyed to a suitable treatment and disposal facility. Systems that dispose of the liquid fraction on site require a soil percolation system, and the amount of liquid that can be disposed of will depend on the system's design and the permeability of the soil.

7.4.4.1   Flushing toilet with septic tank and subsurface soil absorption system

Water is used to flush the waste from a conventional toilet pan into an underground septic tank, which can be placed a considerable distance from the toilet. The septic tank receives the sewage (toilet water and sullage). The solids digest and settle to the bottom of the tank in the same manner as in the settled sewage system. The septic tank therefore provides for storage of sludge. The effluent from the tank can contain pathogenic organisms and must therefore be drained on the site in a subsoil drainage system. The scum and the sludge must be prevented from leaving the septic tank as they could cause permanent damage to the percolation system. It is therefore advisable to inspect the tank at intervals to ascertain the scum and sludge levels. Liquid waste from the kitchen and bathroom can also be drained to the septic tank. Septic tanks should be regarded as a high level of service.

Factors to consider before choosing this option are the following:
- It is relatively expensive and requires a water connection on each stand.
- It requires regular inspection and sludge removal every few years.
- Percolation systems may not be suitable in areas of low soil permeability or high residential density.
- They provide a level of service virtually equivalent to water-borne sanitation.
- The system is hygienic and free of flies.
- The toilet can be inside the dwelling.
- This system has potential for upgrading since the septic tank outlet can easily be connected to a settled sewage system at a future date.
- An adequate, uninterrupted supply of water must be available.

7.4.4.2 Low-flow on-site sanitation systems (LOFLOs)

The terms LOFLOs refers to the group of on-site sanitation systems which use low volumes of water for flushing (less than 2.5 litres per flush). These systems include a pedestal, digestion capacity and soakaway component. They are:
  (1) Aqua-privies;
  (2) Pour-flush toilets and low-flush systems; and
  (3) Low-flow septic tanks.

Aqua-privies:
An aqua-privy is a small, single-compartment septic tank directly under or slightly offset from the pedestal, The excreta drops directly into the tank through a chute which extends 100–150 mm below the surface of the water in the tank. This provides a water seal which must be maintained at all times to prevent odour and keep insects away. The tank must be completely watertight; it may therefore be practical to use a prefabricated tank. The tank must be topped up from time to time with water to compensate for evaporation losses if flushing water is not available. This can be done by placing a washing trough on the outside wall of the superstructure and draining the used water into the tank. The overflow from the tank may contain pathogenic organisms and should therefore run into a soil percolation system (it can also be connected to a settled sewage system at a later stage). The separate handling of grey water is also covered by Winneberger (1974) and Ledin (2000).

Pour-flush toilets and low-flush systems:
Pour-flush toilets use a small amount of water which is poured into the pedestal to effect flushing and to maintain a water seal.
   Low-flush toilets use a manual flushing mechanism to effect removal of excreta.

Low-flow septic tanks:
Low-flow septic tanks are usually manufactured as package units. Factors to consider before choosing these options are the following:
• The excreta are visible to the user in the case of aqua-privies.
• The tanks must be completely watertight.
• The user must top up the water level in the aqua-privy type to compensate for evaporation losses.
• The system is hygienic provided that it is used and maintained correctly.
• It is relatively inexpensive.
• The system can be regarded as a permanent sanitation solution.

- These systems have potential for upgrading since the tanks work in the same way as septic tanks and can thus be connected directly to a settled sewage system.
- The tanks need regular inspection and sludge removal is required from time to time.
- An adequate, uninterrupted supply of water must be available.
- The possibility of using water from a hand basin to flush the toilet, saves water.

Factors in the design of septic tanks are discussed by CSIR (1997), Drews (1986), Mara (1986), Rivett-Carnac (1991) and Winneberger (1984).

7.5 FACTORS AFFECTING CHOICE OF SANITATION

The primary reason for installing a sanitation system in a community is to assist in the maintenance of health and should be seen as only one aspect of a total health programme. The choice of a sanitation system by a community will be influenced by several factors, such as the following:

- The system should not be beyond the technological ability of the community in so far as operation and maintenance is concerned.
- The system should not be beyond the financial ability of the community to meet capital as well as maintenance costs.
- The system should take into account the level of water supply provided and possible problems with sullage management.
- The likelihood of future upgrading should be considered, particularly the level of service of the water supply system.
- The system should operate despite misuse by inexperienced users.
- In a developing area, the system should require as little maintenance as possible.
- The system chosen should take into account the training that can be given to the community, from an operating and maintenance point-of-view.
- The system should be appropriate for the soil conditions.
- The community should be involved to the fullest extent possible in the choice of an appropriate system.
- To foster a spirit of real involvement and ownership, the community should be trained to do as much as possible of the development work themselves.

Local customs should be carefully considered.

Table 7.3. Environmental classification of excreta-related diseases (Mara, 1996)

Category	Environmental transmission features	Major examples of infection	Environmental transmission focus
1. Non-bacterial faeco-oral diseases	Non-latent Low to medium persistence Unable to multiply High infectivity No intermediate host	*Viral:* Hepatitis A, E Rotavirus diarrhoea *Protozoan:* Amoebiasis Crystosporidiasis Giardiasis *Helminthic*: Enterobiasis Hymenolepiasis	Personal Domestic
2. Bacteriael faeco-oral diseases	Non-latent Medium to high persistence Able to multiply Medium to low infectivity No intermediate host	Campylobacteriosis Cholera Pathogenic *Escherichia coli* infection Salmonellosis Shigellosis Typhoid Yersiniosis	Personal Domestic Water Crops
3. Geohelminthiases	Latent Very persistent Unable to multiply No intermediate host Very high infectivity	Ascariasis Hookworm infection Strongyloidiasis Trichuriasis	Peri-domestic Field Crops
4. Taeniases	Latent Persistent Unable to multiply Very high infectivity Cow or pig host	Taeniasis	Peri-domestic Field Fodder crops
5. Water-based helminthiases	Latent Persistent Able to multiply High infectivity Intermediate host(s)	Schistosomiasis Clonorchiasis Fasciolopsiasis	Water Fish Aquatic species or aquatic vegetables
6. Excreta-related insect-vector diseases		Infections in 1-3 transmitted by insects Bancroftian filariasis transmitted by *Culex quinquefasciatus*	Peri-domestic Water
7. Excreta-related rodent-vector diseases		Infection in 1-3 transmitted by rodents Leptospirosis	Peri-domestic Water

Table 7.4. The Kalbermatten (1980a) model for sanitation programme planning

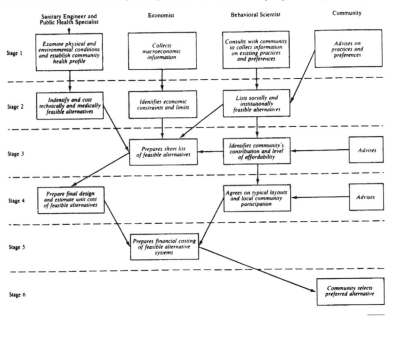

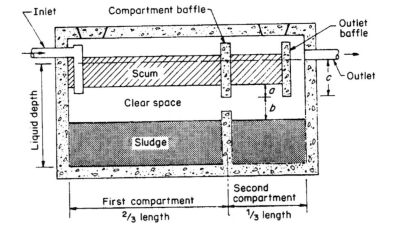

Figure 7.8. Septic tank

Table 7.5. Variations in person load (Henze et al., 2000)

Component	Level
BOD g/(person/d)	15 – 80
COD g/(person/d)	25 – 200
Nitrogen g/(person/d)	2 – 15
Phosphorus g/(person/d)	1 – 3
Wastewater m^3/(person/d)	0.05 – 0.40

Table 7.6. Concentration of pollutants in raw wastewater by separation of toilets and cleantech cooking (assuming: non-P detergents) (Henze et al., 2000)

Wastewater production		250 ℓ/cap/day	160 ℓ/cap/day	80 ℓ/cap/day
COD	g/m^3	130	200	400
BOD	g/m^3	80	125	250
Nitrogen	g/m^3	6	9	19
Phosphorus	g/m^3	1.6	2.5	5

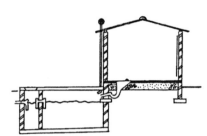

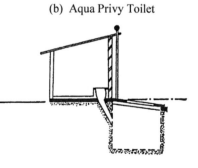

(a) Pour Flush Toilet (PFT)   (b) Aqua Privy Toilet

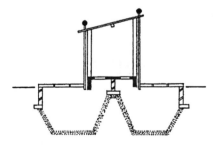

(c) Ventilated Improved Double Pit   (d) Reed Odourless Earch
    Latrine (VIDP)   Closet (ROEC)

Figure 7.9. Types of low cost toilets

Table 7.7. Micro-organisms in wastewater (number/100 mℓ) (Henze and Ledin, 2000)

Type	High	Low
E. Coli	5.10^8	10^6
Coliforms	10^{13}	10^{11}
Cl. perfringens	5.10^4	10^3
Faecal streptococcae	10^8	10^6
Salmonella	300	50
Campylobacter	10^5	5.10^3
Listeria	10^4	5.10^2
Staphyllococus aureus	10^5	5.10^3
Coliphages	5.10^5	10^4
Giardia	10^3	10^2
Roundworms	20	5
Enterovirus	10^4	10^3
Rotavirus	100	20

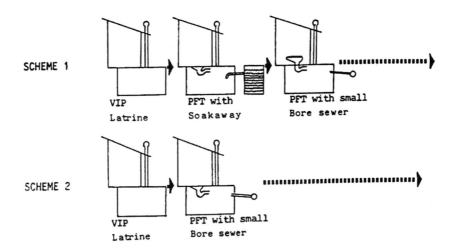

VIP = Ventilated Improved Pit latrine
PFT = Pour Flush Toilet

Figure 7.10. Comparison of on-site sanitation system with examples of phased improvements

7.6 REFERENCES

Alaerts, G.J., Blair, T.L. and Hartvelt, F.J.A. (Eds) (1991) *A Strategy for Water Sector Capacity Building*. IHE Report Series No. 24, International Institute for Hydraulic and Environmental Engineering, Delft.

Batchelor, A. (1994) Recent advances in constructed wetland technology. *Water and Sewage Effluent*. **14**(20), June.

Batchelor, A. and Loots, P. (1995) The performance and cost of waste water treatment systems incorporating constructed wetlands: Results from pilot scale systems. *Natural and Constructed Wetlands for Wastewater Treatment and Reuse*, International Seminar, Italy.

Benefield, Judkins and Parr (1984) *Treatment Plant Hydraulics for Environmental Engineers*. Prentice Hall.

Bester, J.W. and Austin, L.M. (1997) *Building VIPs: Guidelines for the design and construction of domestic ventilated improved pit toilets*. Division of Building Technology, CSIR, Pretoria.

Cairncross, A.M. (1990) Health impacts in developing countries: new evidence and new prospects. *Jnl of the Institution of Water and Enviro Management*, **4**(6), 571-577.

CSIR (Council for Scientific and Industrial Research), Division of Water Technology (1991) *Guidelines on the cost effectiveness of rural water supply and sanitation projects (Part 1)* and *Guidelines on the technology for and management of rural water supply and sanitation projects (Part 2)*. Report 231/1/93, Water Research Commission, Pretoria.

CSIR, Division of Building Technology (1997) *Septic tank effluent drainage systems*. Report No. BOU/R9706, Pretoria.

DWAF (Department of Water Affairs and Forestry) (1994) *Water Supply and Sanitation Policy*. White paper, Water – an indivisible national asset, Pretoria.

Drews, R.J.L.C. (1986) *A Guide to the Use of Septic Tank Systems in South Africa*. CSIR Technical Guide No. K86. National Institute for Water Research, Pretoria.

Fair, G.M., Geyer, J.C. and Okun, D.A. (1968). *Water and Wastewater Engineering*, J. Wiley & Sons, N.Y.

Franceys, R., Pickford, J. and Reed, R. (1992) *A guide to the development of on-site sanitation*. WHO, Geneva, p222.

Henze, M. and Ledin, A. (2000) Types, characteristics and quantities of classic, combined domestic wastewaters. Chap. 4 in *Decentralised Sanitation and Reuse*, ed. Lens, P. and Lettinga, G. IWA Publications, London.

Henze, M., Harremoes, P., La Cour Jansen, J. and Arvin, E. (2000) *Wastewater Treatment – Biological and Chemical Processes*. 3rd ed., Springer Verlag, Berlin.

Kalbermatten, J.M., De Anne, S.J., Gunnerson, C.G. and Mara, D.D. (1980a) *Appropriate sanitation alternatives – A planning and design manual*. World Bank.

Kalbermatten, J.M., De Anne, S.J., Gunnerson, C.G. and Mara, D.D. (1980b).*Appropriate technology for water supply and sanitation*. World Bank.

Ledin, A., Eriksson, E. and Henze, M. (2000) Aspects of groundwater recharge using grey wastewater. Chap. 18 in *Decentralised Sanitation and Reuse*, eds. Lens, P. and Lettinga, G., IWA Publications, London.

Mara, D. (1996) *Low Cost Urban Sanitation*. John Wiley & Sons, Chichester.

Mara, D. amd Sinnatamby, G.S. (1986) Rational design of septic tanks in warm climates. *The Public Health Engineer*, **14**(4), 49-55.

Meiring, P.G.J., Drews, R.J.L.C., van Eck, H. and Stander, G.J. (1968) *A guide to the use of pond systems in South Africa for the purification of raw and partially treated sewage*. CSIR Report WAT 34, Pretoria.

Metcalf and Eddy (1991) *Waste Water Engineering: Treatment, Disposal and Reuse*. McGraw-Hill, New York.

Nichols, P. (1991). *Social Survey Methods: A Field Guide for Development Workers*. Development Guidelines No. 6, OXFAM, Oxford.

Palmer Development Group (1992) *Technical, socio-economic and environmental evaluation of sanitation for developing urban areas in South Africa*. Draft Report No. B3.1, Water Research Commission, Pretoria.

Perkins, R.J. (1989) *Onsite Wastewater Disposal*. Lewis Publishers, Chelsea MI.

Pickford, J. (1995) *Low cost sanitation – A survey of practical experience*. Intermediate Technology Publications.

Rivett-Carnac, J.L. (1991) *Technical guidelines for designing and specifying on-site, low flush, anaerobic digester soakaway sanitation systems*. Appropriate Technology Information.

Serageldin, I., Cohen, M.A. and Leitmann, J. (Eds) (1995) *Enabling Sustainable Community Development*. Environmentally Sustainable Development Proc. Series No. 8., The World Bank, Washington DC.

Shrinivasan, L. (1990) *Tools for the Community Participation: A Manual for Training Trainers in Participatory Techniques*. PROWWESS/United Nations Development Programme, New York.

WISA (Water Institute of Southern Africa) (1988) *Manual on the Design of Small Sewage Works*. 1st Ed.

WHO (World Health Organization) (1994).*Operation and Maintenance of Urban Water Supply and Sanitation Systems: A Guide for Managers*. Geneva.

Winblad, E. Ed. (1998) *Ecological Sanitation*. SIDA, Stockholm,

Winneberger, J.H.T. (Ed.) (1974) *Manual of Greywater Treatment Practices*. Ann Arbor Science, Ann Arbor, MI.

Winneberger, J.H.T. (1984) *Septic Tank Systems: A Consultant's Toolkit – Vol. 1: Subsurface Disposal of Septic Tank Effluents; Vol. 2: The Septic Tank*. Butterworth Publishers, Stoneham MA.

8

Stormwater drainage

8.1 INTRODUCTION

Conventional stormwater drainage requires the removal of excess water from roads and other surfaces. A gravitational network leads water to rivers or lakes. A major step in the design process is the estimation of drain sizes. To do this, the engineer needs to estimate discharge rates. The methods described in this section are elementary methods of establishing design flows. Using locally applicable rainfall data, the engineer is able to select a storm intensity and convert this to a runoff rate for a particular catchment (Jones, 1971).

The methods described first are based on certain restricting assumptions, the main one being that any catchment has a unique time of concentration equal to the travel time down the catchment. The methodology culminates in the so-called rational method. This expression is the modern version of a number of earlier formulae. Despite their limitations, the methods are reputed to yield reasonable answers (Ardis et al, 1969; Schaake, 1967). The rational method in particular is simple to apply and it is easy to visualize the reasoning behind the formula. Although it only yields an initial design, subsequent refinement by more sophisticated methods and computer modelling are always available. Time area methods are based on the Rational method and linear hydrology, but break the catchment into a number of sub-catchments.

8.2 THE RATIONAL METHOD

The rational formula was proposed by an Irish engineer, Mulvaney, in 1851. It was first adopted in the United State of America by Kuichling in 1889, and in England by Lloyd-Davies in 1905. Lloyd-Davies used the equation in conjunction with an empirical equation for excess rain to yield a relationship between catchment area and runoff rate. The rational equation is

$$Q = CiA \qquad\qquad (8.1)$$

Where: Q = the flow rate
           i = rainfall intensity
           A = surface area of the catchment, all in compatible units.

Table 8.1. Rational Coefficient C

URBAN CATCHMENTS			
General description	C	Surface	
City	0.7 – 0.9	Asphalt paving	0.7 – 0.9
Suburban business	0.5 – 0.7	Roofs	0.7 – 0.9
Industrial	0.5 – 0.9	Lawn heavy soil, + 7° slope	0.25 – 0.35
Residential multi-units	0.6 – 0.7	2 – 7° slope	0.18 – 0.22
Housing estates	0.4 – 0.6	– 2° slope	0.13 – 0.17
Bungalows	0.3 – 0.5	Lawn sandy soil, + 7° slope	0.15 – 0.2
Park, cemeteries	0.1 – 0.3	2 – 7° slope	0.10 – 0.15
		– 2° slope	0.05 – 0.10
Frequency factor:			
Recurrence interval	*Multiplier*		
2 – 10 years	1.0		
25 years	1.1		
50 years	1.2		
100 years	1.25		
RURAL CATCHMENTS (less than 10 km^2)			
Ground cover	Basic factor	Corrections: Add or subtract	
Bare surface	0.40	Slope <5%: – 0.05	
Grassland	0.35	Slope > 10%: + 0.05	
Cultivated land	0.30	Recurrence interval <20 yr: – 0.05	
Timber	0.18	Recurrence interval > 50 yr: + 0.05	
		Mean annual precipitation < 600 mm: – 0.03	
		Mean annual precipitation > 900 mm: + 0.03	

Thus if A is in square metres and i is in metres per second, then Q is in cubic metres per second. C is a dimensionless coefficient normally less than unity. Thus C is the proportion of precipitation rate which contributes to peak runoff rate. Values of C for selected catchment characteristics are indicated in Table 8.1.

The hydrograph shape may be compiled from a simple rectangular model of the catchment (Fig. 8.1). If the rain continues indefinitely, the runoff will eventually equal the excess rainfall rate multiplied by the catchment area for flow balance. Initially the runoff will increase as more and more of the catchment contributes. Thus at any time t, the length of catchment contributing is x and the runoff rate is $CiAx/L$ where L is the total catchment length.

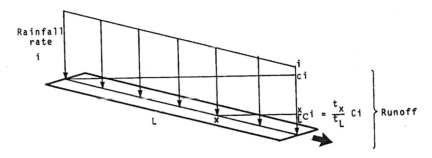

Figure 8.1. Distribution of rainfall and flow down a simple model catchment

If the concentration time is independent of the discharge rate then,

$$x/L = t_x/t_L \qquad (8.2)$$

where t_x is the concentration time over the length x. The hydrograph will therefore increase linearly as depicted in Fig. 8.2 until the entire area is contributing. Runoff will diminish after the rain stops. t_d is the duration of the storm. Then assuming a constant flow velocity, the tail of the hydrograph will fall over a time t_c again where $t_c = t_L =$ the concentration time of the catchment.

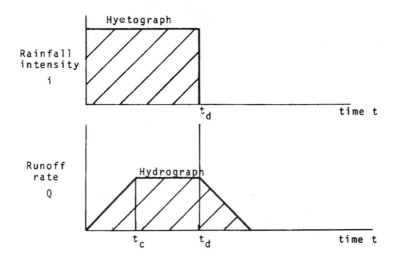

Figure 8.2. Rainfall and runoff versus time for Fig. 8.1

If the storm stopped at time t_c, then the hydrograph would be triangular with a base equal to $2t_c$ (Fig. 8.3). Thus the area under the triangular hydrograph is $CiAt_c$. This indicates that C represents the ratio of the volume of runoff to volume of precipitation, as well as the ratio of peak runoff rate to precipitation rate.

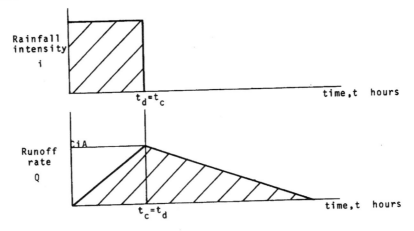

Figure 8.3. Rainfall and runoff for design storm

It is generally accepted, however, that the falling hydrograph tail has a duration exceeding t_c and it may even exceed $2t_c$. Then C does not represent the ratio of volumes. The longer recession limb implies an acceptance that the overland flow velocity reduces as the depth of flow reduces, and casts doubts on the reasoning behind the rational method.

C accounts for initial losses due to depression storage as well as infiltration during the runoff process. It implicitly accounts for the hydrodynamics of the runoff process whereby the runoff from throughout the catchment flows down to the mouth where the discharge Q is to be computed.

It includes the relationship between the recurrence interval of a storm and the recurrence interval of the runoff. For the way to use the formula is to select a storm of known frequency and compute the corresponding runoff assuming the same recurrence interval is applicable. Thus antecedent moisture conditions in the catchment, storm distribution and hydrograph shape are disregarded in deciding C. It is possible that different C values apply to different storms, but the C's listed here are those found to apply to representative design storms, i.e. of the order of 10 year recurrence interval. Thus the 10 year recurrence interval runoff rate is computed from the 10 year recurrence interval storm using the given C.

Variations in storm distribution in time and space are not accounted for. The effective duration of a storm to use in the intensity-duration relationship may not be the total storm duration. Rainfall intensity may vary during a storm and the duration over which the intensity average the design figure may be only a fraction of the total storm. Whether the design intensity occurs at the beginning or end of the storm will influence the antecedent moisture conditions which should affect C.

C should theoretically increase with rainfall intensity if losses are independent of intensity. Once the initial fraction is used to replenish initial abstractions, the balance occurs as runoff and the proportion of runoff to total precipitation increases the bigger the storm.

Rossmiller (1980) proposed the following empirical equation for estimating C from a variety of variables:

$$C = 7.7 \times 10^{-7} CN^3 R^{0.05} (0.01CN)^{-6S^{0.2}} (0.001CN)^{1.48(0.15-I)} (IM/2+0.5)^{0.7} \quad (8.3)$$

where: R is the recurrence interval in years,
       S is the land slope in percent,
       I is the rainfall intensity in inches per hour,
       IM is the fraction of watershed which is impervious, and
       CN is the SCS curve number.

Now for any selected storm recurrence interval, rainfall intensity reduces with storm duration and conversely increases the shorter the storm, in a manner which can be described generally by an equation of the for form:

$$i = \frac{a}{(b+t_d)}c \qquad (8.4)$$

where: i is the average rainfall intensity,
t_d is the storm duration, and
a, b and c are constants.

It is therefore apparent that the storm which will result in maximum runoff rate should be as short as possible, subject to equilibrium being attained, i.e. for maximum runoff intensity

$$t_c = t_d \qquad (8.5)$$

Table 8.2. Formulae for time of concentration for overland flow

Name	Formulae for t_c	Comments
Kerby	$3.03\left[\dfrac{rL^{1.5}}{H^{0.5}}\right]^{0.467}$	$L < 0.4$ km $r = 0.02$ (smooth pavement) 0.1 (bare packed soil) 0.3 (poor grass or bare) 0.4 (average grass) 0.8 (dense grass, timber)
SCS	$\left\{\dfrac{0.87L^3}{H}\right\}^{0.385}$	
Bransby Williams	$\dfrac{0.96L^{1.2}}{H^{0.2}A^{0.1}}$	
Izzard	$\dfrac{\left(0.024i^{0.33}+878k/i^{0.67}\right)L^{0.67}}{\left(CH^{0.5}\right)^{0.67}}$	
Airport	$3.64(1.1-C)L^{0.83}/H^{0.33}$	C = rational coefficient
Kinematic	$58N^{0.6}L^{0.9}/i_e^{0.4}H^{0.3}$	N = Manning roughness

A = area, km^2
H = elevation difference, metres
I = rainfall rate, mm/h. Subscript e refers to excess (runoff)
L = length of catchment, km
c_c = concentration time, hours

There are many empirical methods for establishing the time of concentration of a catchment. Various formulae in use are summarized in Table 8.2.

The formulae apply to specific types of surface, and use of an inapplicable formula should be avoided. Values of the constants in the equations are also indicated. Where compound areas are involved, the concentration time may be estimated by summating the concentration times over individual areas in series. Where the rational method is applied to compound catchments, the formula may be written as

$$Q_p = i \sum_{j=1}^{m} C_j a_j \qquad (8.6)$$

where

$$A = \sum_{j=1}^{m} a_j \qquad (8.7)$$

8.3 LLOYD-DAVIES METHOD

Lloyd-Davies in 1905 published a paper on an approach very similar to that of the more modern rational method. It was in fact Lloyd-Davies who proposed that the storm which produces the greatest runoff of all storms of the same frequency, is the one with a duration equal to the concentration time of the catchment. It was assumed that the concentration time was equal to the travel time from the top end of the catchment to the point at which flow is to be determined. He went further in assuming the concentration time was a function of catchment area.

A relationship between excess rainfall intensity and storm duration was also produced. This could be expressed in terms of the Birmingham formula

$$i = \frac{40}{20 + t} \qquad (8.8)$$

where: i is in inches per hour, and
         t is the concentration time in minutes

The formula was varied slightly by others for storm duration less than 20 minutes. The formula was for rainfall in England specifically, for an acceptable recurrence interval storm, which varies from twice a year for short storms to about once in 15 months for storms over one hour duration. It is interesting to note that the early British formulae, and even some modern English approaches, allow for runoff off impermeable surfaces only. 100 percent loss is assumed on permeable surfaces and the formulae are quoted as being applicable to stated percentage impermeable surfaces. In fact, the runoff formulae went so far as to give runoff directly in many instances (such as the Birmingham formula above). Thus incorporated in the formula is a percentage imperviousness, a storm intensity-duration relationship and a recurrence interval. The formulae are thus more of historical interest than for application The rational method allows for many of the earlier shortcomings. Nevertheless we are also indebted to Lloyd-Davies for development of the step method of computation of drain sizes.

8.4 STEP METHOD

Although the rational method yields a design flow rate at the mouth of a catchment, it does not provide sufficient date to design the individual drains in a catchment. In fact, none of the empirical equations for time of concentration are applicable to built-up areas with impermeable surfaces, artificial channels and circular drains. The concentration time in built-up areas is reduced due to the higher volume of runoff, the smoother surfaces and canalization.

In an effort to account for the flow time through each drain in the accumulation of flow, a step-by-step method was evolved in England by Lloyd-Davies (1905). The runoff from impermeable areas is accounted for, although earlier applications ignored runoff from pervious areas. The runoff from any catchment could be accounted for by applying individual runoff coefficients to each sub-area.

The method in common with the rational method uses a basic assumption contrary to hydrodynamic principles. This is that the concentration time can be estimated from the travel time for full flow down the drains. In fact, during flow concentration, flow rates will be less than the maximum, and flow velocities will be correspondingly lower for the full pipe case. Also at design flow for any pipe, pipes upstream will not be at full flow, as they are designed for a storm of shorter duration and consequently greater intensity. This is offset by another misunderstanding. Water does not need to travel the full drainage system length before an effective equilibrium is attained. In fact, reaction time of the system an be faster than flow time, as it is more a function of wave speed than water speed.

Nevertheless, the Lloyd-Davies step method yields results of satisfactory engineering accuracy. It has been found to overestimate peak flow rates for pipes over 600 mm diameter, but in order to improve on the method, more sophisticated mathematics is required (for example, the kinematic method). The Lloyd-Davies method is relatively simple to apply and drains may be sized in a systematic manner. The procedure is set out in tabular form, and calculations proceed from the top drain to successively lower drains (e.g. ASCE, 1969). Generally it was common to use a rainfall intensity-duration relationship of the form

$$i = \frac{a}{b + t_c} \qquad (8.9)$$

This is not a necessity for application of the method. The computations are set out with the aid of an example (Fig. 8.4) in Table 8.3.

The steps in the computations are shown on the following page (the numbers refer to the columns in Table 8.3):

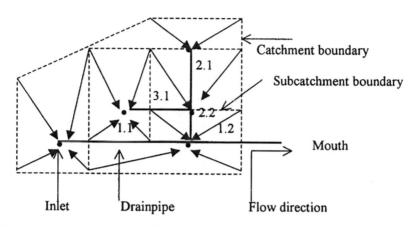

Figure 8.4. Catchment example for Lloyd-Davies step computations

Table 8.3. Step computations

1	2	3	4	5	6	7
Pipe No.	Contrib. area $(A(\text{m}^2))$	Coeff. C	Effect. area CA	Total ΣCA	Length (m)	Slope S
1.1	15 000	0.5	7 500	7 500	180	0.01
2.1	5 000	0.5	2 500	2 500	90	0.02
3.1	8 000	0.7	5 600	5 600	80	0.01
2.2	7 000	0.6	4 200	12 300	50	0.008
1.2	14 000	0.3	4 200	24 000	100	0.0025

8	9	10	11	12	13
Conc. time $t_c(\text{s})$	* i = (see eq. below)	$Q = i\Sigma CA$ m^3/s	Dia. mm	Vel. V m/s	Incr. $\Delta t_c = LV$
180	64	0.480	480	2.7	67
120	69	0.173	250	1.7	33
120	69	0.389	440	1.5	30
153	66	0.817	620	1.8	18
247	59	1.417	950	1.9	53

$$* \quad i = \frac{0.052}{630 + t_c} x 10^{-6} \ (\text{m/s})$$

(1) Mark pipe numbers in a plan, proceeding from the top pipe of each leg. In this catchment there are three levels of subdivision of the drains.

(2) From the contour plan and demarcated subcatchments planimeter is the area contributing to each pipe. If the inflow is along the length of the pipe it may be taken as into the head for simplicity. Alternatively (see White, 1978) it may be fed in along the pipe, in which case the design storm duration depends on that pipe diameter. The pipe diameter must therefore be determined by trial in such case.

(3) The proportion of runoff for each subcatchment must be estimated. The C is similar to the C in the rational formula.

(4) The effective contributing area is CA.

(5) Add the effective areas down to the pipe in question.

(6) Measure pipe lengths from the layout plan.

(7) Establish the gradient from contours. In the case of adverse ground slopes or minor drains, the minimum gradient may be dictated by minimum flushing velocity. In this case, a trial and error method may be required.

(8) The concentration time for upper pipes is based on the time of entry. This varies from 2 to 4 minutes for urban catchments, but may be larger for overland flow. In that case, it must be established as in the rational method or the kinematic method. For lower pipes, it is necessary to compare alternative feeders and select the feeder resulting in maximum concentration time. Thus for pipe 2.2, the concentration time down route 2.1 is 153 seconds (120 + 33) whereas for route 3.1, it is 150 seconds.

(9) The rainfall intensity is assumed to obey the relationship $i = a/(b + t_d)$ where a and b are constants, i is in m/s ant t_d is the storm duration, assumed equal to travel time for maximum runoff rate.

(10) Establish the peak discharge rate by multiplying i by the total effective contributing area.

(11) Pipe diameter may be selected from a flow versus head loss chart, which will also yield flow velocity for (12). It is unlikely that the full-flow pipe diameter so yielded will be a standard commercially available pipe size. In such cases, the nearest larger standard pipe size is selected, and the pipe may run part-full. Theoretically the flow velocity will then be different, but it is often conservative to utilize the full-pipe velocity.

(12) The flow velocity at design discharge can be read from pipe charts or calculated.

(13) The increment in travel time is now calculated by dividing the length of the pipe just determined by flow velocity. This increment is added to the travel time down preceding pipes to obtain travel time to the next lower pipe.

8.5 TIME-AREA DIAGRAM AND ISOCHRONAL METHODS

If travel time is assumed to be independent of storm intensity, then every point in the catchment will have a unique travel time to the mouth or point of discharge. In fact, one could plot isochrones on a catchment map as illustrated in Fig. 8.5 An isochrone in this context is a line of constant travel time. On a simple plane isochrones may be equidistant. Where conduits and overland flow are involved, water velocity down conduits is generally faster than over land, so the isochrones may exhibit anomalies at conduits.

For design purposes it is sufficient to mark isochrones along the conduits proceeding from the outer extremities and marking down each drain. After reaching an intersection, one can correlate isochrones on each leg meeting at the intersection. By joining points of equal travel time, one establishes isochrones. Ultimately, the entire catchment is thus demarcated into time zones.

One could then plot a graph of area contributing to the flow at the mouth of a catchment against time. As time passed, so more and more runoff from further up-basin would reach the mouth, until the entire catchment area was contributing. The time-area diagram shows the rate of build-up in contributing area (which is assumed proportional to flow) during the entire storm. It is a massed area curve. If area is multiplied by excess rain intensity, one obtains a massed flow. The slope of such a curve is proportional to flow rate. The storm intensity, however, also affects the runoff rate, and this is a function of duration. It is therefore necessary to consider the storm intensity-duration relationship together with the time-area graph and this is what the tangent method (Watkins, 1962).

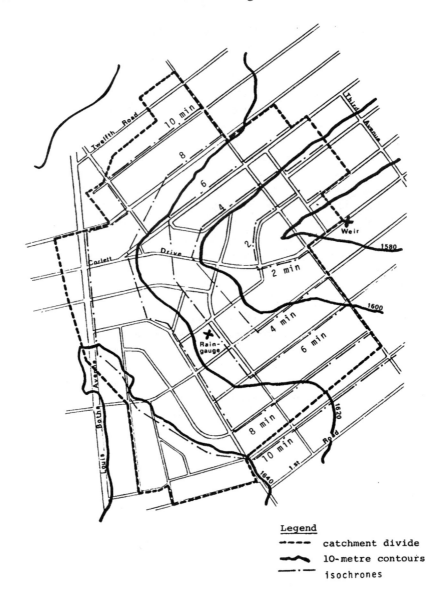

Figure 8.5. Contour plan with isochrones

8.6 TANGENT METHOD AND MODIFICATIONS

The step method outlined will yield a successively larger concentration time and hence longer design storm, for successive pipes, i.e. it assumes the entire basin must contribute for the maximum runoff. This is not necessarily so, as some outlying areas of the catchment may not contribute significantly to the area, but could nevertheless add to the travel time down the catchment. For odd catchment shapes, it will be shown that the rate of runoff can exceed that calculated using the Lloyd-Davies method. This stems from the time-area diagram which uses the steepest segment to define the effective contributing area for maximum runoff intensity.

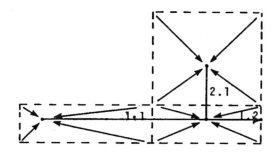

Figure 8.6. Catchment example for tangent method

The application of the tangent method to a time-area diagram will be demonstrated with the example in Table 8.4. The example is illustrated in Figure 8.6. Before constructing the time-area graph, it is necessary to do the steps in the Lloyd-Davies calculations (Table 8.4).

In Figure 8.7 are plotted the various contributing areas, starting on the time axis at the time at which they start to contribute and building up to the full value over the time of entry for that area. Thus the area contributing to drain 1.2 will reach the mouth first. The flow from this area will start at $t = 17s$, the travel time down the drain.

Table 8.4. Data for tangent method example

1	2	3	4	5	6	7
Pipe No.	Contr. area (m²)	Runoff coeff. C	Effect. area (m²)	Total ΣCA	Length (m)	Slope (m/m)
1.1	4 000	0.3	1 200	1 200	200	0.005
2.1	12 000	0.4	4 800	4 800	50	0.010
1.2	3 000	0.5	1 500	7 500	50	0.008

8	9	10	11	12	13	14
Time of entry (s)	Conc. time (s)	* $i =$ (see below)	$Q = i\Sigma CA$ m³/s	Pipe Dia. mm	Vel. m/s	Incr. $\Delta t_c = L/V$ (s)
140	140	217	0.261	430	1.9	117
120	120	238	1.140	670	3.2	16
120	257	144	1.081	690	2.9	17

$$* \ i = \frac{0.05}{90+t} x10^{-6} \ m/s$$

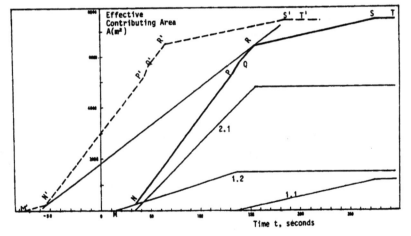

Figure 8.7. Time-area diagram and tangent solution

The build-up time for each drain is its time of entry (or overland flow concentration time). If inflow were over the entire length of drain, outflow would begin immediately and build up over $t_e + t_L$ where t_e is the time of entry and t_L is the travel time for that drain. Tracing back through the drainage system, each contributing area is plotted, commencing at its time of arrival at the mouth. Each contributing hydrograph (or in this case area-graph) will rise over its time of entry and then become horizontal, implying equilibrium is reach for that subcatchment. Each line is lagged by its travel time to the mouth. The total area graph is the sum of the individual area graphs, in the correct time positions (line OMNPQRST). It will be noted that portions of the area graph are steeper than others, in particular over the time during which the lateral contributions arrive. This fact points to the possibility that the peak discharge may result from a storm over the duration of the steeply rising portion of the area graph and not for the full concentration time of the basin. The shorter storm will be more intense and this may more than offset the reduction in area by omitting the flatter portions of the curve. If one assumes a storm intensity-duration relationship of the form

$$i = a/(b + t_d) \tag{8.9}$$

then $$Q = iA = aA = aA/(b + t_d) \tag{8.10}$$

Hence $$Q/a = A/(b + t_d) \tag{8.11}$$

Thus discharge Q is proportional to the slope of a line with a vertical to horizontal slope of A to $(b + t_d)$. A line drawn from $-b$ on the $A = 0$ axis in Figure 8.7 to point S on the area graph will have a slope proportional to the runoff from the entire catchment. Runoff from a portion of the catchment may, however, produce the peak flow. This will be represented by a line not originating at the base of the area diagram.

Now to find the worst (maximum) rate of runoff, it is necessary to find the steepest possible tangent. This is done most easily by drawing another time-area line a distance b in the time axis direction before the original time-area line. (Line 0' N' P' Q' R' S' T' in Figure 8.7). Now draw in a straight line tangential to the convex down part of this curve and also tangential to the convex upward part of the original curve (line N' R). This indicates that the maximum runoff will be associated with a storm of duration $t_d = t_R - t_{N'} - 6 = 154 - (-54) - 90 = 118s$. The corresponding contributing area is $\Delta A = A_R - A_N = 6450 - 200 = 6$ 250 m^2, and the runoff is $Q = \Delta A a / (b + t_d) = 6\,250 \times 0.05 / (90 + 118) = 1.5$ m^3/s.

An alternative method is as follows: Instead of drawing one tangent line, two parallel lines spaced b apart on the time scale are required. For maximum discharge, the left hand line should be tangential to the convex upward part of the curve as before. The right hand lower line should be tangential to the convex downward part of the area-time curve. To find the points of tangency resulting in maximum slope, it is necessary to keep the right hand line tangential to the convex downward part of the curve and gradually increase the slope from the horizontal until the parallel line spaced *b* to the left is also tangential to the convex upward part of the curve. At this stage, the maximum slope is achieved. The corresponding discharge Q is equal to the line slope multiplied by $a\Delta A$ / $(b + t_d)$ where ΔA is the contributing area between the two points of tangency, i.e. the vertical distance, and the design storm duration t_d is the horizontal distance between the same points.

The tangent method has been modified to allow for storm intensity-duration relationships not of the form suggested previously. Formulae of the following more general form have received recent recognition:

$$i = \frac{a}{(b+t_d)^c} \tag{8.12}$$

This type of formula produces curved tangent curves. Escritt (1972) proposed that transparent overlays be prepared on which a number of tangent curves of equal runoff be drawn. The method is time-consuming and not recommended. The extra effort is seldom worthwhile. If necessary, the method could be replaced by numerical methods.

The tangent methods are generally time consuming and not recommended for all catchments. An inspection of the plan should reveal whether there are outlying or sparse areas not likely to contribute to the runoff but which would influence the concentration time of the basin.

Escritt suggests that a modified rational method yields reasonable results for small catchments. He recommends taking a rainfall intensity independent of storm duration for durations of less than 15 minutes in England. Thus for a three-year recurrence interval storm in England, he suggests a storm intensity of 1 inch (25 millimetres) per hour. This procedure does away with the need for time of concentration calculations. It is also stated that it eliminates the necessity of doing a time-area graph.

In general in order to do numerically what the tangent method does in locating the critical contributing area, it is necessary to resort to a trial and error method. Once travel times are established for all points in the drainage system, a storm of a selected duration, equal to or less than the longest travel time is selected. By multiplying Ci by the area between any two isochrones spaced t_d apart, a runoff rate is yielded. This is repeated for different isochrones and for different storm durations. The worst storm is then selected from the results.

8.7 KINEMATIC METHOD

Time area methods do not consider the true hydraulics of runoff. They do not allow for higher rainfall rates causing more rapid concentration of flow and shorter storm durations, which in turn result in higher rates of runoff. Hydraulic methods, e.g. the hydrodynamic equations allow for many effects as indicated by the equations of runoff (see Stephenson and Meadows, 1986) for the derivations and simplifications of the equations):

Continuity: $\dfrac{\partial q}{\partial x} + \dfrac{\partial y}{\partial t} = i_e$ (8.13)

Dynamics: $\dfrac{\partial y}{\partial x} + \dfrac{v}{g}\dfrac{\partial v}{\partial x} + \dfrac{1}{g}\dfrac{\partial v}{\partial t} + S_f = S_o$ (8.14)

  (1)   (2)   (3)   (4)   (5)

where: y = water depth
   v = water velocity
   q = flow per unit width
   i_e = excess rainfall rate $i - f$
   f = infiltration and loss rate
   x = distance in direction of flow
   t = time
   g = gravitational acceleration

Term (1) is a backwater term (non-uniform flow) as well as term (2). Term (3) is the unsteady component. In the case of overland flow, terms (1) to (3) are negligible so the friction gradient S_f is equal to the bed slope S_o. Hence from Manning equation $q = \alpha y^{5/3}$ in SI units, where $\alpha = \sqrt{S_o}/n$ and n is the Manning's roughness, near 0.1 for shallow overland flow.

In general terms, $q = \alpha y^m$ (8.15)

8.7.1 Time of concentration for a plain

At any time to before equilibrium (t_c) to the right of x in Fig. 8.8, $y = i_e t$, and to the left of x, $\dfrac{\partial q}{\partial x} = i_e$.

Equilibrium is reached when $x = L$, and $y_L = i_e t_c$

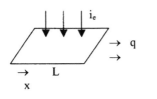

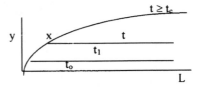

Figure 8.8. Sloping plain　　　　　　　　Figure 8.9. Water profile along plain

And from (8.15), this is $y_L = (q_L/\alpha)^{1/m}$,

where $q_L = e_e L$

Hence　　　　　$t_c = (L/\alpha i_e^{m-1})^{1/m}$　　　　　　　　(8.16)

And before $t = t_c$, $q_L = \alpha(i_e t)^m$　　　　　　　　　(8.17)

The shape of the hydrograph is thus as in Figure 8.10:

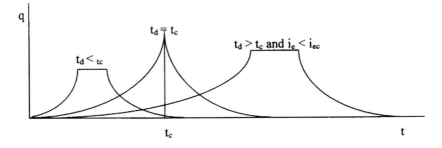

Figure 8.10. Hydrograph shapes for different storm durations

Since the concentration time of a catchment is a function of excess rainfall intensity, it is necessary to solve for q as follows:

From:　　　　　$i_e = i - f$

　　　　　　　　$t_c = (L/\alpha i_e^{m-1})^{1/m}$

　　　　　　　　$i = \dfrac{a}{b + t_d}$

　　　　　　　　$t_d = t_c$

Hence　　　　　$i_e = \dfrac{a}{b + (L/\alpha i_e^{m-1})^{1/m}} - f$　　　　　(8.18)

which can be solved for i_e

and $q = i_e L$

It is found that only very short plains reach equilibrium when time of concentration can be calculated thus. For plains in general, the peak occurs before $t = t_c$ since the rainfall is more intense and it is necessary to solve for the maximum runoff rate and corresponding storm duration as follows:

$$q = \alpha(i_e t)^m$$

$$= \alpha\left(\frac{a}{b+t_d} - f\right)^m t_d^{\,m} \tag{8.19}$$

q can be plotted against t_d to find maximum q. Such graphs have been generalized and $q/\alpha a^m$ has been plotted against t_d for selected values of the dimensionless loss parameters f/a and u/a, where u is an initial abstraction (loss) (Stephenson and Meadows, 1986).

This method has application to long and permeable catchments, where maximum flood peaks are for storms shorter than the theoretical equilibrium time. The results stem from the fact that concentration time is a function of excess rainfall rate, and infiltration rate is not proportional to rainfall rate as per the rational method.

8.8 REFERENCES

ASCE (American Society of Civil Engineers) (1969) *Design and Construction of Sanitary and Storm Sewers,* 332pp.

Ardis, C.V., Dueker, K.J. and Lenz, A.T. (1969) Storm drainage practice of 32 cities, *Proc. ASCE,* **95**(HY1), 6365, p383-408, Jan.

Escritt, L.B. (1972) *Sewerage and Sewage Disposal.* Macdonald and Evans, London, 494pp.

Jones, D.E. (1971) Where is urban hydrology practiced today? *Proc. ASCE,* **97**(HY2), 7917, p257-264, Feb.

Lloyd-Davies, D.E. (1905) The elimination of storm water from sewerage systems. *Min. Proc. Inst. Civ. Engineers,* **164**(2), p41-67.

Rossmiller, R.L. (1980) The rational formula revisisted. *Proc. Int. Symposium on Urban Storm Runoff.* University of Kentucky, Lexington.

Schaake, J.C., Geyer, J.C. and Knapp, J.W. (1967) Experimental examination of the Rational method. *Proc. ASCE,* **93**(HY6), 5607, p353-370, Nov.

Stephenson, D. and Meadows, M. (1986) *Kinematic Hydrology and Modelling,* Elsevier, Amsterdam, 250pp.

Watkins, L.H. (1962) *The Design of Urban Sewer Systems.* Road Research Tech. Paper 55, HMSO.

White, J.B. (1978) *Wastewater Engineering,* Edward Arnold, London.

9

Stormwater management

9.1 DESIGN ALTERNATIVES

Urban development is spreading over more and more of the earth's surface. The problems associated with urbanization are compounded as the density and extent of development proceeds (Schneider, 1975). The effect of particular concern is the elimination of most natural processes and their replacement by man-made streamlined procedures. One such system is the water cycle. Excess rain is no longer free to flow overland and meander along unlined channels. Instead, precipitation is on roofs or concrete or bitumen pavements, and it washes off conveying pollution created by mankind. Stormwater drains replace streams. They intensify runoff and destroy nature's balance. Channels flow more strongly in times of flood. Erosion and deposition occur. Natural self-purification processes such as re-oxygenation may be destroyed as the ecology is affected. Ground water is starved due to increased surface runoff. Vegetation, dust problems and the habitat may be affected. These factors demand a thorough environmental study in parallel with town planning and design of the infrastructure.

Civilization has focussed attention on the urban system. The convenience of
central facilities, mass transport systems and easy trade, have encouraged a
concentration at nodes we call towns or cities. The resulting disadvantages, such
as pollution and elimination of natural fields and streams, follow because man's
ambition exceeds his desire for a balanced life. Many of the problems are
unavoidable except at extreme expense. It is no use blaming the engineer for
problems which manifest. The engineer is able to solve problems at a minimum
of cost, but must work within a budget. The municipal councillor or national
politician is also limited in his abilities and budget. He must balance the ballot
against the fulfilment of ideals.

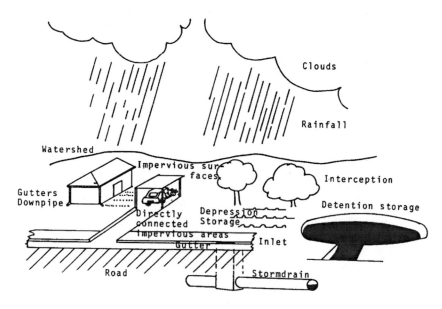

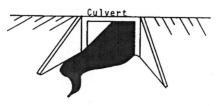

Figure 9.1. Urban drainage process

This chapter is aimed at the drainage engineer. It provides ideas and technology to compromise between limited budgets and best solutions. It presents design methodology for evaluating a best design to be achieved for stormwater drainage. Stormwater design objectives have changed over the years. The engineer used to attempt to remove stormwater as rapidly as possible. Nowadays the consequences of downstream flooding and alteration of the water cycle's regime are recognised. Methods of managing the flows or managing with them are developed (Stephenson, 1981) and most computer drainage models facilitate the management of flow quantity and quality.

The science of urban drainage has received considerable attention in recent years, especially in the United States. Legislation has forced engineers to think carefully about the drainage process. As a result, there have developed numerous research groups and mathematical models for simulating the runoff process.

The components of stormwater systems (Fig. 9.1) can be split into two groups: (Yen, 1978). One is the surfaces, basins, groundcover, gutters and inlets which are evident to all. The other is the major component from the engineer's point of view: namely, the drains, controls, underpasses, treatment or holding works, and final discharge rates and effluent quality.

9.2 STORMWATER MANAGEMENT PRACTICES

The construction of a stormwater drainage system is not the only way to avoid flooding or pollution. The day-to-day operation or management of the catchment will have an important bearing on runoff quantity and quality. On-site detention, retention and regular cleaning could relieve the drains of a considerable load. A reasonable management policy will be assumed at the time of design. Failure to maintain this programme could result in exceedance of the capacity of the system. On the other hand, improved management could alleviate the load on an underdesigned system, or enable more intensive development to take place in the catchment. Control measures may be structural (e.g. diversion, storage or channel improvements) or policy (e.g. insurance, flood warning systems or building control regulations for flood zones). Trotta et al (1977) described an automatically controlled drainage system.

Pollution control is the most obvious result of catchment management. Street sweeping, efficient refuse disposal, discharge monitoring and treatment of runoff are some of the possibilities.

Runoff control is a more difficult and more recent practice. Stormwater retention, groundwater recharge, provision of rough surfaces to retard flow and disconnection of impervious areas, are all logical practices but not explicitly required until recently in the United States and Europe. The methods are still not practiced in many countries.

Table 9.1 summarizes some practices and their accomplishment. The efficiencies and costs are discussed by Wanielista (1979).

Table 9.1. Stormwater management practices

Purpose	Method	Reason
Peak flow rate attenuation	Storm monitoring	Flood prediction
	Detention monitoring	Flood routing
	Channel storage	Flood routing
	Gravel surfaces	Retardation
	Rooftop storage	Routing and lag
	Parking lot storage	Routing and lag
	Disconnected impervious areas	Infiltration and attenuation
Runoff volume reduction	Retention storage	Removal of flow
	Diversion	Subtraction of flow
	Soakaways	Infiltration
	Basin recharge	Increase groundwater
	Infiltration	Flow reduction
	French drains	Seepage
	Swales	Retard flow, infiltration
	Porous pavements	Infiltration
	Contour ploughing	Infiltration
Provision for flooding	Insurance	Compensation
	Building control in flood zone	Limit damage
	Flood warning	Evacuation or diversion
Catastrophe aversion	Evacuation	Structural failure
	Sandbagging	High water levels
	Emergency overflows	Water flow control
	Weir strengthening	Dangerous flood levels
	Water tanks	Polluted water supplies
Erosion control	Berms	Settling
	Vegetation	Stabilization, retardation
	Rockfill	Flow control
	Mulching	Runoff control
	Fertilizing	Encourages vegetation
	Sediment removal	Basin renewal
	Settling basins	Catching sediment
	Screen	Detritus
	Centrifuge	Separation
	Contour ploughing	Surface storage
Pollution control	Street sweeping	Catching solids
	Street vacuuming	Catching fines
	Street flushing	Total removal
	Street de-icing	Ice removal
	Catching first flush	Most concentrated
	Refuse removal	Avoidance of pollution
	Storage	Settling
	Aeration	Biochemical oxidation

Purpose	Method	Reason
	Chemicals	Neutralization, precipitation
	Comminuters	Grinding large solids
	Flotation	Scum, emulsion, oil
	Legislation	Enforcement of standards
	Summons or fines	Discouragement
	Waste dump isolation	Runoff detention
	Grassing street verges	Catching fines, scum
	Fertilization methods	Minimization of washoff
	Land disposal	Removal of recoveries
	Treatment (screens, vortex, filter)	Separation of pollutants
	BMP (Best Management Practice)	Effluent quality, cost effectiveness
	Combined sewer overflow	Economy, ease of installation
	Real-time control	Minimize releases

Many of the systems for flow reduction must be incorporated at design stage. These include means of retarding the concentration time (increased surface roughness and detention basins), methods of catching part of the volume of runoff (diversion systems and retention basins), and means of reducing excess runoff (percolation basins, catchment cultivation and restricted capacity drains) (Carcich et al, 1974) combined with surface channels.

The system or combination of systems to adopt for any particular catchment will depend on the catchment characteristics, such as topography, soil type and cover, climate (rainfall and evaporation pattern) desired risk and the consequences of flooding. It should be recalled that interference with the runoff process complicates the relationship between storm recurrence interval and the recurrence interval of a failure of the stormwater system to do its duty. This aspect was studied by Kamedulski and McCuen (1978).

In order to at least reduce the flood flows from developed areas to the figures before man-made development proceeded, it is useful to understand the runoff process and its assessment. Urbanization reduces the average permeability of the ground by the construction of pavements and buildings. The concentration time is reduced due to the increased runoff intensity, smoother surfaces and man-made channels. The design storm is therefore a shorter, more intense storm than that resulting in maximum development. Natural basins or depressions may be levelled, thereby increasing excess runoff even further.

9.2.1 Safety factors

Any design or structure will normally be constructed with certain safety margins. This is not the same as the risk of overtopping or exceeding the capacity of the system. A storm of a certain recurrence interval will be selected from economic considerations using probability theory as outlined in a later section (Walesh and Videkovich, 1978). The present consideration is the margin of safety on top of the estimated design figure.

In the case of structures such as pipes, manholes, kerbs or weirs, the factor of safety with respect to strength is established routinely by designing according to a structural code of practice. Thus design stresses may be 50 percent of yield stress of steel, or 30 percent of the crushing strength of concrete. Hydraulic designs usually have little such margin. Thus freeboards may be calculated from wave height formula and not an arbitrary additional depth. In stability calculations for dams or weirs, a reasonably severe loading condition with extreme uplift and minimal resistance factors, should be allowed. In view of the dire consequences of failing of such a structure by overturning, the hydraulic engineer should consider applying safety factors in addition to designing for a high recurrence interval hydrological event.

The acceptance of mathematical models for hydrological and hydraulic analysis of a drainage system may lull the design engineer into a false sense of security. Impressive sensitivity studies and verification runs by computer may indicate reasonable margins of safety, but they may remain unknowns in the input data, the programming assumptions or in the interpretation of output. The engineer who is responsible for the drawings should therefore continue to apply normal safety factors based on judgement and the consequences of a failure.

Increasing legal action against drainage engineers in the United States of America has highlighted the need for precautions in design. Hopefully this will not result in excessive conservatism and increase in costs. The balance between economy of design and consequences of failure must not be imposed on the designer or constructor but on the responsible authority or its insurers. It is good practice to inform the affected public of design risks, flood levels and about insurance.

Where design alternatives exist which have similar costs it is sensible to select the least-risk system. Thus an open channel has usually a higher margin of safety against flooding than a closed conduit. This is because the capacity increases rapidly with increased depth in a channel. Adjacent low-lying parking lots and parks, even if not designed detention ponds, could, in an emergency, serve as such. Thus the drainage system should be integrated with the town planning.

9.3 DETENTION AND RETENTION PONDS

To compensate for runoff intensification due to urbanization, stormwater could be stored in man-made basins within the catchment. The ponds can be sited in non-essential areas, e.g. parks, the storage may be provided in depressions, which can subsequently be drained (or water permitted to percolate or evaporate in isolated situations, termed dry basins). Alternatively, the storage may be in channel freeboard or ornamental ponds or recreational lakes (referred to as wet ponds).

Drains could be led directly into the basins, in which case all initial flow would b caught before the drain overflowed and water filled the downstream drains. In this situation, the end-weir arrangement may induce a routing effect by the basin. Thus in addition to retention, the upper levels of the pond would act as a detention system. Detention is the temporary storage of runoff such as in freeboards, while retention involves the permanent diversion into evaporation or seepage ponds, i.e. flow is not returned to the drainage system. A comparison of different methods was made by Poertner (1978). Detention ponds are often referred to as onstream, or outlet controlled, but they are not necessarily so. Offstream ponds are usually inlet controlled.

The pond could be off-channel and water could be diverted from the drainage system. Here an unrestricted inlet such as a channel with a drop into the pond, could be constructed in which case the pond would fill during the initial stage of the storm. In such a situation, the pond may not affect the peak flow much in the case of extreme event storms. Alternatively, the inlet could be designed to divert flow only above a certain minimum to an off-channel pond. Such a device could be a side channel weir. The rate of diversion would increase with increasing discharge in the basin. Various control devices for influencing the stage-discharge characteristics of pond outlets were described by Hall and Hockin (1980).

Retention can be at the outfall of the catchment, in which case a large-scale pond is often required such as a lake or in a park. The retention may be provided along the drains, or it may be at the head of the drains (on site). The latter may obviate the requirements of a single large dam, and instead small ponding due to uneven surfaces, curbs, vegetation, terraces and rooftops, may be all that is required. On steep slopes, deep channels with cross weirs may be used.

The difference between retention and detention storage is illustrated in Figs. 9.2 to 9.5. (The reason for the difference in time to peak for high and low return periods will become clear if the section on kinematic hydrology in Chapter 8 has been studied.)

The method of design of retention basins is to select a design risk and plot the corresponding design hydrograph. Select a design discharge rate and indicate this on the hydrograph. The area under the hydrograph above the design discharge in the volume of storage required. It should be borne in mind that the critical storm duration is not that associated with the most intense storm. It will be considerably longer for the maximum volume of runoff. The critical storm duration depends on the pond design and must be determined by trial.

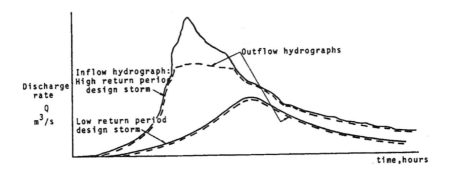

Figure 9.2. Effect of off-channel retention storage on runoff, assuming an inlet weir to pond off drain

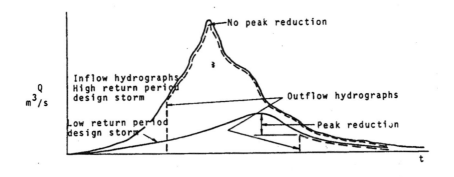

Figure 9.3. Effect of in-line retention storage on runoff, e.g. dam

For instream detention storage, the volume required may be estimated by drawing a straight line from the start of the inflow hydrograph to a point on the recession limb equal to the desired peak outflow rate. The area above the line below the hydrograph is an indication of the pond volume. Flood routing using numerical or graphical methods (e.g. Wilson, 1974) must be performed for the ultimate design. Approximate mathematical methods have also been proposed (Sarginson, 1973). It may be necessary to correct for the storage-discharge characteristics of the inlet to the pond, which means the cut-off line is not horizontal. In the case of in-line retention storage, the volume under the design hydrograph is stored.

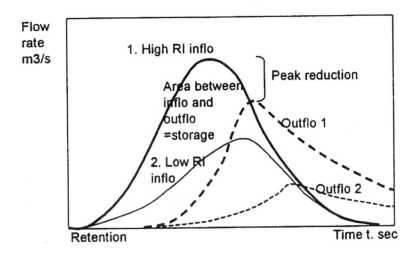

Figure 9.4. Effect of detention storage on hydrograph (weir type outflow)

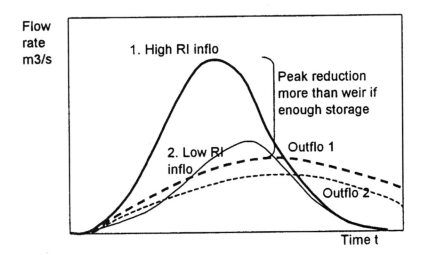

Figure 9.5. Effect of detention storage on hydrogaph (orifice type outflow)

For detention storage, alternative storage-discharge characteristics will have to be tried until a satisfactory compromise for a range of design hydrographs is achieved. A reduction will be achieved on all recurrence interval design storms.

To determine the critical inflow duration for an instream detention pond with controlled discharge, Wright-McLaughlin (1969) proposed a graphical method. The ordinates of the intensity-duration curve are multiplied by the storm duration and runoff coefficient, C. The resulting ordinates are plotted in a mass-flow curve of volume versus time. Now for any discharge (outflow) rate which plots as a straight line on the same plot, the storage required is the maximum difference between the massed inflow and outflow curves.

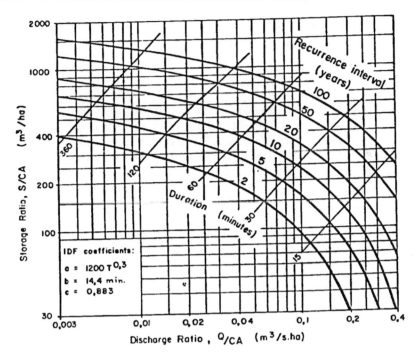

Figure 9.6. Example curves for the preliminary design of storage ponds

For an intensity-duration relationship such as

$$i = \frac{a}{(b+t)^c} \qquad (9.1)$$

an analytical solution for maximum storage is possible, Here i is storm intensity, t is storm duration and a, b, and c are constants for any locality and storm recurrence interval T. Now storage required is:

$$S = \frac{CAat}{(b+t)^c} - Qt \qquad (9.2)$$

For maximum S, dS/dt = 0   (9.3)

Hence $Q/CA = a\{b + t - ct\} / (b + t)^{c+1}$ (9.4)

Thus one can plot S/CA versus Q/CA with t as a parameter. Figure 9.6 is such a plot prepared by Watson (1981) for $a = 1200T^{0.3}$ (mm/h), b = 14.4 min and c = 0.883. From this chart, one is able to calculate the critical storm duration and storage S for any desired outflow rate Q, provided CA and T are known.

9.4 PERCOLATION BASINS

In theory, the ground can provide storage capacity equal to any storm which could be anticipated. The volume of storage per unit area is Dn, where D is the depth to the water table and n is the soil porosity. n is the ratio of volume of voids between soil particles to total volume, and is usually between 0.3 and 0.4, irrespective of soil particle size. Thus 1 m of soil could contain at least 300 mm of rainfall provided it could be absorbed sufficiently rapidly. Unfortunately, the permeability of the soil and impermeable cover usually limits the rate of infiltration. The rate of seepage per unit area is

$$\overline{v} = ki \qquad (9.5)$$

where $\overline{v}$ is the apparent seepage velocity (flow rate per unit area), k is the permeability, which may be as low as 10^{-9} m/s for impermeable clays and is affected by partial saturation (or binding). For granular soils it may be approximated by the equation

$$k = gd^2 n/800v \qquad (9.6)$$

where d is particle size, and v is the kinematic viscosity of water. Thus for 1 mm particle, $k = 9.8 \times 10^{-6} \times 0.3/800 \times 10^{-6} = 0.004$ m/s. This is greater than any rainfall rate. The hydraulic gradient i can reach a maximum value of unity. The actual rate of penetration of water is

$$v = k/n \qquad (9.7)$$

Thus the depth required to store p mm of rain is p/n and the time to infiltrate it is p/k.

Factors affecting the theoretical percolation will include the initial moisture content, which is water suspended on soil particles by surface tension, and this may be anything up to the full porosity for fine clays, although it is lower for coarse granular soils.

Air will also have to be released from the aquifer as water permeates down. The upflowing air will tend to suspend the water permeating downwards and may cause airlocks or impermeable barriers. Perched water tables may also form on the slightest lense of impermeable material. Attention should be paid to the drainage of the aquifer subsequent to saturation. The drainage rate will increase as the water table is raised and this may result in unexpected springs, marshes, soil erosion or even embankment instability.

9.4.1 Effect of holding on water quality

The retardation of escaping water by basins or seepage pits will reduce the rate of reoxygenation. In the case of wastewaters, or polluted runoff, this may result in obnoxious smells, or affect the ecosystem. If the water turns anaerobic, this problem is severe.

Oxygenation plays an important part in natural purification processes by reducing bio-degradable material, emulsifying solutions or oxidizing pollutants. The rate of oxygen absorption depends not only on the water surface area exposed, but also on the water depth and more particularly on the energy input. Turbulence in flowing water is a natural mixing device and considerably improves aeration. Stagnant water will lack this.

On the other hand, if treatment is required before the water can be discharged into natural water bodies, the storage pond may be the place for it (Wipple, 1979). Aerators, skimmers or sediment removal facilities may be constructed in ponds.

It must be realized that a pond will act as a natural silt trap and the ground may become unusable after a storm. The same adverse effect applies to floating pollution, such as oils, or to debris such as broken glass, tin cans, etc. The use of recreation fields for storage is therefore open to question.

9.4.2 On-site detention

Where practicable, property developers and owners should be encouraged to minimize direct runoff by directing gutters and downpipes to pervious or planted areas. The detention is thus spread over a large area and the depth of water is not at all inconvenient. Where the entire property is paved or covered, holding basins, porous layers or restricted capacity stormwater discharge pipes should be considered. In many countries, legislation is sufficiently advanced to force owners to discharge at a rate not greater than the virgin land would. Economic incentive to provide even greater control may be difficult though.

Such flow attenuation also minimizes soil erosion and may reduce the concentration of pollutants such as may settle or float or even have time to react. The concept of gutterless roofs used in some countries, actually retards runoff if runoff from the roof is direct onto the garden. Although gutters detain runoff, they concentrate it with a resulting net increase in peak flow.

9.5 STRUCTURES TO CONTAIN RUNOFF

9.5.1 Parking-lot storage

Parking lots offer one of the most convenient areas for ponding of water in densely built, commercial or industrial areas. By careful grading or dishing of the area, the ponds can be confined to isolated areas which are rarely used except at peak shopping hours.

Porous verges (Fig. 9.7) may be constructed adjacent to the parked area to absorb or convey the surplus runoff. They will also arrest sheet flow and retard flow into drains.

9.5.2 Rooftop detention

Flat roofs may be constructed with parapets to contain precipitation. Although the idea offers an otherwise unused area, it may increase the cost of construction considerably. Special attention will have to be paid to waterproofing. Additional loading must be allowed for in the structural design. Well designed control inlets to downpipes are required. These must also be covered by grates to minimize blockage by debris.

9.5.3 Combined sewers

The economics and hazards of combining stormwater and sewage are discussed in Chapter 6, but the question of attenuation effect on stormwater ingress is of interest here. There is generally a limit on the stormwater capacity of sewers, whether or not it is the intention to convey stormwater. Consequently, there is often backup which attenuates downstream flows. Alternatively, ways of handling stormwater in sewers may attenuate peak flows. Holding ponds and skimming weirs are designed to reduce peak flows into wastewater treatment works. Thus the hydraulics and economics are advantageous. But effluent quality and the environment suffer.

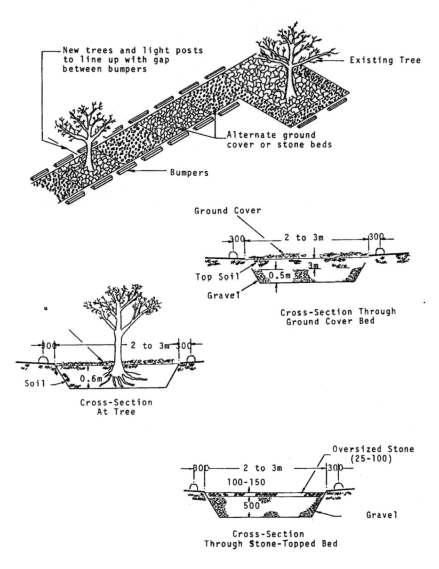

Figure 9.7. Details of median strip for parking lot drainage

9.6 OVERLAND AND CHANNEL RETARDATION

Although detention storage has been proposed as one solution to the attenuation of floods, this is in fact an artificial way, and suffers a number of disadvantages. Not least is the fact that valuable space is required to store floods. The factor of risk becomes increasingly difficult to assess. Whereas storage may be provided for a flood of a certain recurrence interval (the so-called design flood), what about greater floods? The attenuation effect is certainly not proportional to the flow rate, and in fact the reduction in the peak may be negligible in the case of larger floods than the design flood. There is also the question of storm duration to consider. The design storm duration for detention storage is invariably greater than for the channel design storm. The relationship between storage capacity and risk is therefore complex.

A better solution in many cases is to provide channel storage. This is a form of detention storage. Channel storage is affected by decreasing the flow velocity. This again has two effects. It increases the concentration time of the catchment, thereby reducing the design inflow, since storm intensity is known to reduce with duration for any particular recurrence interval. The channel storage also provides a way of holding back water and so reducing the peak discharge rate lower down.

One way to reduce flow velocity is to roughen the channel perimeter or adding bends. The cross sectional area required for any discharge is therefore increased, but certainly not in inverse proportion to the flow velocity. This is because the discharge rates actually reduce through the collecting system if the channel system is so designed to retard flow.

The flow at any section for any storm intensity could be obtained by routing excess rain down the system. But a simpler approach would be to estimate the concentration time, or time to reach equilibrium for any storm, and design the channel to convey the excess runoff corresponding to that storm. The kinematic approach could be used to estimate concentration time at any section. Flow retardation can be carefully controlled and easily designed if rockfill is used as a channel lining.

9.6.1 Dual drainage

Most urban drainage systems are designed to cope with a limited flood, called the minor system. Major floods surcharge underground drains and flow down roads. The flow stream thus created is shallow and slow flowing, i.e. the hydrograph is attenuated. Care should however be taken that the overflow of kerbs is negligible, and road layouts should be such that there is a continuous down gradient until a river is reached.

9.7 FLOOD MANAGEMENT

It is generally the peak of the flood which is of interest as that is the flow associated with the worst flooding along river banks. However, it is not only the peak of the hydrograph but also the volume of runoff over a length of time that affects the flood levels. When we look at the effects of storage in attenuating or routing floods, the volume of water under the hydrograph is of as much concern as the peak of the hydrograph, for it is the volume of water which fills the reservoir prior to it overtopping. In fact the discharge over the spillway or through an outlet from a reservoir is more a function of the storage volume in the reservoir than it is of the peak rate of flow into the reservoir.

With regard to storm drainage design, i.e. the sizing of man-made conduits which affects the water depth in natural channels, the original policy was to remove the storm water as quickly as possible, thereby alleviating local flooding problems. This could be taken to an extreme in many instances as the drainage channels, pipes, culverts and gutters were all designed to be as hydraulically efficient as possible. That is, the water velocities were increased over natural runoff velocities. Impervious cover due to urbanization not only increased the rate of runoff but also the volume of runoff. There was therefore less infiltration and the water resources, in particular the ground water regime of the catchment, were affected. Nevertheless by removing the stormwater as rapidly as possible, there was less flooding on roads and the cross sectional area of man-made conduits was minimized, hence the construction cost was minimized, particularly in the upper reaches of an urban catchment. However, downstream the flood flow rates increased over the natural flows as a result of the rapid concentration of water in man-made conduits.

It is generally the case that for the rapid concentration of water due to urbanization, the time to equilibrium between storm input and discharge is reduced, and therefore it is a shorter duration storm which becomes more critical. Shorter duration storms are associated with higher intensities, so in fact it is a higher intensity storm, and therefore a higher rate of discharge, which is associated with the rapid drainage of the system. This effect is compounded with the increased runoff creating even bigger floods downstream of the man-made drainage system and therefore higher rates of flow and deeper waters or greater flooding once the floods reach natural waterways. The waterways may then be unable to cope with the increased floods, with the result that the streams overflow onto their banks and cause damage previously rarely encountered.

With the realization that floods are not absolute and can be affected by the catchment and man, one can investigate which factors affect the magnitudes of floods. These include:

- *Rainfall*: intensity, duration, movement, spatial and time variation
- *Other sources of water*: snowmelt, dew, hail, groundwater
- *Man induced floods*: dam break, pumping, conduit burst
- *Topography*: slopes, overland, flow lengths
- *Catchment*: cover, roughness, reservoirs, temperature, shape
- *Ground*: antecedent moisture, permeability, capillarity, absorption, porosity, depth
- *Channel*: network, lengths, depths, cross sections, gradients
- *Vegetation*: foliage, height, root system
- *Man made*: cover, buildings, usage, drains

Many of the factors are inherent or cannot be affected by man. These could include topography and soil type. Others can be affected to differing degrees. For even rainfall can be affected by thermal currents from cities, or denudation of forests. The greatest effects of man are probably on the surface system, thereby speeding up runoff and creating bigger floods. Reduction of vegetation, construction of pavements and roofs, add to the volume of runoff. Increased runoff in turn creates erosion, deepens channels, reduces groundwater penetration and thereby accelerates catchment deterioration until the exaggeration effect is irreversible, even if planning policies change.

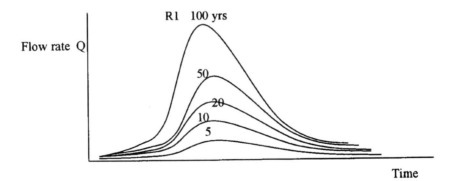

Figure 9.8. Hydrographs for critical storm peak

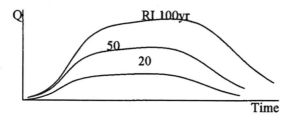

Figure 9.9. Hydrographs producing maximum volume

With the realisation of the effects of rapid drainage, the possibility of stormwater detention in order to attenuate the flood became popular. That is, storage basins were built along waterways in order to retard the flood and therefore reduce downstream flow rates, if not to original virgin catchment conditions then at least to an economic effect. Detention storage involves a temporary retaining of the water by means of a dam wall which would subsequently spill and release water downstream, i.e. after the reservoir behind the dam had filled, therefore lopping off a proportion of the peak and also attenuating the hydrograph due to routing.

There could be dangers due to higher floods than the design flood, as the detention storage would tend to catch the initial flows and once full would have a reduced routing effect. The same or even more so would be the case for retention storage, which is the abstraction of water from the river channel and dissipation by means of infiltration or evaporation. Again, once the reservoir is full, the routing effect on the hydrograph becomes negligible.

The same effect can be obtained by encouraging flood plain storage or channel storage, that is, by roughening the channels by means of rocks, gabions or vegetation, the water is retarded and will tend to be stored in the channel and on the banks and that also has a routing effect.

However, the more water which is stored along the waterway, the greater the area which will be inundated, and the average flood band would be wider than for an unattenuated flood. Thus, whereas the storage may act advantageously for smaller and more frequent storms, when extreme floods occur, the storage merely retards the flow to spread it over a greater width and cause greater flooding. Therefore, a proper flood management policy must go with detention storage construction.

The construction of dams and other obstructions such as culverts and bridges should have environmental studies to be aware of the flooding problems and other effects due to the backing up of water. There are a number of ways of reducing this detrimental effect including:

Water volume management in reservoirs: By releasing water from the reservoir prior to the flood hydrograph arriving, it is possible to minimize flood backwater upstream while maintaining the largest possible volume to absorb the flood hydrograph. However, some form of prediction of the inflowing hydrograph is required in order to release the correct volume of water. Too great a rate of release could cause a bigger flood downstream than actually would occur into the reservoir and the operator of the reservoir could be liable for damages. On the other hand, too small a release would minimize the routing effect. It is possible to predict rainfall in order to estimate the release and time available for releasing water from the reservoir using climate models. In the case of larger catchments, the rainfall may occur days or weeks ahead of the flood peak. That is, there is enough time to model the rainfall/runoff process and predict the flows and water levels in the river ahead of the flood. This would be real-time operation rather than projection into the future as the flood flows and water levels can be predicted as accurately as the hydraulic models are able to operate. Alternative management practices include warning systems to minimize damage and danger to life.

Flood release gates, i.e. gates on dams, also offer a way of attenuating floods but to a limited extent because in the long run the more storage the greater the backup and the lesser the effect on large floods. In fact, the net effect for abnormal floods is to increase the flood width upstream of the dam. Such gates can be operated automatically by floats, electrical sensors or by PLC (Programmed Logic Control) or manually. Radial gates which require a small lifting force, and collapsible/inflatable rubber cylinders have been used.

As a last resort, *fuse plug-type spillways* can be used. These are designed to collapse under extreme flows in order to prevent damage to the dam. However, in the extreme they are unlikely to reduce downstream flooding much and in fact may increase the flood due to a rapid increase in outflow from the dam. This type of mechanism includes gravitational structures such as blocks or collapsing walls and spring loaded gates.

It will be observed from Figure 9.4 that the attenuation effect of a reservoir with limited storage can be small, if not negligible, if a bigger flood occurs than designed for. This would be the case particularly if the outlet were an orifice with low outflow flexibility. An overflow spillway would be more inclined to act proportionally better for larger inflow.

The use of retention storage (abstraction of water) could be problematic for extreme floods. Once the storage was full it would pass the full flow rate. So it may only be the front of the hydrograph which is caught for big floods, and the peak may flow past unhindered. A wide spectrum of floods needs to be considered in designing flood management storage.

One can reach the optimum relationship between storage cost and outflow conduit cost by comparing a number of storage capacities. This could be facilitated using a generalized relationship such as Figure 9.6. This graph was prepared using generalized regional storm depth-recurrence interval data and assuming different outflow rates. All data is rendered dimensionless by dividing by the catchment effective area. Outflow rate was assumed constant during storm.

To use such a chart for any selected Recurrence Interval storm, one selects an acceptable discharge, divides by the catchment area times runoff proportion and reads off the required storage ratio. Multiply this by C x A where A is catchment area in hectares and C is runoff coefficient, to get storage in cubic metres.

9.8 RESERVOIR ROUTING METHODS

- Routing as a function of storage in the reservoir: This could be an increasing discharge over a spillway (weir flow) or an attenuated outflow such as through an orifice.
- Passive release: This includes spillways where outflow is a function of water level in the reservoir. Automatic gate operation would be included as passive release.
- An active operating rule would be based on the inflow as well as the storage in the reservoir so that there is some anticipatory attenuation. The inflow could be treated in one of three ways:
  (1) The routing could be done as a function of the measured inflow rate. This could be particularly useful if the inflow were measured a distance upstream of the dam so there was time enough to make a decision as to water to release from the dam.
  (2) Based on predicted inflow. That is, rainfall data could be telemetered to a computer and the river flows modelled using a deterministic model. Alternatively, climatic models may predict the rainfall rate even further ahead of time.
  In either of these cases, a sufficiently long time is made available between the calculated inflow and the actual inflow to enable the release to be optimised. Again, the object would be to release at a rate less than the peak inflow rate.
  (3) Probabilistic basis. If a probability analysis of previous inflows is made, then the probability of different rates of inflow and volumes could be made and an optimum operating rule devised based on these. However, there is no anticipatory component in this and individual storms may not be optimised, i.e. it is only over a long period of time that an average damage function can be minimized.

Method	Example	Limitations	Problems	Hydrographs:
				Flow rate
Absorption	Infiltration	Area total.	Town planning, Directly connected impervious areas	
Abstraction	Pumping or bypass	Rate of abstraction	Offchannel storage	
Proportion diversion	Weir	Cost	Backup	
Retention	Storage	Volume	Longer storms Bigger storms	
Detention	Storage dam Channel storage	Volume Depth	High flow Extra flood width	
Routing	River reach	Channel storage	Limited efficiency	
Flood plains	Wetlands	Area	Pollution	
Dual drainage	Pipe and road	Flood	Flooding capacity	
Gate operation	Barrages	Mechanical	Over-release	
Warning system	Telemetry	Electrical	Human	
				Time

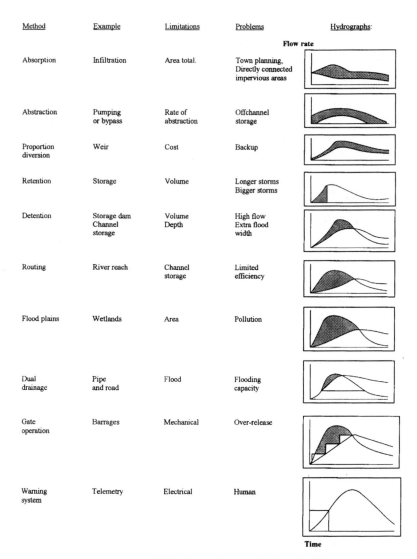

Figure 9.10. Summary of flood routing methods and effects

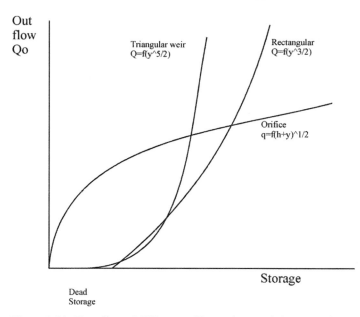

Figure 9.11. The effect of different spillway characteristics on outflow and storage

Whatever management practice is adopted, the flood spectrum will change and a new flood frequency probability diagram will result.

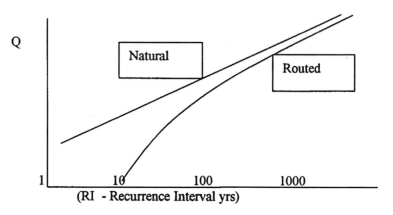

Figure 9.12. Effect of routing on flood spectrum.

Table 9.2. Optimization of bridge flood recurrence interval

A	B	C	D	E	F	G
Recurrence interval yrs	Flood m^3/s	Bridge cost $/an	Damage cost $	Probability $y = 1/A$	Probable damage cost	Total cost C+D+E
2	47	2,700 000	2,000 000	0.5	1,000 000	3,700 00
5	53	2,800 000	2,000 000	0.2	400 000	3,200 000
10	65	2,900 000	2,000 000	0.1	200 000	3,100 000
20	80	3,100 000	2,000 000	0.05	100 000	3,200 000
50	110	3,300 000	2,000 000	0.02	40 000	3,340 000
100	150	3,600 000	2,000 000	0.01	20 000	3,620 000
200	200	4,200 000	2,000 000	0.005	10 000	4,210 000

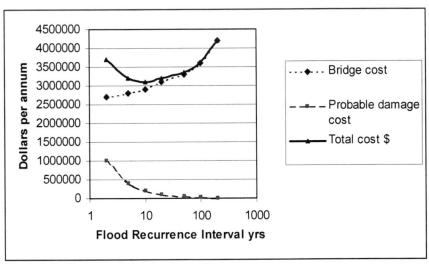

Figure 9.13. Optimum design flood

9.9 FLOOD RISK ANALYSIS

The most appropriate way of selecting a design flood for a structure is with probability analysis. This could apply to a bridge, culvert or spillway. The bigger the recurrence interval of the design flood at a site, the bigger the flood and therefore the higher the cost of the structure. On the other hand the risk of overtopping diminishes and the likely damage is reduced due to lower cost of repairs and danger. The bridge will be damaged if the design flood is exceeded. In Table 9.2, the cost of a bridge over a river is converted to an annual figure, and given for different flood recurrence intervals. The corresponding damage costs are also estimated, i.e. the cost of repairing the bridge and compensating for losses incurred. These are multiplied by the probability of them occurring in any year and added to the annualised cost of the bridge construction. The optimum design flow recurrence interval is where the total cost is a minimum, in this case 10 years. In Chapter 13, the theory of risk is taken further.

9.10   FLOOD PLAIN MANAGEMENT

The flood width on the banks of the river is a function of the flood magnitude represented by the recurrence interval as well as the physical factors, such as bank development and land use. There could be regular flooding at successively higher water levels. There is an area which will be inundated regularly, i.e. every few years, and this is referred to as the waterway. No development should be permitted in the waterway as there is a real and frequent danger to life and property.

At a higher level, such as the 1 in 25 year flood, there is still a risk of flooding which could endanger property and life, but it does not occur frequently enough to be a nuisance. The area within the boundaries of this flood would be called the flood zone, which comprises the waterway and the flood plains.

There is an area outside this flood zone at a higher level which could be flooded even less frequently and this area is referred to as the flood fringe. Figure 9.14 illustrates the various zones. Above the 100 year flood line, the possibility of flooding is so remote, it is seldom accounted for in planning, but it should be borne in mind in important developments, e.g. dangerous depositories.

If artificial flood banks are built along the river the flood plain can be confined. Then the question of what recurrence interval to design for is even more important for overtopping could be catastrophic. And it is necessary to detain the side inflow until the main river has subsided.

9.10.1 Hazards associated with flooding

The reason that development of private and public buildings is undesirable within flood areas is not only due to the fact that it endangers lives and property but also because the buildings obstruct the flow and cause backing up of the river. The concern regarding this damage is due to a number of reasons:

If the river is very deep and flows fast, it could wash away buildings and property and also drown people. At successively lower flows corresponding to lower water velocities and shallower depths, the water becomes safer and at a low depth it may not be life endangering, but it could still result in damage to property. However, shallow depths can be countered by protective measures which could be instituted if there was a warning of an impending flood.

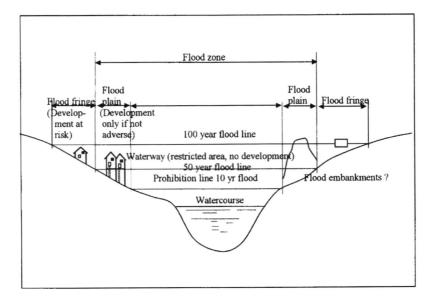

Figure 9.14. Flood plain zones

In other less dangerous situations, very shallow flooding on roads could be minimized by vehicles avoiding those roads or travelling very slowly. The danger or hazard associated with flow is generally expressed in terms of a diagram such as the flood hazard diagram below. This type of diagram has been adopted by various flood management organizations such as Minnesota in the United States of America (1983) and New South Wales in Australia (1986). The ranking of the hazard was simplified by Stephenson and Furumele (2001) in studies for Eastern Gauteng and a hazard risk index was proposed. Similar studies were done on rural rivers (Du Plessis, 2000; and Adriaans, 2001).

Another point to consider is the risk, or probability, of the hazard occurring. Thus, if the hazard is only likely to occur with a probability of 1% in any one year, it may be tolerable, but if it occurs frequently, e.g. at least once every year, it may be intolerable. This is the risk factor and it can be evaluated in terms of the recurrence interval of the flood. Thus a high risk index could be associated with floods occurring more often than once in 20 years and an intermediate risk index for floods occurring with a recurrence interval between 20 and 100 years, and low risk index for events which occurred less frequently than once in 100 years.

The hazard index and the risk index could be combined by multiplying them to give a hazard risk index which is an indicator of desirability of development within the floodway or flood fringe. Figure 9.15 below shows the results. A hazard risk index of 2 or less is considered acceptable.

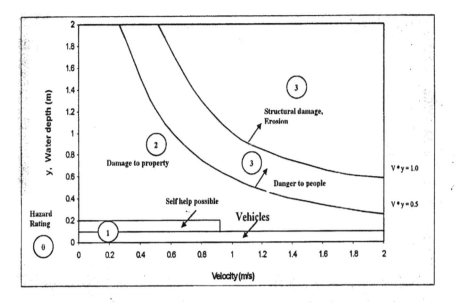

Figure 9.15. Flood hazard diagram

9.11   INTEGRATED FLOOD PLAIN MANAGEMENT

It is thus apparent that a more flexible approach than previously adopted is possible for planning development along rivers. In fact, an integrated management approach is considered at this stage. This implies that a trade off could be made between flooding and prohibiting development within the flood zone. The cost of not permitting development in the flood zone will increase over the years as attractive river bank land becomes scarcer. An economic balance between the average hazard cost and the availability of land is needed.

This will require a proactive approach. That is, a flexible approach within guidelines whereby developers may negotiate to develop within the flood fringe or even the flood zone. The developer may be able to physically design his development such that it results in negligible adverse effects but on the other hand has high social benefits. Alternatively, he may be able to contribute economically to the enhancement of the banks in return for being permitted to develop within an approved area. The principles of integrated management can be applied at this stage, and a detailed plan will emerge as more data is gathered. It would also be desirable to benchmark the proposed integrated flood plain management approach with other practices. International attention paid to integrated flood plain management is now producing useful guidelines and an international comparison would be of use.

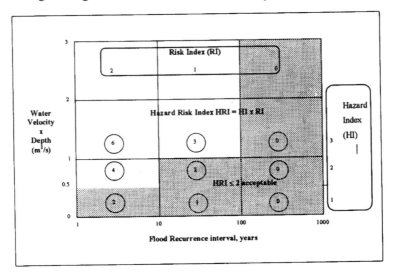

Figure 9.16. Hazard Risk Indices

The decision as to level of development on the banks of a river and how to close the river depends on what the effect of the development will be on the flood, as well as what the effect of the flood is on development. That is, will construction obstruct the flow of water, thereby affecting water levels upstream of the development? Also,

- What the effect of development will be on flood levels
- What a flood will do to the development
- How the value of the new and existing development is affected
- What the damage cost would be
- What pollution threat there is to the water
- The risk of adverse effects

9.11.1 Channel confinement

When land is particularly valuable, or the cost of flooding plains excessive, river channels have been confined between embankments. These earth embankments may be designed to contain a 100 year flood, but overtopping could have severe consequences.

The problem of detaining lateral inflow can be handled by gates. The critical storm duration of the lateral catchments is usually shorter than for the main channel and storage can be provided until the main channel subsides.

9.11.2 Anti-flooding devices

Many mechanical devices are available for use in drainage systems to prevent flow reversal. In gravity drainage systems, check valves and external flap gates have sometimes been used, but neither type is fully suitable for this function.

External flap gates (often referred to as tide flaps) are most commonly installed at the downstream ends of surface water drains to prevent flow entering them when river or tide levels are high. The externally hinged flap, usually of metal, plastic or rubber, can be attached to the end of a pipe to a headwall or to an internal wall of a manhole. Rubber duckbill devices that squeeze flat when the downstream pressure exceeds the upstream pressure are also included in this category. External flap gates are less suitable for use within gravity drainage systems because head differences may not be large enough for them to seal positively if the flow contains grit rag or faecal solids.

The anti-flooding devices selected need to have low head-loss characteristics and a good resistance to blockage. The method of closure that prevents backflow can take a variety of forms, including flap gates, gate valves or ball valves, either singly or in combination.

Inflatable rubber cylinders are also used in urban flood management systems.

9.12 REFERENCES

Adriaans, R. (2001) Development control in the floodplain, a key feature of river management. *IMIESA*, **26**(1), Jan, 22-23.

Carcich, I.G., Hetling, L.J. and Farrell, R.P. (1974) Pressure sewer demonstration. *Proc. ASCE*, **100**(EE1), 10315, 25-40 (Feb).

Du Plessis, L.A. (2000) A new and unique approach of flood disaster management. *S.A. Water Bulletin*, **26**(5), Sept., 16-19.

Hall, M.J. and Hockin, D.L. (1980) Guide to the design of storage ponds for flood control in partly urbanized catchment area. *CIRIA*, Tech. Note 100, London.

Kamedulski, G.E. and McCuen, R.H. (1978) Comparison of stormwater detention policies. *Proc. ASCE Conf. Verification of Mathematical and Physical Models in Hydraulic Engineering*, Maryland.

Minnesota Floodplain Act amended (1983).

New South Wales (1986) *Flood Development Manual*, Sydney.

Poertner, H.G. (1978) Stormwater detention and flow attenuation. In *Storm Sewer System Design*, Yen, B.C. (Ed.), Univ. of Illinois.

Sarginson, E.J. (1973) Flood control in reservoirs and storage ponds. *J. Hydrol.* **19**, 351-359. Also discussion by West, M.J.H. (1974), 23, 67-71.

Schneider, W.J. (1975) Aspects of hydrological effects of urbanization. *Proc. ASCE,* **101**(HY5), 11301, 449-468.

Stephenson, D. (1981) *Stormwater Hydrology and Drainage.* Elsevier, Amsterdam.

Stephenson, D. and Furumele, M. (2001) A hazard-risk index for urban flooding. *IAHR Congress,* Theme A: Development planning and management of surface and ground water resources, Beijing, Sept., 413-418.

Transport and Roads Research Laboratory (1976) A guide for engineers to the design of storm sewer systems. *Road Note 35, HMSO,* London.

Trotta, P.D., Labadie, J.W. and Grigg, N.S. (1977) Automatic control for urban stormwater. *Proc. ASCE,* **103**(HY12), 13396, 1443-1459, Dec.

Walesh, S.G. and Videkovich, R.M. (1978) Urbanization: damage effects. *Proc. ASCE,* 104 (HY2), 141-155, Feb.

Wanielista, M.P. (1979) *Stormwater Management, Quantity and Quality.* Ann Arbor Science, Ann Arbor, Michigan, 383 pp.

Watson, M.D. (1981) *Sizing of urban flood control ponds.* Hydrological Research Unit, Univ. of the Witwatersrand, Johannesburg.

Wilson, E.M. (1974) *Engineering Hydrology,* 2nd Ed. Macmillan, London.

Wipple, W. (1979) Dual-purpose detention basins. *Proc. ASCE,* **105**(WR2), 403-412.

Wright-McLaughlin, Engineers (1969) *Urban storm Drainage Criteria Manual,* Denver.

Yen, B.C. (1978) *Storm Sewer System Design.* Dept. Civil Engineering, Univ. of Illinois, 282 pp.

10

Drainage structures

10.1   HYDRAULICS OF BRIDGES

Bridges, culverts, causeways and fords are constructed by engineers to get traffic across waterways. A single span over a full channel width would not interfere with the flow in the channel. Economics and structural limitations usually require the bridge length to be less than the water surface width at maximum flow. The restriction on width and opening height often has the effect of backing up the water upstream of the bridge. The backwater thus created floods additional land upstream. A compromise between bridge opening and flooded area can often be achieved on an economic basis. The question of what risk of overtopping to design for is handled in Chapter 13. Most of the time the roadway is far above the water level and the opening is only partially used.

In the case of bridges, the control is usually at the entrance to the channel constriction, so streamlining the approach flow can increase the hydraulic capacity of the opening. The local reduction in width will cause higher water velocities than normal, with the result that the scour regime is affected and local scour in the vicinity of the bridge is likely unless some form of bed and bank protection is employed (Laursen, 1962; Kindsvater, 1957).

Beyond the constriction, the flow expands again to the full cross section of the channel. The flow expands after the constriction at about 5° from the centre line on each side. There is dead water on the downstream side of the embankments, and even some circulation which dissipates energy. The energy loss through the constriction is a function of the velocity head difference in the constriction and downstream. The water level, in the case of subcritical flow, is controlled further downstream. It may be normal level if there is a long uniform channel downstream. Figure 10.1 shows the flow pattern between two encroaching embankments (US Dept. Transport, 1978).

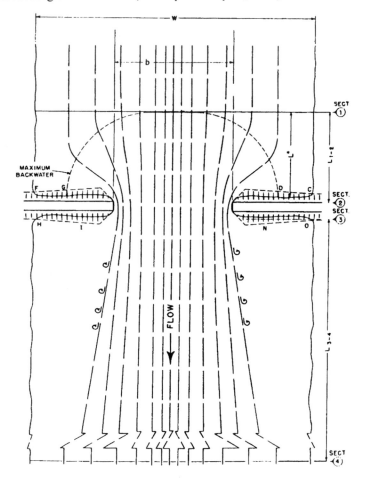

Figure 10.1. Flow lines for normal bridge crossing.

10.1.1 Flow through gap

The velocity through the trapezoidal shape formed by two facing embankments across a river gap obeys a relationship of the form

$$v = C\sqrt{2g(y_1 - y_2)} \qquad (10.1)$$

and flow
$$Q = CA_2\sqrt{2g(y_1 - y_2)} \qquad (10.2)$$

where the depth y_1 is upstream of the gap and y_2 is in the gap as indicated in Figure 10.2. In the case of drowned flow, y_2 is taken as the depth downstream or y_3 above bed level in the gap. The value of the coefficient C was found by Naylor (1976) to vary from 0.75 to 1.09 with a mean of 0.9. In the case of supercritical flow through the gap y_2 should be replaced by the critical depth y_c.

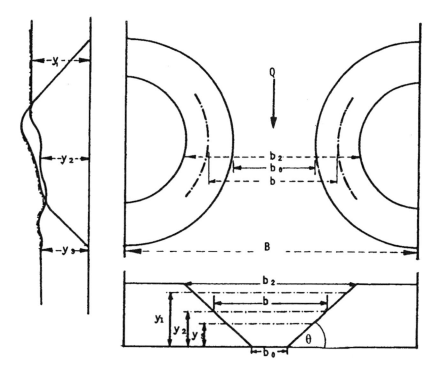

Figure 10.2. End-tipped embankment.

The critical depth in a trapezoidal section is given by

$$y_c / y_1 = 0.4\left[1 - 1.5p + \sqrt{1 + 2p + 2.25p^2}\right] \qquad (10.3)$$

where $p = b_0 \, (tan \, \theta)/2y_1$ and b_0 is the bottom width at the control section in the gap and θ is the angle of the embankment. The value of y_c/y_1 varies between 0.67 for wide gaps and 0.80 for triangular gaps. Thus for a triangular gap with free flow,

$$Q = 0.26\sqrt{2g} \; y_1^{5\,2} / tan\theta \qquad (10.4)$$

For the triangular shaped gap, it is possible to solve for the inside slope for stability of a granular or rockfill surfacing. Eliminating v and y_1 from Equations 10.1 to 10.4 and combining with an expression for stable stone size d, we get an expression for θ in terms of d and Q (Stephenson, 1979).

$$\frac{(tan\theta)^{2\,5}}{cos\,\theta\sqrt{tan^2\,\phi - tan^2\,\theta}} = \frac{8.25dg^{1\,5}(S-1)}{Q^{2\,5}} \qquad (10.5)$$

This equation may be solved by trial and error or iterative techniques for θ.

10.1.2 Surface profile

There may be one of three types of flow through a bridge waterway. The corresponding water surface profiles are depicted in Figure 10.3 and described below:

(1) If the water surface is above critical depth at every section, the flow is subcritical (type I flow). This is the condition normally encountered in practice and the calculation procedures following generally refer to this type of flow.

(2) The flow depth may pass through critical in the constriction. Under these conditions, the water depth upstream becomes independent of downstream conditions. If the depth passes through critical in the constriction, but not below critical depth downstream, it is referred to as a type IIA flow. If the flow depth drops below downstream critical depth, it is referred to as a type IIB flow. In this case, a hydraulic jump will occur below the constriction if the downstream depth is above the critical depth.

(3) If normal flow in the channel is supercritical, the water level in the constriction will rise as illustrated in Fig. 10.3 (III). Undulations of the water surface will probably occur and waves may occur upstream and downstream. No backwater in the normal sense will occur.

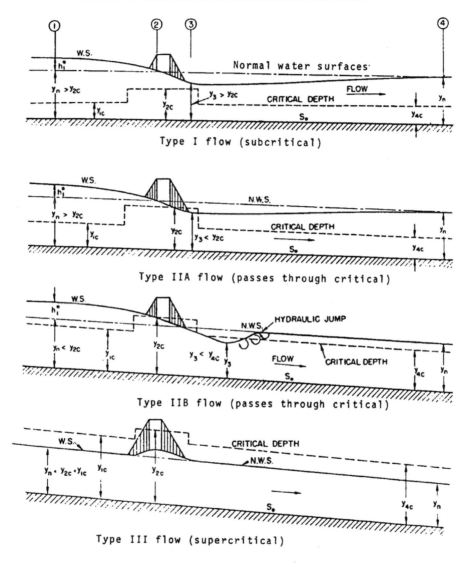

Figure 10.3. Flow profiles past embankments.

The head loss due to the constriction will be designated h_b and this is assumed to be given by an expression of the form

$$h_b = K\ \alpha_2\ v_2^2\ /\ 2g \qquad (10.6)$$

where v_2 is the average velocity at cross section 2 (the constriction) for water level at normal depth for the river section and α is a velocity energy coefficient, yielded by an integration across the section of qv:

$$\alpha = \dfrac{\sum_i q_i v_i^2}{QV^2} \qquad (10.7)$$

Q is the total discharge and V is the mean velocity across the section.

$$h_b = y_1 - y_4 = \dfrac{\alpha_4 v_4^2}{2g} - \dfrac{\alpha_1 v_1^2}{2g} + \dfrac{K\alpha_2 v_2^2}{2g} \qquad (10.8)$$

Now, since sections 1 and 4 are essentially the same, $\alpha_1 = \alpha_4$, and by continuity $A_1 v = A_2 v_2 = A_4 v_4$. Therefore

$$h_b = \dfrac{K\alpha_2 v_2^2}{2g} + \alpha_1 \left\{ \left(\dfrac{A_2}{A_4}\right)^2 - \left(\dfrac{A_2}{A_1}\right)^2 \right\} \dfrac{V_2^2}{2g} \qquad (10.9)$$

It should be noted that $y_1 - y_4$ is not the difference in water levels; it represents the build-up in water level (or backwater) upstream of the bridge. In addition, there will be friction head losses due to normal flow.

The backwater head loss coefficient K_b for flow normal to a symmetrical restriction may be read from Fig. 10.4. Here M is the bridge opening ratio,

$$Q_b\ /\ (Q_a + Q_b + Q_c) \qquad (10.10)$$

where Q_b is flow which pass through the same section as the bridge opening without the bridge there.

Since A_1 is not known until h_b has been determined, it is necessary to estimate h_b initially from

$$h_b = K\alpha_2 v_2^2 / 2g \qquad\qquad (10.11)$$

The value of A in Eq. 10.9 can then be determined.

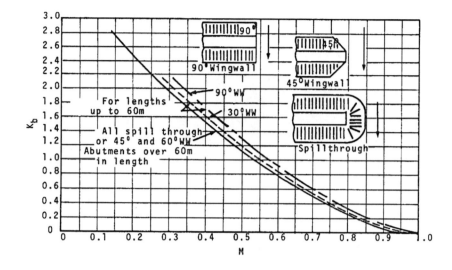

Figure 10.4. Backwater coefficient base curves (subcritical flow).

The backwater head loss is also affected by:

(1) The number, size, shape and orientation of piers in the constriction.
(2) The eccentricity of the bridge in the river section.
(3) Skewness of the bridge relative to the direction of the river.

The influence of these effects on K is given in Figures 10.5 to 10.7. Thus the K to use should be

$$K = K_b + \Delta K_p\, \Delta K_e\, \Delta K_s \qquad\qquad (10.12)$$

In the case of skew openings, the projected width of opening and not the total width of opening should be employed to determine M.

It should be noted that the results here are based on the assumption of one-dimensional flow. Laursen (1970) indicates that lateral flow can significantly increase the backwater effect.

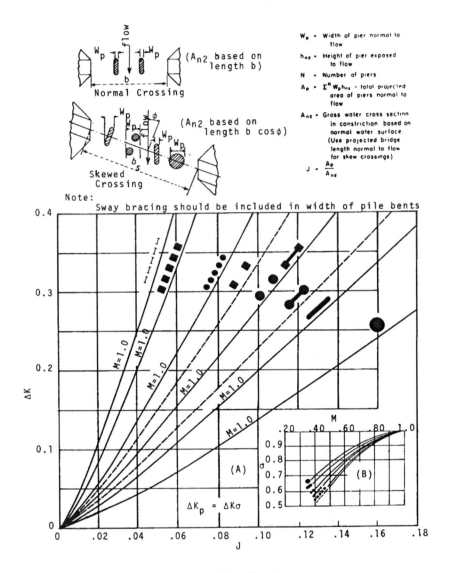

Figure 10.5. Increment on backwater coefficient for piers.

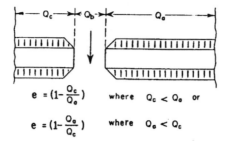

$$e = \left(1 - \frac{Q_c}{Q_a}\right) \quad \text{where} \quad Q_c < Q_a \quad \text{or}$$

$$e = \left(1 - \frac{Q_a}{Q_c}\right) \quad \text{where} \quad Q_a < Q_c$$

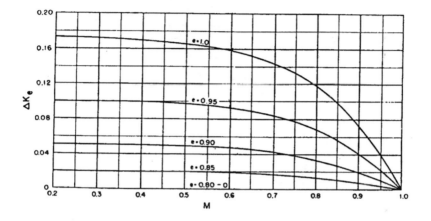

Figure 10.6. Increment on backwater coefficient for eccentricity.

Eccentric and skew channel approaches are unavoidable in some cases as road alignment is dictated by town layout, topography and economics. However, a higher deck is the consequence, as well as additional flooding upstream. Water velocities are also higher through constricted openings, resulting in more erosion unless banks are protected. Nowadays sloping embankments are not economic, as concrete walls can be built steeper even vertical. Reinforced Earth and Loffelstene offer structured concrete bank protection. The use of large opening culvert sections as bridges is another solution and these can be obtained with splayed approaches, high soffit faces and lined floors.

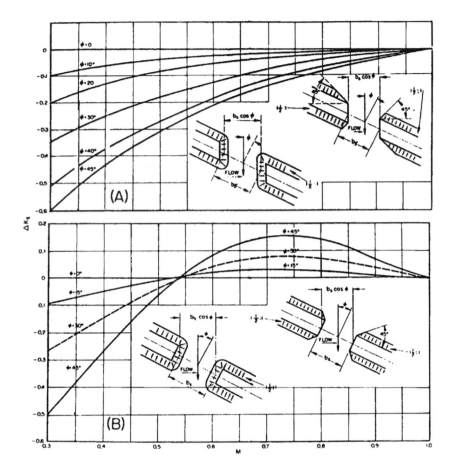

Figure 10.7. Increment on backwater coefficient for skew.

10.1.3 Drop in water level

The difference between the water level upstream and downstream of the bridge embankment is not the same as the backwater. The water level in the restriction is difficult to evaluate theoretically and it was investigated by model testing. Figure 10.8 presents the resulting data. To use the curve, compute and contraction ratio M and read off D_b the differential level ratio where $D_b = h_b / (h_b + h_3)$. Now with the previously computed backwater for a normal crossing, H_b, compute h₃ the drop in level.

$$h_3 = h_b\left(\frac{1}{D_b} - 1\right) \qquad (10.13)$$

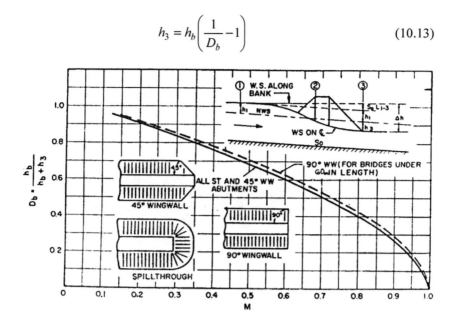

Figure 10.9. Differential water level ratio base curves.

The water level downstream of the construction is h_3 below normal level. With piers, it was found that although backwater h_b increased, h_3 remained as for a construction with no piers.

In the case of eccentric or skew crossing one adds the additional backwater Δh_c or h_s to h_b and determines h_3 from

$$h_3 = \left(h_b + \Delta h_c \ \text{or} \ \Delta h_s\right)\left(\frac{1}{D_b} - 1\right) \qquad (10.14)$$

where D_b is obtained as before.

Now the total difference in water level across the embankment is

$$\Delta h = h_3 + h_1 + S_o L_{1\text{-}3} \qquad (10.15)$$

where h_1 is the total backwater allowing for piers or eccentricity, S is the bed slope and L is the distance from section 1 to 3.

10.1.4 Complex structures

In the case of two bridges in close proximity, the backwater effect is not necessarily the sum of the individual effects. The close the two bridges, the nearer the resulting effect approaches that of one bridge. The backwater was found to increase by 30% to 50% for a distance between two identical bridges varying from three to ten times the embankment width at waterline level in the direction of flow. It would be wise to model the system in a hydraulics laboratory in order to confirm the water levels where complex bridge structures are contemplated.

Where scour of the bed is possible under the bridge (Laursen, 1962), the backwater effect may be reduced due to the reduced velocity through the constriction. Spur dykes (Fig. 10.9) have been found to assist greatly in reducing scour where it is likely to be a problem. Diving currents beside steep banks and around piers have been known to cause extensive damage to foundations.

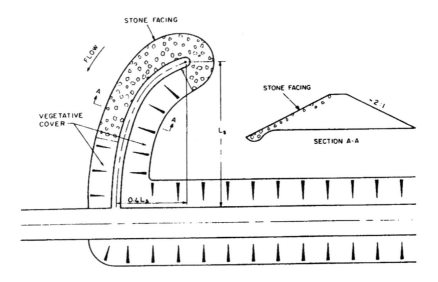

Figure 10.9. Plan and cross section of spur dyke.

10.1.5 Flow over an embankment

Flow conditions under a submerged bridge are similar to those through a culvert with inlet control. Flow may also occur over the top of the embankment and bridge. This is a case of a broad crested weir.

The depth/discharge relationship over a broad crested weir such as an embankment is more difficult to analyze than for a sharp crest on account of the unknown position at which critical depth occurs, and the problem of evaluating energy losses. The hydraulics of flow over a broad crested weir may be studied using momentum principles and neglecting friction. Equating the net force on the water body between sections 1 and 2 in Fig. 10.10 to change in momentum

$$wy_1^2/2 - wy_2^2/2 - wh(y_1 - h/2) = (wq/g)(q/y_2 - q/y_1) \qquad (10.16)$$

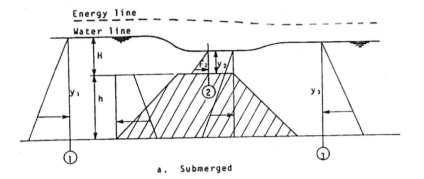

a.   Submerged

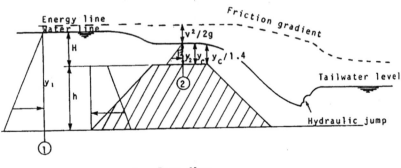

b.   Free flow

Figure 10.10. Broad crested weir wave.

Solving for q, the flow per unit width of crest,

$$q = \sqrt{\left[y_1^2 - y_2^2 - h(2y_1 - h)\right]gy_1 y_2 / 2(y_1 - y_2)} \qquad (10.17)$$

Chow (1959) indicates that experiments have proved that

$$y_2 = (y_1 - h) / 2 \qquad (10.18)$$

and that
$$q = 0.612\sqrt{g}\left[\frac{y_1}{y_1 + h}\right]^{1/2}(y_1 - h)^{3/2} \qquad (10.19)$$

or
$$q = C\sqrt{g}\ H^{3/2} \qquad (10.20)$$

where $H = y_1 - h$ (Fig. 10.10). Over the maximum range of h from zero $to\ y_1$, C could vary from 0.612 to 0.432. From observations, it is found to vary from 0.54 for low sill height to 0.47 for a high sill weir.

The previous theory applied to a weir or sill with the tailwater level above or below the critical depth over the weir. If the tailwater is lowered below the critical depth level, the depth over the sill will fall until it reaches critical depth. At this stage, the specific energy of the flowing water, $y + v^2/2g$, is a minimum, and the critical depth is given by

$$y_c = \sqrt[3]{q^2 / g} \qquad (10.21)$$

When the tailwater drops below the critical depth over the sill, it no longer affects the flow conditions over the sill. Actually the critical depth for a free overflow occurs a little way upstream (about $3y_c$) from the crest. The depth at the crest is less than y_c on account of the non-parallel flow. Depth is found for a free drop crest to be $y_c/1.4$, so that the flow in terms of the depth over the crest y_o is

$$Q = 1.65\sqrt{g}\ y_o^{3/2} \qquad (10.22)$$

10.1.6 Inundation of bridge

If upstream water level rises to above the soffit of the bridge, flow conditions may alter. If the water touches the upstream face, orifice flow may result instead of free flow. Discharge is then proportional to the square root of the head and not the head to the power of 3/2. In this case, the upstream water level rises considerably in order to achieve an increase in discharge when compared with free surface discharge. Inundation of the roadway is likely. Other problems also arise if this type of flow occurs. Damage to the superstructure by floating objects is possible. The opening may become blocked more easily by floating debris. There may occur uplift under the superstructure.

10.1.7 Erosion due to overflow

It is possible to estimate the maximum height of embankment to avoid erosion by flow over the crest. The permissible scour velocity over the crest may be estimated from the equation

$$d = \frac{0.25v^2}{g(S-1)\cos\theta(\tan\phi - \tan\theta)}$$

(10.23)

where d is the erosive particle size.

For a flat horizontal crest, with ϕ equal to 35°, this gives the permissible velocity over the crest as

$$v_2 = 1.6\sqrt{dg(S-1)}$$

(10.24)

10.2 CULVERT HYDRAULICS

A culvert is defined here as a structure for conveying stormwater under an embankment. A culvert or a bridge would be constructed over a natural river or man-made channel to assist traffic to pass over the waterway.

The design of culverts to convey stormwater under roads or embankments has been the subject of considerable research and considerable misunderstanding. The difficulty invariably arises in connection with the point of control – either inlet or outlet control is usually the case. However, in order to appreciate the problem it is necessary to start a step earlier in the design process. That is, to understand why a culvert is a control structure at all. For this we need to consider the aspects of economics and risk.

10.2.1 Economic design

The cost of a culvert is much greater than the cost of the equivalent length of channel. The culvert will have to be designed and built to resist high earth and superimposed loads, both vertical and lateral. The cross sectional area of a culvert is invariably smaller than that of the water in the channel at flood flow in order to reduce the cost of the culvert.

A culvert also has a larger wetter perimeter than a channel as it is closed on top. The head loss and average energy gradient through the culvert is therefore steeper than in the channel without the culvert. If the channel bed is prefixed at a subcritical gradient, the only way this steepening of the hydraulic gradient through the culvert can occur is by raising the headwater level above the normal depth. This causes a backwater in the channel upstream of the embankment. Head is gained by reducing the friction loss in the upstream channel. Inlet conditions into the culvert then control the discharge through the culvert.

If the channel is at a supercritical bed gradient, the culvert will probably be installed at a flatter grade, with the result that the water level will fall towards the discharge end and may reach critical depth. This condition gives rise to outlet control conditions.

Outlet control is more likely to occur for culverts in defined streams or channels. Inlet control is more likely in the case of an embankment across a catchment which collects water towards the culvert crossing, Flow is thereby concentrated at the inlet whereas the outlet will be free.

10.2.2 Principle of controls

An hydraulic device is said to control flow if it limits the flow of water which would otherwise exceed that flow with the prevailing upstream and downstream conditions. If the river flow is specified then neglecting backwater storage, the head will adjust across the control section until the inflow equals the discharge.

Flow can be controlled from either the upstream side or the downstream side depending on whether flow is supercritical or subcritical respectively. The velocities of water relative to that of an hydraulic reaction dictate whether flow is supercritical or subcritical. Thus if the flow is supercritical, the water velocity is faster than the velocity of a wave, so that waves cannot pass upstream, and control cannot be effected from downstream. A downstream control or constriction would create a standing wave which may be in the form of an hydraulic jump.

A control from upstream will uniformly affect the downstream flow depth. On the other hand, if the velocity is subcritical waves can travel upstream at a speed faster than the water is flowing, so any control on the flow downstream will back up water until it reaches an equilibrium profile upstream of the control. Flow downstream will be at normal depth.

Supercritical depth occurs when the Froude number, $F = v / \sqrt{gy}$ is greater than unity, i.e. $v > \sqrt{gy}$ where $\sqrt{gy}$ is the celerity of a shallow water wave. If F is less than 1, the flow is subcritical.

The relative gradient of the culvert and channel and the geometry will dictate where the control section is in a culvert section. It can be altered by careful design, and in fact if control can be transferred from the inlet to the outlet, or else if a balanced design is achieved, the possibility of upstream flooding is minimized for any outlet size.

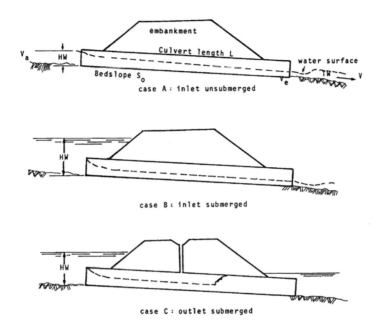

Figure 10.11. Culvert longitudinal sections illustrating inlet control conditions.

10.2.3 Hydraulic profiles

Some of the different water surface profiles with the corresponding control sections, are indicated in Figures 10.11 and 10.12. In the case of inlet control, the tailwater level will be relatively low so that the culvert runs part full for some or all of its length.

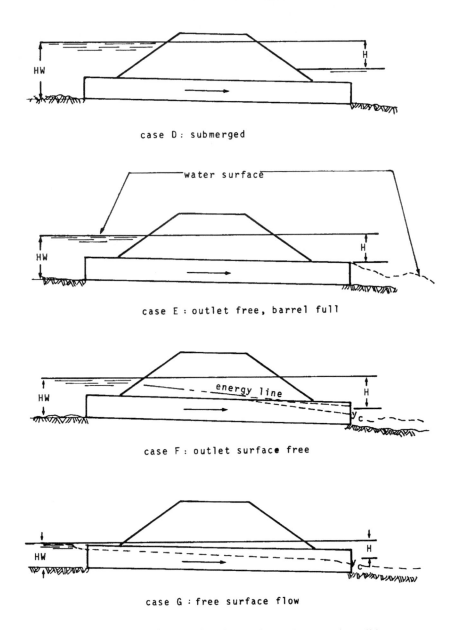

case D: submerged

case E: outlet free, barrel full

case F: outlet surface free

case G: free surface flow

Figure 10.12. Culvert longitudinal sections illustrating outlet control conditions.

The slope of the bed may be supercritical in which case depth will pass through critical at the entrance (case A). It may even occur that the headwater is higher than the barrel soffit without the water touching it if there were an inlet taper. H/D should exceed approximately 1.2 for submergence.. If discharge were higher, the headwater may cover the entrance, in which case the situation would be case B for a low tailwater. Critical depth could be induced at this inlet either by a steep downstream slope, or a high headwater H, creating a high velocity and large contraction.

Case C where a hydraulic jump occurs is possible for a high tailwater. Observe that for case C to be stable, the culvert barrel upstream of the jump would have to be vented. Kalinske and Bliss (1943) indicate a jump would evacuate air at a rate $0.006 \ Q(F_1 - 1)^{1.4}$ where Q is the water discharge and F_1 the upstream Froude number $v_1/\sqrt{gy}_1$.

In each of the inlet control cases the barrel size beyond the inlet could be reduced without affecting the discharge. Conversely if the inlet conditions were improved the capacity of the culvert for any limiting headwater could be increased.

For a tailwater level so high that it drowned the culvert completely, the discharge would be controlled by the difference between entrance and exit water levels. This is a form of outlet control (case D).

Assuming the barrel was reduced in capacity until it limited the flow or increased the headwater, control would transfer to the barrel (but this is classified as one form of outlet control, Case E).

The latter two cases are equivalent to pipe flow, with the head drop being primarily in conduit friction. The slope could be subcritical or supercritical. For relatively long culverts, the inlet end only may be surcharged and the discharge end may run with a free surface. This case (F) will only occur with a low tailwater level and subcritical slope. In some extremes with a flat culvert bed gradient and large cross-section the flow may be free-surface and subcritical the entire length, which is illustrated as case G.

10.2.4 Inlet Design

If the culvert cross sectional area is to be fully utilized or conversely is to be minimized the culvert should run full or nearly full. In the case of low tailwater levels or steep gradients, this may be a problem. It was indicated that for these cases the control is often at the inlet. Careful attention is therefore necessary in the design of the inlet to ensure minimum contraction of the flow (French, 1969). The objective is to ensure that flow rounds the edges of the inlet with minimum of separation, thereby filling the barrel cross section as much as possible. The discharge coefficient is then maximized.

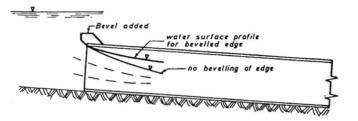

a. BEVELED TOP

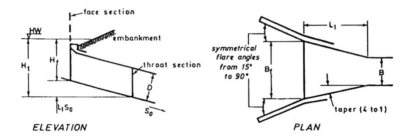

b. SIDE-TAPERED INLET

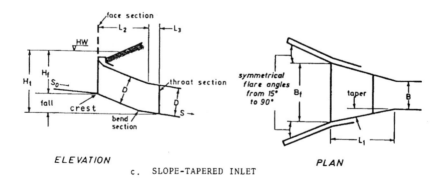

c. SLOPE-TAPERED INLET

Figure 10.13. Inlet details.

The improvement may be obtained with a steep throat, a drop inlet, wing walls, a hood, or just bevelled edges. The shape of the top entrance appears to be the most important, and the bottom or invert the least important, since flow there is horizontal. Thus an inlet meeting the battered embankment is highly conducive to flow concentration and results in a low discharge coefficient.

A taper should be in straight sections for ease of construction and rounded edges are found to have little improvement over plane bevels. Nevertheless, there are shaped precast concrete inlets available for circular culverts in the smaller sizes. It is always good policy to lay pipes with the barrel end facing upstream as this provides something of a transition.

Figure 10.13 illustrates some possible inlet arrangements. The simplest type of improvement is a vertical head-wall on top of the entrance to the culvert in the case of a battered embankment. This eliminates the re-entrant angle. The next step would be to bevel the top of the inlet. The bevel should be at least 10% of the culvert height at 33° to 45° to the axis of the culvert. In the case of skew culverts, the acute approach edge should also be bevelled. This will increase flow by up to 20%.

The second degree of improvement would e to taper the sides of the inlet. A taper angle of 45° (angle measure from the culvert axis) is perhaps the best compromise between hydraulic efficiency and length of approach. This will increase flow 25% to 40% over a square-edged inlet. Associated with side tape is usually a bevelled soffit, or a drop inlet to ensure the soffit height is not the control.

A slope-tapered section (see Fig. 10.13c) is the third degree of improvement. This form of design increases the head of the barrel as well as tapering the inlet and 100 percent improvement in flow is possible.

10.2.5 Inlet control equations for box culverts

The position of the control section in a box culvert will depend on the type of inlet. In the case of composite designs, e.g. with wing walls, a slope taper or a drop inlet, the necessary headwater at each change in section should be determined.

If the inlet control is due to a drop or a narrow entrance and flow is free, then the depth at the entrance is critical and weir flow occurs. This occurs for H/D less than about 1.2. Thus

$$Q = C_B Bg^{1/2} \left(\frac{2}{3} H \right)^{3/2}$$
(10.25)

where B is the width at the point and H is the headwater level above the invert or effective weir crest. Strictly, H is the energy level of the headwater, not the water level, but in most cases the approach velocity is negligible. Values of the discharge coefficient C_B are tabulated in Table. 10.1.

Table 10.1. Discharge coefficients for culverts

Control position	Flow condition	Coefficient	Box culverts		Circular culverts
			Side taper	Slope taper	
Crest	Unsubmerged	C_B	0.92	0.92	
Face	Unsubmerged	C_B	0.77	0.92	
15-26° wingwalls + top bevel		C_C	0.59	0.59	Square edge 0.57
or 26-90° wing, no bevel		C_h	0.84	0.64	Square edge 0.79
26-45° wingwalls + top bevel		C_C	0.64	0.64	Bevel edge 0.65
or 45-90° +top and side bevel		C_h	0.86	0.70	Bevel edge 0.83
Bend	Unsubmerged	C_B	-	-	
	Submerged	C_C	0.80	0.80	
	Submerged	C_h	0.87	0.87	
Throat	Unsubmerged	C_B	1.00	1.00	
	Submerged	C_C	0.94	0.93	
	Submerged	C_h	0.95	0.96	

Where the water touches the soffit, the culvert acts as an orifice. Discharge is related to head according to an equation of the form:

$$Q = C_C BD\{2g(H - C_h D)\}^{1/2} \qquad (10.26)$$

A set of rectangular culverts in parallel can be treated as a single culvert ignoring the dividing walls, for the purpose of selecting wing-walls. The noses of the dividing walls should, however, have a bevel. Practical considerations may limit the side taper on very wide culvert sets.

10.2.6 Circular pipe culverts

A circular cross section has better hydraulic characteristics than a rectangular one. Head losses are lower. The structural resistance is good as arching is induced. A higher headwater may however be needed for any cross sectional area owing to the shape if it is to run full at the entrance. For this reason, the inlets are often rectangular with a subsequent transition to a circular section (see Fig. 10.14).

Where rectangular section inlets are provided, the discharge control equations are similar to those for box culverts. In circular sections, the type of control is generally similar to that for rectangular sections and similar equations apply in the case of submerged flow. The inlet is normally submerged if H/D is greater than 1.25.

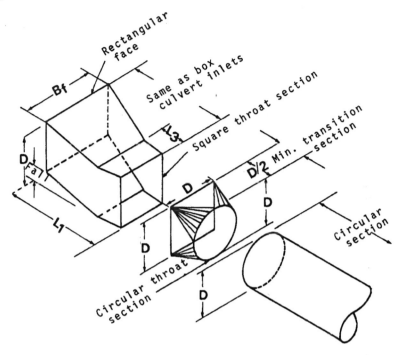

Figure 10.14. Slope-tapered inlet transition for circular pipe.

For free surface discharge, a control section will occur where the depth is critical depth. Direct derivation of an expression for critical depth is difficult for circular conduits, but the principle of minimum specific energy is used to derive the relationship as for rectangular sections. The resulting expression for all shapes is

$$A_c^3 / B = \frac{Q^2}{g} \qquad (10.27)$$

where A_c is the cross-sectional area of flow at critical depth, B is the width of surface and Q is the discharge rate. This expression cannot be solved directly for critical depth y_c as a function of diameter D, and the relationship must be derived numerically. The relationship between discharge and critical specific energy, E_c, can be approximated over the range E_c/D less than 0.8 by the expression

$$Q = 0.48 C_B g^{1/2} D^{5/2} (E_c / D)^{1.9}$$ (10.28)

As a guide, the discharge coefficients C_B in Table 10.2 for box culvert could be used, although Henderson (1966) indicates they are sensitive to slope. There are difficulties in evaluating specific energy or discharge at any particular depth in non-rectangular conduits. Diskin (1962) produced dimensionless charts for use of circular conduits running part full.

For submerged inlets, the discharge equation is the orifice equation

$$Q = C_c A \{2g(H - C_h D)\}^{1/2}$$ (10.29)

Blaisdell (1960) indicated that a considerable improvement in inlet capacity of circular culverts was possible with a hood and vortex suppressor over the entrance.

10.2.7 Outlet Control

The outlet may be free-discharging, in which case the depth in the culvert at the outlet will be critical, or submerged in which case the culvert will flow full. Alternatively the tailwater depth may be above critical depth in the culvert but below the soffit of the culvert. In either case of free surface discharge, the downstream water level is known in which case one can backwater using the direct step method) to determine the point beyond which the culvert will run full.

In all cases where the culvert runs full, the head loss along the culvert can be determined from a friction formula, e.g. Darcy-Weisbach

$$S_f = \frac{\lambda}{4R} \frac{v^2}{2g}$$ (10.30)

where R is the hydraulic radius A/P, and λ is a friction factor which for most culvert cases is the fully developed turbulent factor and is obtainable from a Moody diagram. Alternatively, the Manning resistance equation can be employed. The head losses at the entrance may be evaluated from an equation of the form

$$h_L = K_e \frac{v^2}{2g} \qquad\qquad (10.31)$$

where the coefficient K_e may be determined from Table 10.2.

Table 10.2. Entrance loss coefficients for $v = Q/A$

Outlet control, full or partly full, Entrance head loss $H_e = K_e \dfrac{v^2}{2g}$	Coefficient K_e
Pipe	
Projecting from fill, socket end	0.2
Projecting from fill, square cut end	0.5
Headwall or headwall with wingwalls	
Socket end of pipe or rounded	0.2
Square edge	0.5
Mitered to conform to fill slope	0.7
End-section conforming to fill slope	0.5
Bevelled edges, 22.7° or 45° bevels	0.2
Side- or slope-tapered inlet	0.2
Metal pipe projecting from fill, no headwall	0.9
Reinforced Concrete Box Section	
Headwall parallel to embankment, no wingwalls. Square on 3 edges	0.5
Round 3 edges top radius 1/12 barrel or bevelled 3 sides	0.2
Wingwalls 30° to 75° to barrel. Square edged at crown	0.4
Crown edge rounded to radius 1/12 barrel or bevelled top	0.2
Wingwall 10° to 25° to barrel. Square-edged at crown	0.5
Wingwalls parallel (extension of sides). Square at crown	0.7
Slope- or slope-tapered inlet	0.2

10.2.8 Balanced design

The discharge characteristics as a function of headwater for each section of a culvert will differ with the type of flow. Thus flow into the inlet is usually orifice-type flow (Q proportional to $H^{1/2}$). Barrel control may be similar (Q proportional to $H^{1/2}$), while outlet conditions may be weir flow (Q proportional to $H^{3/2}$). Under different headwaters, different sections may control the flow. It is therefore useful to plot the discharge characteristics of each section on a common chart, such as Figure 10.15. It will be seen that at lower headwater levels, the inlet conditions limit the flow, while at higher heads, the outlet conditions may limit the flow.

The optimum design will be that for which inlet and outlet conditions give a similar discharge (the design flow) for the required maximum headwater permitted. Inlet control curves should be plotted for different inlet configurations. The required inlet configuration for any headwater and barrel size can then be read off the plot.

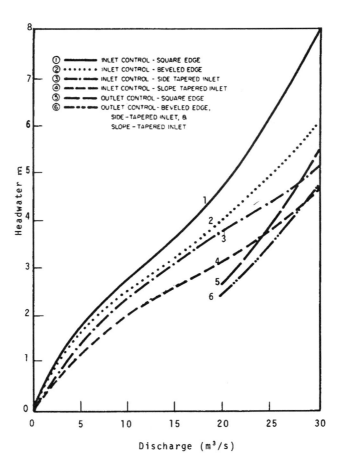

Figure 10.15. Performance curves for single 2m box culvert with alternative inlets.

10.3   REFERENCES

Blaisdell, F.W. (1960) Hood inlet for closed conduit spillways. *Proc. ASCE*, **86**(HY5), p7, May.
Chow, V.T. (1959) *Open Channel Hydraulics*. McGraw Hill, N.Y., 680pp.
Diskin, M.H. (1962) Specific energy in circular channels. *Water Power*, 14, p270.
French, J.L. (1969) Nonenlarged box culverts inlets. *Proc. ASCE*, **95**(HY6), 6928, p2115-2137, Nov.

Henderson, F.M. (1966) *Open Channel Flow*. Macmillan, N.Y., 522pp.

Kalinske, A.A. and Bliss, P.H. (1943) Removal of air from pipelines by flowing water. *Civil Engineering, ASCE*, **13** (10), p480, Oct.

Kindsvater, C.E. (1957) Discussion on flood erosion protection for highway fills. *Trans. ASCE*, **122**, p548.

Laursen, E.M. (1962) Scour at bridge crossings. *Trans. ASCE*, **127**(1), p166.

Laursen, E.M. (1970) Bridge backwater in wide valleys. *Proc. ASCE*, HY4, p1019-1038, April.

Lin, H.K., Bradley, J.N. and Plate, E.J. (1957) *Backwater effects of piers and abutments*. Colorado State Univ, Civil Engineering Section, report CER57HKL10, 364pp, Oct.

Naylor, A.H. (1976) A method for calculating the size of stone needed for closing end tipped rubble banks in rivers. *CIRIA*, Report 60, London, 56pp.

Stephenson, D. (1979) *Rockfill in Hydraulic Engineering*. Elsevier, Amsterdam, 216pp.

USBR (U.S. Bureau of Reclamation) (1987) *Design of Small Dams*. 611pp.

U.S. Department of Transport (1972) Federal Highway Commission, Hydraulic Design of Improved Inlets for Culverts. *Hydraulic Eng. Circular 13*, 172pp.

U.S. Department of Transport (1978) *Hydraulics of Bridge Waterways*. Washington, 111pp.

Yarnell, D.L. (1934) Bridge piers as channel obstructions. *U.S. Dept. Agriculture*, Tech. Bull. No. 442.

11

Asset management

11.1 INTRODUCTION

Due to the fact that water services have lagged behind the manbuilt environment in many ways, a formal method of evaluating the industry is required. Whereas asset management has been employed primarily for evaluation and performance improvement, it has many other uses; the documentation of existing facilities will facilitate improved management and bring out a greater awareness of the value of maintenance.

Programmed maintenance and upgrading will be facilitated and their benefit will readily be seen by users. The handing over of services constructed by BOT (Build, Operate and Transfer) or public organizations will be facilitated and book-keeping will be made easier. Transparency with regard to value of assets and upgrading due to plant maintenance will bring closer communication with users and a greater willingness to participate.

The techniques of water services asset assessment, i.e. location, description and condition determination, will require data capturing where this has been previously avoided but it is seen as a necessary step in the standardization and upgrading of water supply and sanitation services. Computer data processing will facilitate this and also enable easy transfer to the next step, namely economic evaluation and planning for improvements of extensions.

The operation of a water authority should, like any other organization, be run on sound business principles. Even though profitability may not be the sole driver, a sound economic basis is required. A business-like or commercial approach is encouraged, as it gives a sense of competitiveness, or stimulus, to workers, and a sense of pride to investors and consumers. As with many authorities, an autocratic approach is no longer tolerated. Current developments in legislation have created an enabling environment to the water services sector to improve its performance and meet the needs of its consumer or customer base. Central to the functioning of a viable organization is the management of its assets. Knowing what your assets are, condition they are in and maintenance thereof is of key importance to the business of any water services provider and authority. Asset management can be considered as the core activity of a water institution that contributes to viability, providing for good decision making, efficiency and performance.

The water services sector has come from a past which has been very fragmented. The sector had lost the opportunity to keep on par with international developments. The testimony of this is reflected in the failure of the sector to be viable and in meeting current challenges. Preliminary findings indicated that no to minimum asset management was occurring at small water services institutions. Larger municipalities and water boards had fully integrated management systems linked to GIS. The reasons given for this current state in the smaller institutions is the lack of capacity, lack of resource, and poor incentives at being efficient.

It is generally recognised that water engineering infrastructure assets development and rehabilitation are capital intensive industries. In order to maximize the productivity of water engineering assets, a sound understanding of the condition and performance of all assets is needed. Based on an adequate information base, the prioritized long-term investment needs for a water services system development will be much easier to plan for, and existing assets managed in a more efficient way.

11.2   ASSETS

In its broader sense, asset management is the assessment of valuables, investment and the management of them, such that their value is optimally enhanced while at the same time providing a benefit to the owner.

In the water services sector, asset management is now applied to evaluate engineering installations, to ensure good maintenance and provide users (consumers) with the best possible service. The steps in asset management, namely compilation of asset registers, institution of best management practices and reporting, are of value in making the service more efficient and transparent.

Potential areas for applying asset management in the water industry are:

Infrastructure: Often split into surface and subsurface hard assets. These include:

- The source of water
- Treatment works
- Pumping stations
- Pipelines
- Reservoirs
- Distribution systems
- Stores, offices and equipment.

The operational side of the works will receive more attention once the installations are functional. This includes pumping, optimum use of energy, quality control (water quality), planned maintenance, adequate backup (spares, standbys), monitoring (quality control of assets and metering) and technical manpower.

Human Resources: Maintaining an adequate level of trained, experienced and motivated staff, ensures efficient operation. The operational side is of concern in developing communities. They may rely on outside assistance (design and equipment) for installation, but operation and maintenance is an ongoing activity which the community has to manage. Capacity building ensures sustainability and responsibility. Training in technology and accountability may require a sizeable budget, but is the only means of ensuring sustainability.

Financial: Investments, holdings (property, shares) all increase the value of the organization. Balance sheets normally value these more accurately than infrastructure. It also extends to profitability, revenue, debts and taxation, so a true valuation is possible.

The steps in asset management for rationalizing the water industry are essentially the following:

- Compile lists of hard assets, e.g. dams, pipelines, water works, reservoirs, meters.
- Assess non-capital costs, i.e. power, operation, maintenance, administration, financial (interest rates, load periods) and staff.
- Rank assets in order of value, and by order of requiring attention, disposal, supplementation.
- Obtain net worth or organization.
- Describe what asset enhancement is required and worthwhile, e.g. rehabilitation, training, re-investment.

Table 11.1. Water services infrastructure assets in a full water services cycle

Water services system infrastructure assets	Asset function in the water cycle	Water infrastructure major asset components
Water source and associated infrastructure. Water supply infrastructure.	Storage Abstraction Conveyance Purification Pumping Storage	Dam, borehole(s), intake, pumps, motors, valves Canal, tunnel, pipeline, filters, pumps, valves, meters, reservoirs
Distribution network infrastructure.	Conveyance Metering Pressure control Fire-fighting	Pipelines, valves, meters, reservoirs, pumps, motors
Wastewater collection network infrastructure.	Concentration Conveyance Pumping	Sewer pipelines, manholes, pumps, motors, outfall
Wastewater treatment infrastructure.	Stormwater attenuation Screening Treatment Conveyance	Screens, grit tanks, metering flumes, settling and balancing tanks, sludge reactor, pumps, meters
Treated effluent disposal infrastructure.	Evaporation Conveyance Reuse Discharge	Ponds, pipelines, pumps, motors, meters, valves, outlets

Some concepts are relatively new, including risk assessment (structural, hydrological and pollution), benchmarking and reportability.

Asset management requires initial surveys and computation of asset registers, analysis and finally action, to optimally enhance the value of assets.

11.3   BENEFITS OF ASSET MANAGEMENT

Stewart et al (1997) suggest Asset Registers should adequately reflect asset description (e.g. pumps) and measure (e.g. lengths of pipes). Initial registers will be from office records and operators' knowledge, but planned data surveys should follow. This would include field inspections, and measurements, verification, system evaluation and testing. This also applies primarily to older works, for completely new works should initiate good asset registers, information systems and data banks (see Bordogna, 1996; Brittan and Rumsey, 1990).

The registers will be of benefit as follows:
- Better understanding of the core of the business
- Accountability
- Evaluation using an EAV (Equivalent Asset Value)

- Serviceability grading
- Maintenance planning
- Strategy planning
- Unit cost establishment
- Forecasting of investment requirements
- Rationalization
- Deferment of expenditure
- Investment planning
- Operating expenditure matching asset serviceability
- Work scheduling
- Operational efficiency gauging
- Statutory reporting
- Bargaining with regulators
- Raising capital
- Water balance
- Organizational review

The following benefits will also manifest:
- Development of asset registers
- Standardized procedures for comparisons and integration
- Standardized nomenclature
- Common data bases for different departments
- Systematic data collection
- Reduce risk of fraud
- Evaluation of financial position
- Costing of risks and establishing procedures to counter
- Development of maintenance and rehabilitation plans

11.4 BEST PRACTICE

Asset management is a component in the process of best practice. Best practice refers to overall system improvement. To do this requires the following steps:
- Benchmarking to establish norms or standards.
- Asset management to improve quality of assets.
- Water audits to minimize misuse.
- Night flow analysis to detect losses.
- Pressures management to reduce losses and faults.
- Economic analysis to minimize costs.

11.4.1 Reviews of world practices

Relevant international references to asset management associated with the water industry were identified in the UK, Australia and New Zealand. The literature available from the USA refers primarily to infrastructural asset management, methods and techniques in upgrading and rehabilitation of all components of a civil engineering character. Also the asset management terminology encountered in the US literature is rather different to, for example, the UK literature.

The UK experience (OFWAT, 1997) revealed that the first pass infrastructure asset register had to be elementary and statistically based. It was largely to justify the Capital Investment Programme. Some five years later, the first asset register became the Strategic Business Plan of the components and the update produced a refined investment programme and operational programme. It touched core business, annual operating efficiencies, levels of service, guaranteed standards, unit costs and the evolution of corporate frameworks for measuring performance efficiencies and annual reporting and review. The advance asset register is expected to extend to loss control, serviceability and investment planning. Standardization of Asset Registers is important for serviceability and performance grading.

Since hard assets depreciate over time, some form of replacement or rehabilitation may be needed. This has to be financed out of a contingency fund, raised loans or revenue. Water authorities in the UK are required to estimate what expenditure is required for renewal, up to 20 years hence. Performance targets have to be set and methods of indexing the state of the system developed, e.g. number of leaks or percentage loss per annum, or pressures or water quality.

The system developed for planning for future renewal of assets was termed in the UK an Asset Management Plan (AMP). The plan may go so far as to optimize the future developments, (e.g. decide routes or sizes of replacement pipelines), or plan expenditure sequence. Operating and capital costs for alternatives will be summed and discounted to get the best sequencing. Alternative discount rates and rates of return may be considered as well as a probable range of demands.

The UK practice followed Rumsey and Harris (1990) who suggested three steps in estimating long term expenditure:
(1) Establishment of policies and standards.
(2) Collect source data on the nature, condition and performance of assets.
(3) Determining and costing a strategy to rectify problems and meet requirements.

In the UK, the process of asset management is still catching up with the privatisation. Theoretically, a register of assets (especially hard assets) was available since 1989, but the process of surveying and assessing is an ongoing one.

Information technology approaches were investigated by Ballantine and Williams (1997) who suggested models that could be used for technical audits, information technology and standards. The data for these functions will however be peripheral to the main purpose, namely operational efficiency.

With the restructuring of the water sector, foreign interests surfaced in the field of asset management practices, particularly from Australian (1995) and New Zealand (1996) national asset management manuals.

It has been established that past experience in the water industry was typified by:

- Public or government ownership
- Little consumer interaction
- Financing by stocks or government loans
- Reporting to government
- Long-term capital interim planning
- Ad hoc maintenance
- Separate technical and financial departments

In addition:

- Less attention was paid to asset management in the past because systems were newer and failures fewer.
- It was easier to deal personally with problems on smaller scales.
- It was easy to raise capital than for maintenance and operation.
- Staff employment was long-term so experience and knowledge was available in-house.
- Consumers were less aware and less interactive.
- Less access to computers and software for formal procedures.
- Environmental responsibility was not enforced.
- Image was not important because capital was obtained from government
- Reportability was minimal.

The impacts of the above listed constraints initiated changes to the systems. The shortage of government funds and the realization of links with industry could make the system more efficient. An important component of asset management is the upgrading of water and sanitation services to acceptable levels. Many such services have in the past been perceived as inferior in level and standard. There has therefore been a resistance to paying for such services.

11.4.2 United Kingdom

During privatisation of the UK water industry, the UK government and the water industry developed an approach for the management of water assets now known as Asset Management Planning (AMP). This approach was based on the four key economic features of the UK water industry, namely:

- Monopoly in business of water services
- Economic externalities regarding the services provided
- Potential asset neglect due to its life-span and out-of-sight character
- The lack of differentiation in the product

In addition, a dimension which surfaced during the privatisation process, was the sentiment that the government and funding agencies to the water industry were becoming increasingly interested in securing private sector participation to assist with the delivery of services which traditionally had been the responsibility of government at national, provincial and local level.

Subsequent to the UK experience, a big consideration for the water sector internationally was the question of the extent of private involvement in water companies. There are distinct advantages in many instances, such as improved cash flow and more businesslike efficiency. The adequate depreciation plays an important role. The theoretical approach to depreciation of infrastructure assets is illustrated in Figure 11.1.

The question of whether a water services provider can be operated as a truly private or commercial concern may arise. A water services authority/provider may be provided with access to raw water by a national authority, and the cost of this may be beyond its control. The infrastructure (or hard assets) may have been inherited, or dictated by others. Although public accountability and transparency are encouraged, there is no measuring stick to gauge efficiency against.

In the UK, OFWAT (the Office of Water Services) attempted to regulate the water industry economically with regard to standards and procedures, but it can only be compared with other water services providers with different circumstances, or against self-imposed criteria.

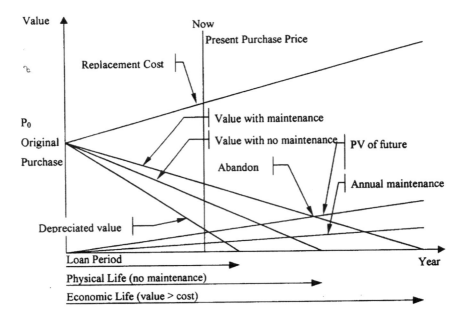

Figure 11.1 Depreciation of infrastructure assets

It is possibly easier to elect at the outset for a new supply organization to plan, build, train and/or operate it privately. This only happens for newly supplied communities, e.g. in developing countries, and they have peculiar problems. Capital may be short, expertise absent and there may be even no motivation. Governments may have to subsidize the services and appoint technical assistance to ensure adequate maintenance and financial management in such cases. The standards set by a national watchdog may not be attainable or applicable. This has since been perfected by determining performance indicators and benchmarking procedures.

The separate management of water supply, sewerage and stormwater was avoided by demarcating business areas largely on a catchment basis. Then pollution is the direct concern of the water supplier. There are still many countries who regard stormwater as the responsibility of the roads department, sewage as the environmental department and water supply as an easily sellable asset. The consolidation of the various facets improves business and technical efficiency.

The current asset management protocols recognized in the UK and internationally are illustrated in Table 11.2.

Table 11.2. Asset management protocols

Management protocol	Public utility	Private utility
Approach to capital	*Capital recovery* – float a bond to cover improvements	*Capital formation* – generate money to cover improvements
Performance	*Focus on delivery* – regulatory and capacity requirements drive decisions	*Focus on performance* – delivery is given; attention is focused on getting the highest return
Risk	*Risk avoidance* – use conservative assumptions and build to withstand low probability events	*Risk averse* – evaluate tradeoffs between risks, costs and outcomes
Procurement	*Complex procurement* – multistep with frequent checks and balances; control costs at each point	*Strategic sourcing* – enrol partners in achieving results; control total solution costs

Source: After Rynowecer and Levin (1999)

11.4.3 Australia and New Zealand

The process of asset management planning (AMP) implementation in Australia and New Zealand started primarily from the reform of local government (i.e. at the municipal council level) and the subsequent re-assessment of a typical council's engineering infrastructure assets (e.g. roads, bridges, water supply systems, parks and gardens, stormwater systems, sewerage systems, buildings, and plant and equipment).

In both countries, new local government legislation as the Council of Australian Government (COAG) of 1994 in Australia and Local Government Amendment Act (No. 3) of 1996 in New Zealand, enabled methods and techniques of basic and advanced asset management planning to be entrenched at the most relevant public level and adopted for the largest infrastructure base.

In Australia, the reform of local government coincided with the National Agenda for Water Reform of 1994. There is a certain similarity with the South African situation regarding two new water acts promulgated in 1997 and 1998. However, both Australia and New Zealand have been implementing water quality guidelines since the late 1990s.

11.4.3.1 The Australian asset management (AM) process

The development of an AM process was primarily accelerated by the requirements for advanced infrastructure technical management fully integrated with accounting standards and federal/state regulations. The key objectives were set as follows:

- AM is recognised as vital by local government (WSAs) and water services providers (WSPs)
- WSAs and WSPs are made aware of the benefits to be gained from AM planning implementation
- AM requirements are looked upon as the catalyst for technical, financial and information needs
- AM programmes give priority to the analysis of life cycle costs

The Institute of Municipal Engineering of Australia approved and published the National Asset Management Manual (NAMM) in October 1994. The NAMM deals with the following aspects:
- Why asset management?
- Choosing an appropriate asset management plan
- Asset management principles and concepts
- "How to" guides

The Australian National Asset Management Manual is available for use by local government authorities and recommended for use by all semi-government and private organizations responsible for the management of infrastructural assets.

11.4.3.2 The New Zealand asset management (AM) process

The development of AM process in New Zealand has been primarily accelerated by the following incentives:
- Ratepayer pressures to enhance water services infrastructure
- Regulatory requirements to meet increased environmental standards
- Re-setting of priorities by the local government authorities to meet community demands
- Public pressures generated creativity in allocating the costs of development

It should be noted that to a large extent the New Zealand approach to the development of asset management benefited from the Australian initiatives. However, the whole process is much more detailed and represents a wider group of stakeholders. The key objectives set out by the NZ National Asset Management Steering Group supported by the Association of Local Government Engineers of NZ, the NZ Local Government Association, the NZ Society of Local Government Managers and the Office of the Auditor-General, were as follows:
- Observe the reform of local government
- Bring about excellence in infrastructure asset management

- Implement a realistic approach to asset management
- Introduce quality management of infrastructure assets
- Develop appropriate infrastructure asset management practices
- Improve the long-term performance of assets at the lowest lifetime cost
- Infrastructure asset management process has to facilitate strategic financial plans which are prepared by local government
- Keep updated developed AM work

The NZ National Asset Management Steering Group published the NZ Infrastructure Asset Management Manual (NZIAMM) in November 1996. The NZIAMM deals with the following aspects:
- The Background - Introducing Asset Management
- The Theory - Asset Management Principles and Concepts
- The Practice - Developing and Using Asset Management Plans
- The Toolbox - Guidelines and Examples.

Between 1996 and 2000, the NZ National Asset Management Steering Group and the Institution of Public Works Engineering Australia worked mutually on the revised Infrastructure Management Manual and are now compiling and presenting a revised Infrastructure Management Manual, New Zealand Edition. The revised IMM provides guidance on:
- Background of IMM (benefits, responsibilities, total AM, LCAM, etc.)
- Implementation tasks (stakeholders, needs analysis, AM plans, auditing AM, organization issues, etc.)
- Asset management techniques (levels of services, demand forecasting, risk assessment and management, optimization, O & M, demand and financial management, etc.)
- Information Systems (asset registers, data capture, GIS systems, etc.)
- Country-specific issues

It should be noted that the revised IMM has been prepared with representation from accountants and auditors, local government engineers and managers, water and waste, gas and electricity managers, parks and recreation managers and transportation managers. The evolution of asset management planning methodology is illustrated in Figure 11.2 below.

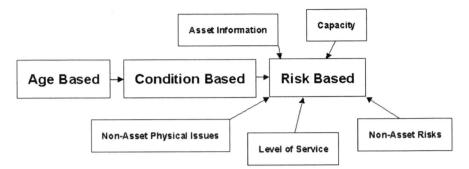

Figure 11.2. Asset management planning methodology

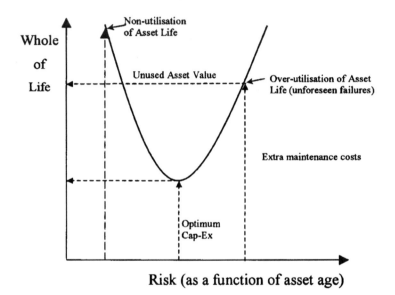

Risk (as a function of asset age)

Figure 11.2 Linkage between risk and life cost of infrastructure (NZWWA, 2000)

In both Australia and New Zealand asset management planning methodology has progressed in recent years and particularly in New Zealand where the principles of advanced asset management are being adopted. Figure 11.3 illustrates the contemporary research level in asset management with regard to risk and life cost management, where WOL COST represents Whole of Life Cost.

11.5   DATA MANAGEMENT

There is sometimes a requirement for national information systems with details being required, particularly with regard to assets. The national information system on water services could form part of a larger system relating to water generally. The public is entitled to access the information and reasonable steps have to be taken to ensure that the information is provided in an accessible format. Water service providers have to make water service development plans or drafts thereof.

A National Information System could be used for development, implementation and monitoring of a national policy on water services, and more specifically to provide information to water service institutions, consumers and the public to enable them to monitor the performance of water services institutions. Whereas government may fund reasonable expenditure incurred in establishing and maintaining the National Information System, they may also charge a reasonable fee for making the information available, presumably to the public.

The water service providers have a duty to report to the Government on assets and liabilities as well as the development planned. They have to report on progress on the development planning each year and this gives an opportunity for updating the list of assets and liabilities.

Contracts and joint ventures may be made with water service providers or water service authorities. In the case of outside organizations participating, financial responsibility is even more important as the rate of return on investment needs to be disclosed. More particularly, the establishment of tariffs will have to be considered such that costs are recovered.

From the point of view of processing National Information gathered from Water Suppliers, there should be large volumes of data to check, respond to and use. Standard forms would simplify such processing by computer. General national statistics could be derived and evaluated. Policy-making would be simplified with a knowledge of the extent of existing (and proposed) water works. Budgets and priorities could be established. On an individual service basis, a regulator or assessor could provide guides and eventually even licences or subsidies.

11.6   METHODOLOGY FOR AMPS

Systematic methods are required to cope with large volumes of information and to ensure transparency in the face of new owners and operators.

The methods which can be used for AMPs (Asset Management Plans) include:
- Surveying: manual, machines, remove images, document scanning.

- Field observations, cameras, tests
- Labour intensive: household, surveys, leak detection, valve inspections, earthworks.
- Computer: GIS, data bases, scanning, image processing, IT (Information Technology), SCADA (Supervisory Control and Data Acquisition), PLC (Programmed Logic Control) and PMC (Plant Monitoring and Control).

The extent of use of these, the scope, and the costs will be assessed and extrapolated. The necessity for and extent of training will be considered.

The activities of water authorities that will use the data include:

- Engineering planning for future installations
- Strategic planning for better operations
- Improvement of standards (quality, pressure, flow, reliability)
- Rehabilitation, disposal, replacement
- Business strategy for future funding
- Cash flow
- Accounting and auditing
- Operation
- Maintenance
- Monitoring
- Loss control
- Safety
- Reporting
- Establishing life cycle costs and viability

11.6.1 Information systems

A key element in efficiency and acceptance of water authorities is the ability to manage data. This includes data collection, storage, retrieval and dissemination. The research project was required to concentrate on the available information base at each of the investigated water services authority/provider. Emphasis given to the information systems was as follows:

Hardware: Volumes of water and rates of flow available
                   Locating pipes, valves and other buried objects
                   Maintenance, repair or replacement, i.e. sizes, fittings
                   Locating leaks, faults
                   Evaluating installations: condition, life, value, replacement
                   Operation - pump and pipe outputs, reservoir storage
                   Responding to complaints
                   Risk of pollution

	Environmental conditions
Financial:	Cash flow forecasts
	Liquidity
	Value for sale
	Necessity to flow loans
	Profitability
	Recovery of debt
	Investment requirements
Human resources:	Salaries, pensions
	Abilities
	Sale of skills
	Capacity to operate and manage
	Training
	Labour unions

11.6.2 Asset management plans

An Asset Management Plan is a programme about capital works and other expenditure associated with maintaining the conditions and performance of the water services authority/provider's assets.

11.6.2.1 What should a typical Asset Management Plan comprise?

- Procedures for preparing and updating the asset management programme
- A statement of the water services authorities (or providers) relevant standards and policies
- A list and description of all asset subsystems, land assets and their use, water supply, water distribution, sewerage, sewage treatment and disposal
- A physical *infrastructure asset register* containing information on the performance and condition of the principal components of each subsystem
- A short-term, say 5-year, relatively detailed investment plan to meet shortfalls in the performance and conditions determined for each subsystem.
- A medium to long-term investment estimate with regard to performance and condition of a whole system considering the likely future demand for services.

11.6.2.2 What should a typical Asset Management Plan deal with?

- It should summarise existing physical infrastructure assets including physical description
- Determine how old they are

- Determine what they are worth now and how much will they cost to be replaced
- Determine the costs of operation and maintenance for available assets
- Assess the structural/hydraulic condition of all relevant assets
- Establish the economic life before possible failure to overloading or deterioration or uneconomic operation
- Determine extent of infrastructure assets capacities compared to demand for services
- Assess the risks associated with infrastructure assets failure or overloading
- Assess the cost of operation and maintenance, both passive and active

Further to information compiled in the infrastructure asset register, a programme should deal with:
- Standard of services that the customer expects (Water Services Authority or Provider decides what standards should apply)
- Determine future services provision to meet adopted standards with regard to procurement, rehabilitation or replacement
- Methods of creating or acquiring new assets
- Methods for disposal of "surplus" assets
- Determine strategies for financing future services provision
- Determine the true cost of services based on available assets
- Establish the Whole-Life-Costing (WLC) for the water services system components (i.e. not only the initial costs and energy consumption during operation, but also the down-time due to premature failure and planned maintenance)
- Determine procedures for economical operating and maintaining of assets under acceptable level of risk
- Propose monitoring procedures for assessing the performance and condition of assets
- Establish risks with regard to public safety and security associated with the ownership of assets
- Evaluate asset management practices and procedures most relevant to the Water Services Authority

International experience gained in recent years indicated that the initial infrastructure asset registers evolved over about a 10-year period into the asset management programme consisting usually of three stages of development:

1st stage: Elementary and statistically based asset register for capital investment.

2nd stage: Refined investment and operational programme – serving as a strategic business plan.

3rd stage: Advanced loss control serviceability and investment planning programme.

11.6.3 Benefits from implementing an Asset Management programme

Development and implementation of an asset management programme can bring several general and specific benefits for a Water Services Authority or Provider. Such programmes would enable them to benefit from the following:

- All assets are identified in a single "asset register" as per water services authority/provider's efforts in compiling it
- All assets will be systematically categorised and allocated specific code, allowing thus for general orientation and electronic monitoring of a specific asset or group of assets (i.e. standardised procedures and nomenclature)
- Detailed valuation of all registered assets will allow for a refined assessment of a WSA's or WSP's financial position
- The programme will set the preconditions for a critical risk assessment procedures
- Available capacity can be logically compared with demands on the water service system
- The programme will provide a background for determining the true system costs and more appropriate customer charges
- The emergency procedures for a system components can be better determined and verified
- The programme enables preparation of a maintenance management plan based on pro-active procedures
- The programme will allow for rehabilitation and replacement strategies to be determined
- The programme provides a vital linkage between financial and technical responsibilities of the top management of a WSA or WSP
- The programme allows for betterment in quality control within a water services system
- The programme will produce common databases for different departments of a WSA or WSP
- The risk of fraud within a WSA or WSP will be reduced.

11.6.4 Types of asset management plans

In principle, the *Basic Asset Management Plan* is designed to meet minimum requirements for services and financial planning. It attends to a basic technical replacement programme and cash flow projections. The Basic Asset Management plan comprises primarily an asset register, operation and maintenance regulations according to designated service levels, simple condition and performance monitoring and reporting. The object of a Basic Asset Management plan is to attend to financial returns and social benefits.

The *Advanced Asset Management Plan* aspires to optimise processes and activities by means of detailed monitoring and activities by means of detailed monitoring and analysis of data on asset condition, performances, lifecycle costs and various options considering improved processes. The main differences between the Basic Asset Management and Advanced Asset Management plans refer to risk management and employment of optimisation techniques to enable evaluating of various options at specific levels of service attended in the Advanced Asset Management Plan.

Taking into consideration the vast diversity of water services, it is practically impossible to determine what would be a *Standard Asset Management Plan* to be recommended for application as both extremes (i.e. basic or rudimentary and advanced plans) exist. It appears that above all, the standard approach in preparation of an asset management plan should attend to the essential steps as follows:

- The levels of services and the extent of the assets controlled would determine at what level (i.e. basic or advanced Asset Management Plan) the water services authority should prepare the Asset Management Plan,
- Currently existing or required asset management processes and data acquisition methods to be established,
- Required level of services and associated performance measures to be confirmed by the community (or consumers),
- The capacity expansion based on the service demand prediction to be balanced with the available resources,
- The options for providing the required level of service, associated risks and trade-offs should be reviewed in-house by the water services authority/provider or in association with a specialised consultant,
- The Whole Life Costing (WLC) of most of the important components of a water services system (i.e. not only the initial costs and energy consumption during the operation but also the downtime due to premature failure and planned maintenance) should be considered in a plan compilation,
- The asset reproductive technical and financial options to be proposed,
- The programme shows how to monitor and upgrade the present asset management plan from one to another stage by introducing more advanced methodology and techniques.

Essentially, an asset management plan developed by a water services authority/provider has to evolve from a "standard approach" designed plan into a plan advancing according to the desired level of services.

The *Total Asset Management Plan* is a comprehensive programme incorporating a substantial improvement in the operation and maintenance of plant and equipment. This totally integrated approach to asset management is primarily used in the manufacturing sector and applies modern principles of operation and maintenance management.

11.7 LIFE CYCLE ASSET MANAGEMENT

The life cycle management component of the Asset Management Plan is, together with an Asset Register, the most important as it outlines what is planned to keep assets managed and operating at the agreed level of service while optimising life cycle costs. Asset life costs determine the total costs of ownership over the life of an asset for the purposes of:

- Evaluating options for the procurement of new assets
- On-going management decision-making throughout the life of an asset
- Benchmarking the actual cost performance of an asset
- Reviewing the process for future design/acquisition decisions

A typical project development is made up of four generic phases, consisting of concept, development, implementation and termination. For different types of project development, these stages are usually broken down into stages specific to the industry or area of project application. This concept is adopted in the life cycle asset management (LCAM) methodology. The LCAM is about all consequences of owning and operating the assets. The principle is illustrated in Figure 11.3, however, also accounting for increases in risk due to the age of a project (or asset).

It must be noted that the asset whole life cycle principle should not be mixed with the Process Analysis and Life Cycle Assessment (PA&LCA) which has been developed primarily as an environmental management tool for product development from a raw material state to the state of product disposal or recycling. The PA&LCA considers the whole material and energy supply chain including treatment and any reuse of wastes.

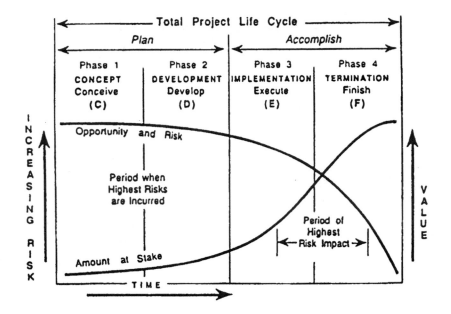

Figure 11.3 Typical life cycle profiles – Risk versus. amount at various stages (Wideman, 1992)

The costs relating to the life cycle of the asset include the following:
- Preliminary investigation and feasibility study costs
- Design and construction costs/or acquisition costs
- Operation and maintenance costs
- Ultimate replacement costs
- Disposal costs
- Depreciation costs

11.7.1 Life cycle costing

To determine these costs, which occur at different stages of the life of an asset, Table 11.3 illustrates the most important cost areas for Asset Life Cycle Management (ALCM) and the life cycle cost components.

All too frequently engineering construction contracts are awarded on the basis of tender price alone. However, the operating costs and life of the works have an important bearing on the best solution. It is often done in Mechanical Engineering practice but seldom in Civil Engineering that operation and maintenance costs are added for tender evaluation. And then after installation, deterioration can affect life and replacement costs. This explains a way of correctly evaluating different proposals. A spreadsheet for lifecycle costing and comparison of different pipe materials is presented to explain the concepts.

Whereas the procedure gives the most economic solution, there could be other constraints affecting the decision. Cash availability may dictate a low capital cost system. This should rarely arise as most capital projects are financed with loans. And the prevailing interest rate enters the formulae used to compute the optimum solution. Then there are taxes to be considered in the case of private installations. Capital may come from taxed profit, whereas operating costs are subtracted before paying tax. In that case an operating cost intensive system may be more economic. In the majority of cases pipelines are public services and tax is omitted in the calculations.

Table 11.3. Cost areas for asset life cycle

Cost component	Life phase	Cost area examples
Capital costs	Planning	Feasibility studies
		Research/concept development
		Programme planning
	Design	Functional design
		Detail design
		Documentation
	Construction	Site and land
		Construction/purchase
		Management
		Quality control
		Commissioning
Recurrent costs	Operations	Operations personnel
		Training
		Energy
	Maintenance	Planned maintenance
		Spares
	Renewal/Rehabilitation	Detailed design
		Construction costs
		Management cost
		Quality control
	Financial	Cost of finance
		Depreciation
Residual value	Disposal	Decommissioning
		Sale value
		Asset disposal/salvage costs
		Site decontamination

Source: NZ Infrastructure Asset Management Manual (1996)

11.7.2 The life of a works

There are alternative definitions of 'life' when it comes to engineering works:
- The Physical life is the time the works exist whether in use or not.
- The Useful life is the time it can be used
- The Service life is the time it can be serviced, e.g. spares obtained or repaired.
- The Safe life is the time in which it can be safely used or is safe to leave in place.
- The Economic life is the time it can be used economically, i.e. before it is economically better to do something else.
- The Engineering life is the time it should be in use i.e. after which it should be closed down whether for economic or safety reasons.
- The Financial life is the time over which it is financed. It may be the period of the loan if a loan is made to pay for the construction. It is also possible to re-float a loan if the works is still serviceable.
- The Payback period is the time to pay off the loan and may not be the same as the original loan period.

The value of the works at any stage is also difficult to quantify. It could be any of:
- The original construction and design cost.
- Inflated cost, i.e. cost of a new works
- Replacement cost, which may be for a substitute.
- Insured value.
- Depreciated value allowing for wear and tear. This depends on level of maintenance kept up.
- Linearly decreased value over the life, whether it be financial or economic life.
- One of the above minus outstanding operating and maintenance costs discounted
- One of the previous plus value of product sold in the case of a non-public owner.

Insurance companies are well aware of the different values and use this knowledge in meeting claims. Unless the insured is able to distinguish between the different values and is aware of the terms in the documentation he may not receive his premium worth.

11.7.3 Economic evaluation

There are many ways of evaluating and optimising assets. The method of discounting should be entered into with care. The standard practice is to convert annual payments to a present value by dividing them by $(1+i)^n$ where i is the interest rate as a fraction and n is the number of intervals from now. n is normally in years. The Present value is the sum of such numbers. Where the annual costs A are the same, the Present Value, PV, becomes

$$A((1+i)^n-1) / i(1+i)^n . \qquad (11.1)$$

The discount rate to use is often taken as the prevailing interest rate paid on loans. It demonstrates a time rate of preference but purely financial. So it assumes no cash flow obstacles or taxation payments. In the long term, interest rates vary so a time average rate would be desirable but the future is uncertain so prevailing rates are generally adopted. The source of funds often affects the interest rate. Development funds may be offered at cheap rates. Private sources generally want higher returns, to cover risk and to ensure profit.

Risk affects the outcome in a number of ways. There may be uncertainty regarding demands, reliability of source, finance or dangers ahead. A probability analysis is a good way of accounting for risk on large projects. Each scenario should be multiplied by its probability to obtain the probable optimum. Short cuts can be made as generally risk reduces the optimum capacity. So a higher interest rate or smaller capacity (shorter life) could be adopted.

When inflation enters into the picture the prices should be escalated at the inflation rate and the discounted. As an approximation the effective discount rate is the interest rate minus the inflation rate.

How to select the optimum solution is the next question. Optimisation methods are available on spreadsheets, and generally the difference between benefits and costs is maximized using these methods. Where a public system is to be optimised the benefits are taken implicitly and the system with minimum cost is sought subject to an acceptable level of supply.

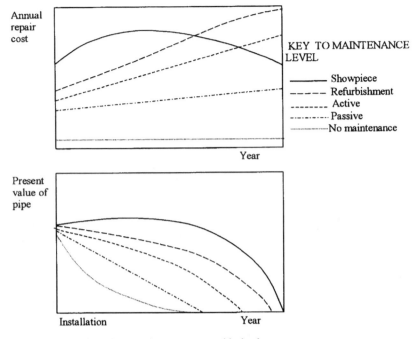

KEY TO MAINTENANCE
LEVEL

——————— Showpiece
— — — — Refurbishment
-------- Active
—·—·—··· Passive
·················No maintenance

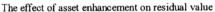

The effect of asset enhancement on residual value

Figure 11.4 Costs and effects of different levels of maintenance

Other criteria for selecting the optimum have occasionally been used. These include the benefit/cost ratio, which is now only used for ranking results optimised prior using the maximum difference principle. Maximum internal rate of return is also a possibility used often in sensitivity studies of the results to interest rate. On national projects other features affect the answers and shadow values are often attached to commodities to allow for intangible or indirect benefits. The benefits of water supply range beyond the price of the water. Improved health, lifestyle and environment result. Less directly, new opportunities arise, e.g. better home life and education, more employment including on the water scheme, and uses of water manifest, e.g. irrigation, hydro power, recreation or fishing. Pipelines generally represent the major assets of a water supply organization. They have many of the components to illustrate the necessity of including running costs as well as capital cost.

11.8   ASSET MANAGEMENT REGISTERS

Infrastructure Asset Management can be defined as the process of managing the creation, acquisition, maintenance, operation, rehabilitation, extension and disposal of the infrastructure assets of an organization, in order to provide a defined level of service in a sustainable and long-term cost-effective manner. This process can only take place if there is adequate data on which to base decisions. Adequate data implies sufficient data that is accurate and up-to-date. The Asset Register is the database that contains all the relevant information on all the infrastructure assets owned by the organization, and it is on this database that the entire Asset Management Plan depends. For this reason, it is vitally important that a well designed asset register is put into place during the implementation of an Infrastructure Asset Management Plan.

Typically, asset registers will differ significantly from one organization to another, as their infrastructure assets and management plans differ, but there will be common elements to all asset registers. There are a number of advantages to be had from designing a standard asset register for a particular sector, such as the water services sector. These advantages include the following:

- Allows consistent comparisons between different organization within the sector
- Simplifies data gathering by central monitoring agencies
- Enhances transparency and public access
- Makes it much easier for institutions that do not yet have an asset register in place to implement one. This is especially useful for smaller institutions that do not have the capacity to design and implement an asset register from the ground up.

11.8.1 National standard for asset registers

As has been discussed earlier in these guidelines, there are a number of advantages and benefits to be obtained from a National Standard for asset registers. The main problem with establishing a national standard is the fact that Water Services Institutions (WSI's) vary a great deal in their size, in the number, value and types of assets under their control, and in many other aspects. It would not be possible to define and implement a perfect National Standard for asset registers, but this could be achieved through an evolutionary process. An initial standard needs to be decided upon and implementation begun. As the advantages and weak points of the standard become evident, it can be refined and improved.

11.8.2 Requirements of an asset register

Because WSIs vary such a great deal, a national standard for assets registers would need to be very flexible. It should not, for instance, prescribe the hardware or software that is used to support it. It should, however, define the types of output required and the methods used for calculating performance indicators, condition and risk assessment gradings and asset valuations. The national standard should also recommend the types of data to record, and present a default identification system for the assets. These will be to assist WSIs in setting up asset registers, but should not be restrictive on those WSIs who have already implemented an asset register.

Some of the requirements of an asset register are the following:

- It should record the details necessary to clearly identify each asset
- It should record a basic set of information that is the same for every asset. This would include the identification, the location, the age, and assessments of the value, performance, condition and risk of the asset.
- It should record, for each type of asset, any information over and above the basic set of information that is necessary to effectively manage that asset. The information that should be recorded is any information for which the value of knowing the information is greater than the cost of obtaining the information.
- It must meet the organization's management, planning, technical and financial needs, as well as any legislative requirements.
- It must be easy to operate and provide quick and accurate access to information, in the form required, to anyone who has a right to that information
- It should facilitate accurate and confident decision making
- It must be secure so as to prevent unauthorized changing of data.

11.8.3 Components of a national standard for asset registers

A National Standard for asset registers would have to be flexible enough to allow the implementation of an asset register by all WSIs, and yet accurate enough to ensure that the benefits of a national standard are achieved. The National Standard should define at least the following:

- Which assets should be included in the asset register
- The minimum set of data that should be recorded for each asset
- A flexible scheme of asset classification and identification
- A standard method of evaluating the condition grading of the asset. This will be different for different classes of assets.

- A methodology of evaluating the performance grading of the asset. As with the condition grading, this will also be different for different classes of assets.
- A standard method for defining the value of an asset. This can be different for different classes of assets. It may also allow the WSI some freedom of choice in the method they use, as long as they use one of the standard methods, and state which method has been used.
- A method for evaluating the risk associated with the asset, and acceptable levels of risk.
- A definition of, and methodology for evaluating levels of service required form the assets.
- A measure of the accuracy of the data recorded for each asset.

11.8.4 Selection of assets on register

It is not feasible, nor is it necessary to record the assets of an organization down to the last pencil and bolt owned by the organization. It is therefore necessary to decide what level of assets will be recorded. The ultimate criterion for deciding is that the value of the information should be greater than the cost of obtaining and maintaining it. This is, however, not easy to establish, because the value of the information depends on the use to which it is put.

The line between which assets are recorded and which are not can be drawn purely on the basis of the value of the asset, for example, any asset with a value of over $100 should be recorded. There are two main problems with this approach. The first is that a standard definition of the value of an asset must be in place before this method can be used. The second problem is that the value of the asset does not necessarily give an accurate indication of the importance of the asset to the organization. Although this method does have problems, it is still probably the easiest, and most accurate, and thus the most commonly used.

11.8.5 Minimum set of information to be recorded

The National Standard for Asset Registers would have to define a minimum set of data that should be recorded for each class of asset. This will have two components, the data that is common to all assets, and the data that is specific to the class of asset.

The data that is common to all assets include the following:
- A unique identifier for the asset. This will be discussed in more detail.
- A name for the asset.
- The location of the asset, such as latitude and longitude.

- The date of acquisition or commissioning of the asset. This is necessary to calculate the age of the asset as well as its condition and value.
- An initial cost. This could be either the original purchase price or an evaluation made at some point in the asset's life.
- The current value of the asset, as well as the methodology used to calculate the current value.
- The current condition of the asset and methodology used to evaluate.
- The current performance grading of the asset and the methodology used.
- The risk currently associated with asset and procedure used to calculate.
- A measure of the accuracy of the data recorded for this asset.

11.8.6 Recording the changes to assets

The asset register contains a picture of the infrastructure assets at an instant in time. However, as time passes the assets will change. Pumps wear, pipes develop cracks, and all manner of changes occur to the assets. In order for the managers of the system to be able to make informed decisions, the information in the database must be kept up-to-data and reflect these changes. This can be done in one of two ways:

- Periodic surveys of the assets, or
- Ongoing capturing of changes whenever they occur.

The advantage of using periodic surveys of assets is that it is not necessary to maintain information about every change that occurs to the asset, but rather the net result of a number of small changes is recorded when the survey is done. The disadvantage is that regular, usually relatively expensive, surveys are required, and that the data in the register is only as current as the date of the last survey.

Ongoing capturing of changes has the advantage that the register is always kept up-to-date. It also has the advantage that trends in the performance or condition of the assets can be tracked and analysed. The main disadvantage is the fact that this required recording a relatively large quantity of data in the register. However, most institutions already record data regarding most changes to their assets such as maintenance or repair work.

An example of an event that indicates a change in the condition of an asset would be a maintenance event. Most institutions already record maintenance event information in the form of a job card and an action report, so no new information would need to be recorded. These are often captured into a Maintenance Management Tool, so to record this information in the asset register would only require linking the Maintenance Management Tool with the Asset Register.

An example of an event that could indicate a change in the performance of an asset is a regular water quality sample. If this shows a deterioration in the water quality, then the performance of the purification plant might need attention. Again, these events are usually recorded by most WSIs, so no new information would need to be recorded, it would just need to be linked into the asset register. Ideally, the asset register would be the central repository for all data regarding the assets and the events that affect them for the entire institution. Different reporting modules would then be used to extract maintenance, or water quality, or billing information, etc.

11.8.7 Capturing the data

As discussed above, the asset register will contain a large amount of data, both on the assets themselves, and on the events that affect them. It is necessary to continuously capture event data, and to regularly capture asset condition data via surveys. These are two basic methods of creating asset registers, depending on the capacity and resources of the WSI.

Entirely manual systems would be applicable for very small WSIs that have neither the capacity nor the technology to use computer based asset registers. Typically, the operator will have a book of forms, in which he will write down the events that are relevant to the system when they occur, as well as the results of the regular surveys that they may conduct.

11.9   SYSTEM DEFINITION

An understanding of the levels of service that an organization should be providing is fundamental for any organization. These define what the organization should be doing, and it is only when these are known that an organization will be able to judge how well it is doing. Ultimately, the objective of asset management planning is to match the level of service provided by an asset with the expectation of the customer. Asset management planning will allow the determination of the relationship between levels of service and the cost of such service. This relationship can then be evaluated in consultation with the customers to determine the level of service for which they are prepared to pay. The appropriate level can be determined by benchmarking.

The first step in determining levels of service is to determine the level of service currently being delivered by the organization. Then next step is to determine what levels of service are sought by the customers. This will then reveal a number of service level gaps.

A national standard for asset registers should give methods for evaluation levels of service, and minimum levels that are required, consistent with

Government policy, the Reconstruction and Development Programme, and customers' expectations, willingness and ability to pay.

11.9.1 Asset identification and classification

A classification system is necessary to break up the organization's assets into a hierarchical system of identification. This system can be based on function, type or location, or multiple systems can be used that cover all three of the above. The system used should:

- Be logical and easily understood.
- Facilitate the collection and recording of data.
- Compliment the financial accounting procedures.
- Recognize the synergies in asset management activities.

An identification system is necessary to assign each asset with a unique identifier. These identifiers must be unique throughout the entire organization, and be used wherever the asset is referred to. Identifiers fall into three major categories:

- Unintelligent – a sequential number that is assigned when the asset is added to the asset register that tells the user nothing about the asset.
- Semi-intelligent – a code that includes some reference to the asset classification. For example, if a functional classification system is used, the code may have two letters to specify that the asset is a pump and a specific type of pump, and then a sequential number assigned when the pump was added to the register.
- Fully intelligent – a code that is made up entirely from the classification system and which allows the user to know immediately which asset is referred to just from the code.

11.10 REFERENCES

Ballantine, S.G. and Williams, M.F. (1997) Network modeling and the improvement of distribution management. *Proc. 3rd Conf. Water Pipeline Systems*. Mech. Eng. Pubs, London.

Bordogna, J. (1996) Civil infrastructure systems: Ensuring their civility. *Jnl. of Infrastructure, Viewpoiont, ASCE*, June.

Brittan, R.J. and Rumsey, P.B. (1990) The role of asset management. *Jnl. of Institution of Water and Environmental Management*, June.

NZIAMM (1996) New Zealand Infrastructure Asset Management Manual, 1st ed., November.

NZWWA (2000) Water 2000. Water Conference and Expo. NZ Water and Wastewater Association, Auckland. Proceedings, March.

OFWAT (Office of Water Services) (1997) Birmingham, U.K.

Rumsey, P.B. and Harris, T.K. (1990) Asset management planning and the estimation of investment needs. In Schilling, K.E. and Porter, E. (Eds) *Urban Water Infrastructure*, Kluwer Academic Publications, 119-131.

Rynowecer, M.B. and Levin, D.E. (1999) Improving performance in water industry. *Water Environment and Technology*, U.K., July.

Stephenson, D., Barta, B. and Marson, N. (2001) *Asset Management for the Water Services Sector in South Africa*. Water Research Commission report no. 897/1/01. Pretoria.

Stewart, A., Gistel, A. and Tait, R. (1997) Asset management planning in the new Scottish Water authorities. *Proc. 3ʳᵈ International Conference on Water Pipeline Systems*. Mechanical Engineering Publications, London.

Water Services Act (1997) *Act No. 108 of 1997*. Government Gazette Vol. 390, No. 18522. Government Printers, Pretoria.

Wideman, M.R. (1992) *Project and program. A guide to managing project risks and opportunities*. Project Management Institute, USA.

12

Privatisation

12.1   INTRODUCTION

Water is a basic commodity and the technology required to deliver it is well known and simple. Yet there are many water companies that have experienced problems in management and operation. There are, on the other hand, many private companies wishing to take over the business. Unfortunately the assets are valuable and the infrastructure has been built up over centuries, so that to take over is not that simple. The processes of abstraction from the source, purification, pumping and distribution and storage, could all be reproduced and improved on but that improved cost effectiveness would be marginal compared with improving the customer end services. That is, obtaining the co-operation of consumers with regard to payment and managing their water consumption effectively is an important business aspect (see Pollit, 1995).

The international market for water is some US$400 billion per year. This may be compared to the market for electrical energy of US$1,000 billion a year. And this is only for urban water consumption, i.e. residential, commercial and industrial, but it includes potable water supply and wastewater collection and treatment.

The market is largely in Western Europe (30%), followed by Asia (28%), North America (25%) and in decreasing order, Eastern Europe (5%), Latin America, Oceania, and Africa. The annual growth in the water market is some 8-10% a year, made up largely from developing countries. Water is becoming an increasingly valuable commodity due to reducing relative resources, pollution and increasing costs of storage (Dinar et al, 1997).

Most water supply companies have been protected by governments and are inefficient and not customer orientated. There is a large loss rate, i.e. about 30% of water supplied is unaccounted-for. Many people still remain unserviced.

Yet there are relatively few organizations able to supply potable water efficiently. Although there are large international companies able to operate water supply companies, they are still met with a high level of distrust in some cases. It should be borne in mind that the water supply industry is capital intensive and the payback time is long. I.e. there are no rapid profits to be made. Most new investment is required where people are least able to pay. For them, water is seen as a right and not as something for which they have to pay or an economic good. The result is that many water companies are in the hands of incompetent or nontransparent government appointed organizations.

The amount of money required to be invested in water supply and water services over ten years is as follows: In Asia, some US$280 billion is required, in Latin America, US$220 billion, in Africa US$80 billion, in the Middle East US$45 billion, in Eastern Europe US$40 billion, and in North America and Western Europe US$35 billion.

Some obvious facts indicate most water companies are under-performing. The unaccounted-for water loss is as high as 80% in some developing countries. In Asia, it averages around 50%, in South America 30%, in Eastern Europe 30%, in Western Europe and the United States, 15-25%.

The percentage of the population not connected to a potable water supply scheme ranges from 58% in Indonesia, down to 20% in South America, while in Western Europe and North America practically all people have access to potable water.

The percentage of the water supply operated by private companies is small. Of the total of the world population of 6 billion, only about 5% are served by private companies. Of this 290 million people, 126 million are in Europe, 72 million in Asia and Oceania, 48 million in North America, 21 million in South America, and 22 million in other countries.

Although it may said that the privatization forces are only starting to push, the fact is that they may have burnt their fingers or are reluctant to embark on high-risk ventures, particularly in the developing world. The models for private ownership are only beginning to emerge and become attractive in some instances. Whereas the past has seen public ownership models, including management contracts, lease and concessions, the change is now to private ownership models or private consultancies, and these are typified by BOOT, joint ownership, i.e. public-private partnership, and decreasing the outright sales.

Limitations in public funds for floating shares has reduced the attractiveness of water utilities to become public. The continued poor performance and comparison with successful private models indicates that private management is likely to become more attractive even to consumers. On the other hand, concerns about monopolization and large populations dependent on private companies are constraints. And, from the private companies point of view, the long payback periods, the considerable risks and the large capital investments are deterrents.

The waves of privatisation in the United Kingdom have been watched by the world. The first wave of privatisation starting in 1973 saw approximately two thousand companies involved in the water industry. The second wave of privatisation after 1989 narrowed the focus to forty companies, i.e. a concentration compared with the earlier diversification. At the same time, the perspective became clearer and the necessity for catchment control, i.e. the integration of wastewater services with water supply, became apparent. Nowadays increased efficiency is being obtained through competition and experience. The creation of a high level of regulation in the United Kingdom has enabled benchmarks to be provided and individual performances to be reviewed, while at the same time cost effectively improving the water companies and their profitability, or at least levelling off and reducing the risk. The time scales and levels of service are becoming clearer after the first wave of privatization. The past models and the trials and failures of previous companies have narrowed down the uncertainty and the scope for opening up unknown management methods. Optimum strategies have only been found by trial and error and experience and those companies staying in the market have gained this, whereas there is little chance for smaller companies to catch up unless they are supported by governments.

The three waves in the privatization can be seen as a learning stage which has been experienced, their integration into the proper business arena and in the foreseeable future, there is likely to be a shift to multi-utility companies. Ways of providing for water requirements in the future will also require a high level of technology including recycling and desalination.

Whereas the big companies in the market at present include Suez, Saur, Thames, Severn, Anglian and Vivendi, there may be amalgamations particularly in view of newer technologies and financial requirements. The amalgamation with the energy business may also bring other players into the water services market (see Roth, 1987).

On the other hand, many previously public companies such as water boards are embarking on public-private partnerships and eventually see themselves as private companies. Whereas they have the skills and facilities, a businesslike approach is yet to be instilled into a previously public servant approach.

Table 12.1. Necessary skills for integrated EPC/Utility companies

	Skill	"Best practice" – examples
1. Megaproject management	Management of big EPC projects • Planning • Execution, especially resource and time management	*Thames Water*: completed "London Main Ring Project" ($4.2 bn) 21 months ahead of schedule below budget
2. Information management	Management and utilization of huge data-bases on customers, machinery and distribution networks	*Severn Trent*: operates distribution networks with radio-based meter reading
3. Risk management	Estimation, reduction or transfer of risks in inter-national operations • Political (changing government) • Environmental (earthquakes, flooding)	*Générale des Eaux, North West Water*: have international experience in other utility industries (gas, tele-communications)
4. Handling of pressure groups	Favourably influencing local politicians, regulators, consumer interest groups, environmental power groups	*British water companies*: have extensive experience with regulators (OFWAT) and consumer groups (Water Watch)
5. Capital access	Access to public and private financing through govern-mental relationships and solid financial rating	*Suez-Lyonnaise des Eaux*: has good relationships with World Bank top management, pools resources in "Asia Water" investment company

12.2   ECONOMIC REFORM

In an attempt to 'economically reform' the management of water in economic terms, Briscoe (1997) of The World Bank sets out lessons learnt. He suggests:

• Change should only be initiated when there is a great need
• A clear strategy and effective information is required for dealing with all interested parties
• Sensitivity is necessary for different institutions and environments
• Allow water markets for effective management
• Start with pilot projects in new environments
• Compensate economically and efficiently third parties affected, e.g. by lowered water tables.

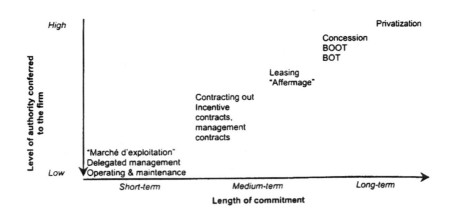

Figure 12.1. The various types of contract (Perrot and Chatelus, 2000)

Urban water is characterized by high supply cost and low opportunity cost (Fig. 12.2). Accordingly, priorities for economic optimization relate rather to the engineering side which dominates the supply cost, rather than the trading of water as an economic commodity. Briscoe (1997) cited examples in India where the reform process floundered because of this inflexibility, yet in California and Australia reform happened.

12.3   CONCESSIONS

The decision on the part of a public authority to "contract out" to a concession or delegate responsibility for an infrastructure project to the private sector, requires three conditions (Bezançon, in Perrot and Chatelus, 2000):

(1) *An appropriate political, legal, fiscal and administrative framework*, which is distinct from an ordinary public procurement contract. The concession is a long-term partnership-building process established between a public authority and a private entity aimed at enabling the latter to invest profitably in a public service or public works related project. In contrast, public procurement merely consists of an acquisition procedure without any consideration paid to either contract period, service operations or level of capital investment.

(2) This authority will be required to follow a specified *contract-award process*, which requires several stages encompassing facility or service design, construction and operations.

(3) *A contract* to bind both parties over the long-term. Such a contract must contain minimum threshold conditions in order to ensure successful

execution of the entire project, assign each party's corresponding set of risks, and protect their vested interests.

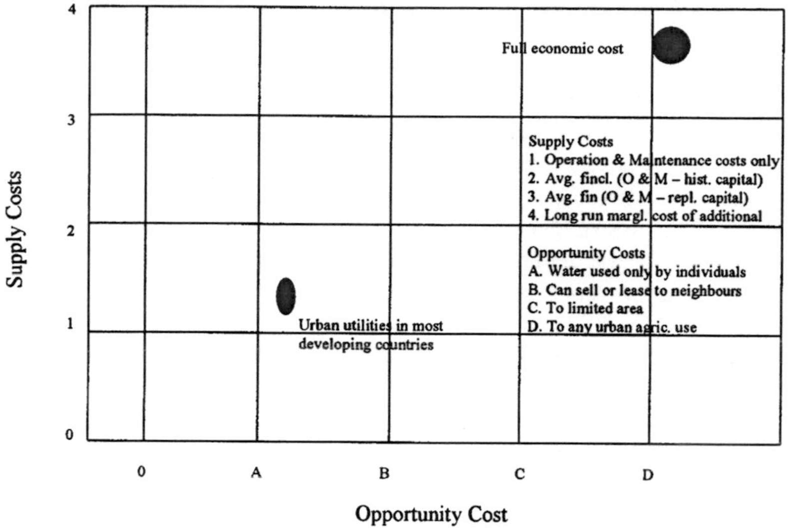

Figure 12.2. Supply cost, opportunity cost and full economic cost for urban water supply (Briscoe, 1997)

A public-private partnership will be completely built around these three basic stages. The concession is a process which allows for a private partner, due to its particular competence and financing power, in the provision of a public service or in the construction and operations of public infrastructure projects over a long time period. This contract, as well as the so-called "public-private partnership" contracts, is dissimilar to a public procurement inasmuch as their complex nature necessitates a different, customized legal framework

The legal framework alone is not sufficient to launch a concession. It is also necessary for the authority or the country backing the project to exhibit a strong pro-partnership stance and fulfil a series of preliminary technical conditions. Without such conditions, long-term capital investment programs could not be implemented.

12.3.1 Definition of a concessionary contract

A concessionary contract is not a basic public procurement contract and differs from the classical contract for the purchase of facilities or services, which is simply awarded to the lowest bidder on the basis of a set of predefined specifications.

A public-private partnership is the outcome of a search for the most appropriate long-term partner that is able to provide the necessary expertise and management capacity to effectively run a public service or facility. This type of contract can be distinguished in broad terms from a public procurement contract by the following characteristics:

- *The total or partial assumption of the project's capital investment program* by the concessionaire within the scope of a long-term contract which affords the time span necessary to both fully depreciate this investment and carryout the contractually-stipulated service obligations;
- *The transfer of public service obligations from the concession-granting entity* to the concessionaire with respect to public liability and the ensuing breakdown of risks between the two parties;
- *The unique, all-encompassing and complex nature of the contract*, including design, financing, construction, maintenance, facility operations, service provision, etc., the magnitude of which necessitates in-depth negotiations between the two parties prior to drawing up the contract;
- *The link established between concessionaire remuneration and service operating performance*, notwithstanding the type of fee collection (e.g. tolls paid by the public authority). Even if the right to financial equilibrium is supported by a considerable body of national jurisprudence, and as the concession-granting authority's participation in project financing or facility operations does not impact contract characteristics, as long as some risk is borne by the concessionaire.

In cases where the concession involves both the construction and operation of a facility as opposed to merely the delegated management of an existing service network, it is commonplace for the project to encompass a long works period prior to service start-up, hence before revenue-generation. This financial situation results from the sizable investment which is required up front, thereby producing project financial curve unique to this type of concessionary activity. On the other hand, a pure service concession typically entails a correlated remuneration as from the contract's effective date.

The concession is defined therefore by a *multiplicity of criteria*. National law must provide for a legal framework dedicated to concessions, as distinct from that set up to accommodate public procurement contracts featuring specific rules for their award and execution. Table 12.2 presents the essential differences between these two contract types.

Table 12.2. Comparison between a public procurement contract and a concession

	Public Procurement	Concessionary Contract
Definition	Provision of supplies, component(s) of a work program, or a service determined by the public authority	Creation of a public facility and management of a public service by a private entity via an agreement negotiated with the public authority
Primary characteristics	Single objective	Multiple objectives
	Short-term	Long-term
The contract	Lack of association with service management	Definite association with service management
The contract awardee	Not granted a public service delegation mission	Has been granted a public service delegation mission
Public characteristics	Supervision of execution of works by the public authority	Supervision of service operations by the concessionaire
	Absence of pre-financing, co-financing or financing of the works on the part of the contracted builder	Pre-financing, co-financing and financing of the facility on the part of the concessionaire
	Zero capital investment from the contracted firm	Capital investment contributed by the concessionaire
	No freedom in service or facility design accorded	Freedom accorded in service/ facility design
	Contract devoid of any service creation or organization function ("secondary" contract)	Contract instituting and organizing the service specified by the public authority ("primary" contract)
	The contracted firm is not the project developer	The concessionaire is the project developer
	Lack of contract management leeway granted to the firm	Contract management leeway granted to the concessionaire
	No long-term hold of public domain	Generally associated with a long-term hold of public domain
	Absence of joint Construction-Management-Maintenance responsibility	Long-term responsibility assigned to the concessionaire

The fundamental feature of any concessionary approach is reliance on a longstanding political commitment, not only for the project in question but also for concessionary formulae applied to other projects within the same country. Failure on the part of the public authority to fulfil its commitments may have negative repercussions on a country's international relations and on other public-private partnership projects. To prevent this from occurring, it is necessary to establish a legal framework for this type of contract.

Many countries with limited experience in the use of public-private partnerships, including those in Western Europe, have sent up specialized working groups for concessionary contracts either within the Finance Ministry or at a high level of government. Such steps have allowed a standardized methodology for awarding concessionary contracts by capitalizing on cumulative experience at a national level. Such working groups are composed of specialists and external consultants, combining a wide variety of skills and providing guidance to public authorities on how to draw up concessionary contracts.

12.4   FORMS OF PARTNERSHIP EVALUATED

This section describes each form of PPP that has been evaluated in the spreadsheet package. Specific characteristics from the descriptions have been identified and weightings have been given to each characteristic according to the level of correspondence between the characteristic and the form of partnership.

12.4.1 Full Privatisation

Under this arrangement the municipality sells off all its assets to a private company. The private company legally owns the entire water and sanitation system and has legal responsibility to provide, operate and maintain services and collect revenue on a permanent basis. The municipality would assume a regulatory role in terms of monitoring water quality, adequacy of services and tariff setting. In many countries, no water and sanitation services have been fully privatised. There is a significant opposition from several stakeholders to full privatisation.

12.4.2 Concession

This form of partnership entails the municipality transferring the ownership of all its water and sanitation assets to a private company for a significant time period, usually between twenty and thirty years. Under the contractual agreement the private company would have to operate and maintain the entire system. It would also be required to invest significant amounts of money in expanding the services. The municipality would regulate the performance of the concessionaire and also approve tariff structures. At the end of the contract the private company has to return the assets to the municipality in acceptable state.

The concessionaire carries out all the capital investment, operates the resulting service and is remunerated primarily through service fees paid by users. The facilities are to be handed over to the oversight public authority at the end of the contract period.

12.4.3 Lease Contract

Under this arrangement a private company leases the entire water and sanitation system from the municipality for a period typically of 20 years. The private company would operate and maintain all the systems. Depending on the arrangement, revenue may be collected by the municipality or the lessee, and the profit being shared between the two parties. Under a lease contract the lessee never owns the asset and never invests any money in new infrastructure. At the end of the contract the lessee has to return the assets to the municipality in god working order. The municipality still has a regulatory role to play.

The Lessee manages either the facilities side or the operations side of the project and is remunerated primarily through service fees paid by users. Equivalent to the Anglo-Saxon Management Contract (plus a "success fee").

12.4.4 Management Contract

A management contract entails a private management team working with the existing municipal staff. Typically of five years duration, the management team would implement programs to improve cost recovery and control, asset management, staff training and systems efficiency. While a management contract should result in improved municipal revenue and efficiency, the responsibility to provide services ultimately still rests with the municipality.

12.4.5 Service Contract

A service contract would be entered into with a private company for a specific municipal task. Service contracts, normally two years long, are commonly used in meter reading and solid waste collection. They can also be utilized in more technical functions like the operation of purification and wastewater treatment works. Service contracts represent a good means of obtaining skills which the municipality does not posses themselves. The municipality needs good management skills and needs a financially sound footing in order to implement effective service contracts.

12.4.6 Corporatisation

Corporatisation is the process of ring fencing and registering a certain municipal function (for example water supply and sanitation) as a private company. While the municipality would still wholly own the new company, corporatisation serves to formalize the relationship between service provider (private company) and service authority (the municipality). This enables the private company to deal more easily with other private sector companies and reduces the political interference from the town council. Corporatisation often is a step taken by a municipality before it enters into other forms of PPP.

12.4.7 Public-Public Partnerships

A public-public partnership is a partnership between two public organizations, where one organization offers its services to the other under some sort of contractual arrangement. While it is not likely that this sort of partnership will occur at present between municipalities in developing countries owing to the difficult state that the majority of municipalities are in, there is very real possibility of public-public partnerships between municipalities and water boards.

12.4.8 BOOT and BOT Projects

BOOT refers to Build Operate Own and Transfer. BOT refers to Build, Operate Transfer. BOOT and BOT arrangements are usually project specific. They are utilized as a means of financing a new development. Under one of these arrangements a private company would supply the capital for a project, implement the project and operate the project for a certain time period, usually twenty to thirty years. Upon expiry of the contract the assets would be transferred to the municipality and the municipality would assume operation of the system.

BOOT (Build, Own, Operate, Transfer): Within a project financing framework, several private entities – via a series of contracts – finance, build, own and operate an infrastructure facility designed to accommodate the set of needs established by the public authority, which in many instances acts as guarantor for the project. Ownership of the facility is to be transferred to the public authority upon expiration of the contract.

BOT (Build, Operate, Transfer): A variant of the BOOT contract, whereby facility ownership is transferred to the public authority once the building phase has been completed.

BOO (Build, Own, Operate): A variant of the BOOT contract, whereby facility ownership remains in the hands of the private investors beyond the contract's expiration.

Reverse BOOT: The public authority finances and builds the facility, and then confers service operations to a private firm which assumes ownership as the building phase winds down.

12.4.9 Municipal Debt Issuance

Municipal Debt Issuance is purely a mechanism of obtaining capital through selling public bonds. Currently in developing countries there is little prospect for municipalities obtaining capital through a municipal debt issuance due to bad credit ratings, and lack of institutional capacity.

12.4.10 Private Consultants

Private consultants would be called in to work on specific specialized technical problems encountered at a municipality. The municipality would benefit immediately from the work done by the consultant but if the situation changed then they would have to seek the consultant's expertise again.

12.5   MUNICIPAL CHARACTERISTICS

The following characteristics have all being identified as being critical in the decision of selecting a PPP. These characteristics have been identified in consultation with the Municipal Infrastructure Investment Unit's Scope of Work Guidelines and also from the characteristics of the various forms of PPP (see also Lyonnaise des Eaux, 1998).

Municipal characteristics considered:
- Ability to reduce liabilities
- Ability to improve cost recovery
- Ability to improve cost control
- Ability to raise capital
- Ability to improve infrastructure maintenance
- Ability to improve customer service
- Ability to improve water quality
- Ability to improve system efficiency
- Ability to improve staff ability
- Ability to adjust staff numbers
- Ability to improve motivation levels

A spreadsheet package can establish, from information entered by the user, an index between one and ten for each of the above-mentioned characteristics. The information entered relates to consumer demographics and service levels; asset age, condition, and value; income and expenses; municipal debts; capital availability; water quality; municipal staff and customer service levels.

In evaluating alternative strategies towards better management, the following must emerge:
- Cost of immediately required water and sanitation infrastructure.
- Capital expenditure required in the maintenance and replacement of existing assets over the next twenty years.
- Income and expenditure estimates
- Municipal debt ratio
- Level of consumer payments
- Unit costs and charges for water and sanitation

Furthermore there are graphical outputs comparing the demand for capital, and capital available for the next ten years and a comparison of income and expenditure over the next ten years.

The organizational spreadsheet outputs the following information:
- Index of condition of infrastructure
- Overall service efficiency
- Index of water quality
- Metered losses in water and sanitation systems
- Index of staff ability in terms of experience, level of training and performance
- Employee consumer ratio
- Level of staff motivation

The final outputs are two tables. The first table is an ordered list of the problems in the municipality with an index indicating the severity of each problem. The second output is an ordered list of forms of PPP that would best address the problems identified in the analysis.

The program provides a useful tool for municipalities to analyse their current situation and identify problems and obtain an indication of the severity of those problems. It also opens the door to the possibilities that PPPs pose for addressing the problems facing municipalities. It also serves as a tool for educating municipal staff about PPPs.

From the researched municipalities it is clear that municipalities are generally not on a secure financial and organizational footing. Municipalities exist with a lot of uncertainty in terms of capital funding, profitability and sustainability of services and stability of tenure for municipal staff. Due to the large number of organizational, structural and jurisdictional changes that public services have undergone in the last eight years a significant proportion of municipalities are struggling to meet the demands put on them. PPPs represent a way of reworking the organizations such that can meet the demands put on them by consumers and new legislation.

Municipalities could look to PPPs in order to address and solve many of the problems that they are currently facing. Municipal councilors and managers need to understand what a PPP can offer, consumers and the town council. A PPP will provide services which are more efficient and more reliable but at no extra cost to consumers. A PPP represents an opportunity for the municipal council to demonstrate that it is at the forefront of service provision and that it is genuinely committed to sustainable service provision. A PPP also represents a constant income source for the town council. Councilors and consumers need to understand that water and sanitation are essential services, which have to be provided in a sustainable manner, and the only way to do that is though consumers paying for the full and recurrent costs of service provision.

The question of degree of involvement of private enterprise requires regular revision. Cases have recently occurred where private industry has walked away from previous commitments. The risk and long-term profitability may not have been correctly gauged.

Similarly, there is a great suspicion in developing countries as to the true motivation of private companies, accompanied by a lack of grasp of international financial dealings. More appropriate public-private partnerships may have to evolve. Participation by poorer communities in non-financial terms, e.g. labour input and local risk transfer, are possible solutions. The risk communities are prepared to tolerate depends on the cost of risk aversion and the alternatives, e.g. heavy demand reductions in times of shortfall, wealth sharing and capacity building would equally make private sharing more palatable.

12.6   PRIVATE AND PUBLIC INSTITUTIONAL ROLES

Constitutional rights in South Africa require that priority be given to providing all people with adequate water supply and sanitation services. It is estimated that in order to provide all South Africans with basic services R100 billion (USD10 billion) is needed in infrastructure development over the next ten years. (Jackson and Hlahla, 1999)

It is the duty of the municipality to provide basic services first, to all consumers before it provides higher-level services to those who can afford it. However, due to inequalities of the past, consumers' attitudes, inexperienced municipal staff and new legislation, South Africa experienced a critical lack of financial and institutional sustainability at municipal level.

Since 1994, numerous water and sanitation projects have been implemented in South Africa. Due to consumers unwillingness to pay, inability to pay and vandalism several of these projects are no longer functional. Consumers unwillingness to pay for services traditionally related to politically motivated consumer boycotts, however nowadays consumers do not pay for services upon claims of poor quality of services, inadequate maintenance and inaccurate billing. Low levels of payment for services can be attributed to the level of services not being in line with what consumers can afford and a lack of community involvement.

The South African Water Services Act of 1997 and the Municipal Systems Act 2000 both create an enabling environment for the establishment of Public Private Partnerships in municipal service provision. The Water Services Act differentiates between water service providers (WSP) and water service authorities (WSA). Through this differentiation there is allowance for a WSP to be anyone who is capable and authorised (by the WSA) to provide water services. The Municipal Systems act allows for the provision of services through internal of external mechanisms. It also introduces the concept of Municipal Service Partnerships (MSPs). Preference is placed on seeking to enter into an MSP with a public partner first and private partnerships are only to be considered as a final option.

12.6.1 Case study – Balfour municipality

The town of Balfour is situated in the south-western corner of Mpumulanga, South Africa. The town is approximately one hundred kilometres east of Johannesburg. The municipality's jurisdiction covers an area of twenty square kilometres. Balfour has an estimated population of sixty thousand. Balfour Municipality's water and sanitation department has a staff of twenty-four people. Water is supplied via pipeline from a purification works at Fortuna Dam situated twenty kilometres from the town. The older parts of the town utilize suction tanks while the recently provided services in Siyathemba have full waterborne sanitation and utilize a wastewater treatment works.

12.6.2 Analysis of municipal situation

The following graphs and tables detail data as to the municipality's situation. This information is then transformed into values which are then analysed in order to identify major municipal problems and decide on a suitable form of Public Private Partnership.

12.6.3 Immediate capital expenditure requirements

Taking into account the current level of service and income of consumers, a demand for new and upgraded services has been calculated. The following levels of services have been assumed for each income category.

Table 12.3 Affordability of services

Household Income	New LOS considered affordable	
	Water	Sanitation
Less than R500 p.m.	Public standpipe	VIP
From R500 to $1 200 p.m.	Yard tap	VIP
More than R1 200 p.m.	In house	Full waterborne
(R1 = US$ 0.12 in 2003)		

In order to provide adequate and affordable water and sanitation services to consumers on the basis established in Table 12.3, the following demand for funds has been calculated.

Table 12.4. Immediate funds required to upgrade services – Balfour

Summary total cost of new and upgraded infrastructure	
Water – Local infrastructure	R705 051
Water – Bulk infrastructure	R1 724 712
Sanitation – Local infrastructure	R4 601 583
Sanitation – Bulk infrastructure	R0
TOTAL immediately required capital expenditure	R7 031 345
(R = US$ 0.12, 2003)	

In reality this spending will not occur immediately, but most likely over a period of five years. Hence, the annual capital expenditure over the next five years has been calculated to be R1 406 269.

Generally the level of service in Balfour is high with a large majority of consumers having received new water and sanitation services since 1994. The few consumers who require upgrading to their services are those who can afford a higher level of service, and those consumers who have been utilizing suction tanks for several years and now require full waterborne sanitation. The total number of consumers requiring new or upgraded services is 1367, this represents 22% of consumers. Currently there are no consumers without any services. The above-mentioned figures represent upgrades to existing services only.

12.6.4 Age, condition, value relationship of existing assets

The age, condition and original cost of all the assets within Balfour's Water and Sanitation Department has been used to calculate an expected life and current value of all assets. This information has been plotted on a time series (Figure 12.3) which shows the required capital expenditure, for the next 20 years in order to maintain and replace the existing assets. Monetary values are all expressed as a present value.

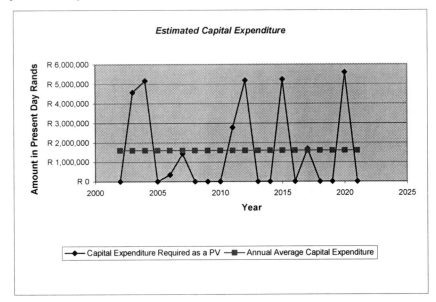

Figure 12.3. Estimated capital expenditure over the next 20 years – Balfour

To solve the problems permanently, Balfour should implement an asset management system in order to constantly monitor the operation, condition, and maintenance of its assets. Should a system of such a nature be implemented, all maintenance and replacement expenditure on assets would be predicted, timed and budgeted for. As a result assets life spans will be maximized and financial management optimised.

12.6.5 Capital available

The capital that is available and expected to be obtained in the foreseeable future has been reduced to an annual present value, so that it can be compared to the annual demand for capital. Capital is sourced mostly through Consolidated Municipal Infrastructure Program (CMIP) Grants or the District municipality may obtain a loan in order to supply the capital needs of Balfour. Balfour Municipality currently does not borrow funds against its balance sheet.

Table 12.5. Capital expenditure demand and availability – Balfour

Source of funds	Amount as a present value	No. of years funds required/available
Immediate new and upgraded capital requirements	R1 406 269	5
Maintenance and replacement capital expenditure	R1 599 428	20
Weighted Average Annual Capital Expenditure required	R1 950 995	20
Average Annual Capital Available	R1 822 352	5

Table 12.5 summarizes the Annual Average Capital Expense and the Annual Average Capital Available and from Table 12.5, it can be seen that over the following five years there is a deficit of R1 128 634 per year. This deficit represents approximately seven percent of annual capital expenditure. This is a relatively minor deficit which could easily be rectified with an internal loan, direct or cross subsidy.

Figure 12.4 displays the information from Table 12.5 as a time series over a ten-year period.

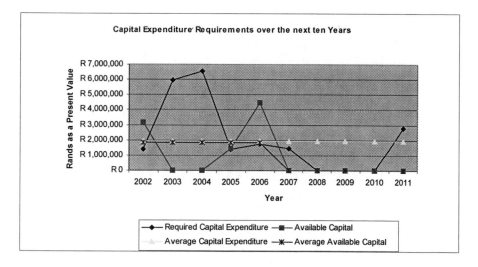

Figure 12.4. Capital expenditure and availability over 10 years – Balfour

At present there is only sufficient information to predict available capital for the next five years; hence the available capital on Figure 12.4 is zero after 2007. The required capital expenditure however remains just under two million Rand per year for twenty years. Over the next five years there is an annual shortfall, between capital required and capital available, of approximately one hundred thousand Rand per year, as seen on Figure 12.3. This is a point of concern and emphasizes the water and sanitation departments need to ensure long-term financial sustainability of service provision either through accurate consumer charges of continued grant funding.

12.6.6 Municipal income and expenses

Table 12.6 summarizes the actual income received and expenses paid by the water and sanitation department. This table shows that the water and sanitation department is running at an annual deficit of R779 792. This represents 20.4% of annual expenses. This is a significant loss which is currently covered by direct and cross subsidies within Balfour Municipality. The apparent anomaly in the distribution of the annual deficit between the water department and sanitation department is due to the very high percentage of unpaid sewage bills (73 percent) compared with unpaid water bills (28 percent).

Table 12.6. Annual income and expenses – Balfour

	Amount in last financial year	Percentage attribu-table to water	Percentage attribu-table to sanitation
Annual Income	R3 035 450	65.5%	34.5%
Annual Expenses	R3 815 242	60.1%	39.9%
Annual Surplus (Deficit)	(R799 792)	39.0%	61.0%

Balfour's water and sanitation department has no debts, i.e. it is generally debt adverse. Most capital expenditure is made through grant funding.

12.6.7 Cost of service provision and level of cost recovery

Currently Balfour Water and Sanitation Department runs at a 20.4 % loss. This is due to the fact that 73% of sewage bills and 28% of water bills are never paid. If 95% of consumers paid their bills the water and sanitation department would make an annual profit of R2 474 567 that is a 64.9% profit.

If this scenario did occur then the water and sanitation department would be able to fund the required capital expenditure of R1 950 995 per year and still make an annual profit of R523 572. This represents a profit of 12.64 % on total billing.

Table 12.7 shows the total cost of provision of services, the average charge to consumers and the profit per kilolitre of services provided.

Table 12.7. Per kilolitre costs, charges and profit of services – Balfour

	Water Supply	Sanitation
Cost of service provision per kilolitre	R1.51	R1.31
Average per kilolitre charge to customer	R1.93	R3.50
Profit per kilolitre	R0.42	R2.19

12.6.8 Water and sanitation staff

Currently there are 24 people employed in the water and sanitation departments at Balfour. There are 6135 households who utilize water and sanitation services. Hence the consumer-employee ratio is 256:1. The recommended consumer employee ratio is 275:1. Hence Balfour should attempt to decrease its staff quota by two employees.

Staff ability were analysed on three fronts, namely: experience, level of training and performance.

12.6.9 System efficiency, quality and service

The average condition of infrastructure at Balfour is poor. This is due to insufficient scheduled maintenance and low levels of staff technical capacity. The poor condition of the infrastructure is reflected in the losses in the water and sanitation systems (see Table 12.8), however it is not reflected in the low level of consumer complaints (see Table 12.9).

The average number of complaints is below 'normal' values. This is complemented by the good to moderate reaction times to complaints within the water and sanitation department.

Table 12.8. Water losses in water and sewage networks – Balfour

	Water Supply	Sewage System
Percentage loss in system	5.14%	5.36%

Table 12.9. Average number of monthly complaints – Balfour

	Number of complaints per 1000 consumers per month	Prescribed value
Water supply	3.26	4
Sanitation	1.63	6

Low levels of water quality both at the purification and the wastewater treatment ends of the system are of particular concern. The purification works is currently being operated above its design capacity. This in addition to the age of the system can account for the frequent violations in water quality standards. The wastewater treatment works struggles continually to maintain wastewater standards; this is due to the large volumes of substandard effluent from the abattoir.

12.6.10 Results of analysis

The spreadsheet package described by Richardson (2002) assigned values, from 1 to 10, to the problems identified in the water and sanitation department. Table 12.10 lists the problems requiring attention in order of severity (a value of 10 being most severe). This list of problems is in agreement with the discussion in the previous section. Cost recovery and cost control are major problems at Balfour. If these problems were addressed then there would be significant profit generated within the department which could be used to address capital needs. Low levels of infrastructure maintenance can be attributed to the poor financial state of the department and the low skills of staff.

From Figure 12.5 it can be seen that the average severity of problems at Balfour is just below seven. This is a moderate to high value and indicates that overall the water and sanitation department needs particular attention and careful management in order to address the serious problems it is facing.

Table 12.10. Ordered list of factors requiring attention within water and sanitation department – Balfour

Severity	Municipal factor
10.00	Need to improve cost recovery
10.00	Need to improve cost control
10.00	Need for capital
10.00	Need to improve infrastructure maintenance
8.00	Need to improve water quality
7.25	Need to improve motivation
6.22	Need to improve customer service
5.39	Need to improve system efficiency
3.97	Need to improve staff ability
3.52	Need to adjust staff numbers
0.00	Need to reduce liabilities

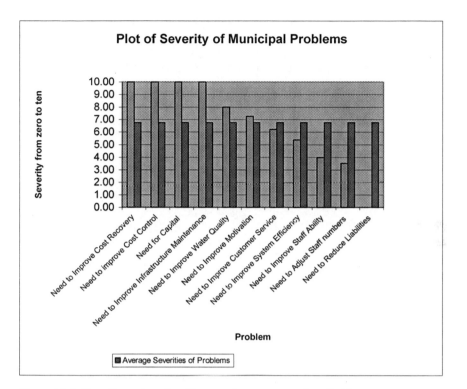

Figure 12.5. Plot of severity of problems and average severity – Balfour

12.7   SELECTION OF BEST FORM OF PUBLIC PRIVATE PARTNERSHIP

The output for the selection of the most appropriate form of public private partnership is in Table 12.11. The table lists the forms of PPP from most to least appropriate. The corresponding index values are listed from smallest to largest.

The spreadsheet package identified a management contract as the most appropriate form of public private partnership. A management contract of approximately five years duration would serve to initially address the problems of cost recovery and cost control. The need for capital would initially remain a problem but as the profitability of the water and sanitation department is restored, off balance sheet borrowing could be obtained. Improving infrastructure maintenance and water quality is where an external management team would implement maintenance programs and teach management staff how to effectively manage infrastructure and implement mechanisms to improve water quality. A management contract would generate a small amount of income for the town council. The other options available are a lease contract and a concession.

Table 12.11. Order list of most appropriate forms of PPPs – Balfour

	Ordered best form of PPP
29.14	Management Contract
32.64	Lease Contract
34.64	Concession
35.36	Built Operate Transfer (BOT)
35.64	Full Privatisation
36.58	Private Consultants
37.36	Build Operate Train Transfer) (BOTT)
37.86	Service Contract
38.37	Public-Public Partnership
45.31	Corporatisation
62.36	Municipal Debt Issuance

A lease contract would probable run for between fifteen and twenty years. A lease contract would address all the same issues as a management contract but there would be little focus on the transfer of management skills from the private partner to water and sanitation department. A lease contract could also generate significant income for the town council which could be spent on other infrastructure needs within the community.

A concession contract, lasting between twenty and thirty years, would represent a complete and long-term overhaul of the water and sanitation department. There would be significant transfer of skills and training of staff. There would also be the potential for significant capital expenditure within the municipal jurisdiction. There would be a moderate amount of revenue generated for the town council.

Other forms of PPP would not be viable to implement at Balfour.

12.8   REFERENCES

Briscoe, J. (1997) Managing water as an economic good: rules for reformers. *Water Supply*, **15**(4), 153-172, Blackwell Science Ltd.
Dinar, A., Rosegrant, M.W. and Meinzen, Dick R. (1997) *Water Allocation Mechanisms.* World Bank Policy Research Working Paper 1779.
Jackson, B.M. and Hlahla, M. (1999) South Africa's infrastructure service delivery needs: the role and challenge for public private partnerships. *Development Southern Africa*, **16**(4).

Lyonnaise des Eaux (1998) *Range of Options for Private Sector Involvement in Municipal Water and Wastewater Service.* Paris.

OFWAT (Office of Water Services) (1997) Birmingham, U.K.

Perrot, J-Y and Chatelus. G. (Eds) (2000) *Financing of Major Infrastructure and Public Service Projects: Public-Private Partnership.* Presses de l'école nationale des ponts et chaussées.

Pollit, C. (1995) Managing Techniques for the Public Sector, In *Pulpit and Practice*, Eds. G Pales and D. Savoie, Kingston McGill Queen Univ. Press.

Richardson, A.E. (2002) *Decision support system for optimizing strategies for the financing and implementation of municipal water services.* MSc(Eng) thesis, University of the Witwatersrand.

Roth, G. (1987) *The Private Provision of Public Services in Developing Countries.* EDI series in Economic Development, published for the World Bank Oxford Univ. Press.

13

Probability and risk

13.1 HYDROLOGICAL UNCERTAINTY

Storms and floods are unpredictable. The magnitude of floods cannot be calculated in advance, although a statistical assessment of the likelihood is possible. The question arises as to what discharge rate to design any drainage structure for. The design flow may seldom, if ever, occur during the life of the structure. Fortunately, the flow rate at other times will usually be less than the capacity of the system. For extreme events, a drain may overflow. This may cause damage or inconvenience and is to be avoided. The decision as to what discharge to design a structure for is usually based on an economic risk analysis. The cost of a larger structure is balanced against the probably cost of damage due to larger floods than the structure can accommodate.

The probability of different storm magnitudes and durations occurring must be obtained from a statistical analysis. The ideal situation for a hydrological analysis would be one where a continuous flow record over many years was available. A direct analysis of peak flows could then yield probabilities of different flow rates at the site.

Unfortunately adequate flow records are seldom available at a selected site, or even anywhere in the catchment in question. Even if there were historical records, it is likely that the catchment has undergone, or will undergo, changes in surface cover and drainage patterns. The hydrologist will therefore need to resort to rainfall records, and from these synthesize the necessary runoff pattern at the site of the proposed drain, culvert or conduit. Regional or local rainfall records are usually come by and are independent of development in the basin, There may, however, be a change in the method of recording at some stage.

Many standard hydrological techniques, such as the rational method, the Lloyd-Davies method and the tangent method, are based on the assumption that the probability of a storm of a particular duration is the same as that of the runoff computed using the method. Any discrepancy in the correspondence is supposedly built into the runoff coefficient. The assumption is suspect as there are many variables likely to affect runoff besides the storm intensity. These include antecedent moisture conditions in the catchment, including the state of the surface or retention storage and soil saturation. The storm distribution in space and its variation in time will also affect the peak rate of runoff. The length of record will also affect the distribution pattern, especially if recurrence intervals are extrapolated. Beard (1978) indicated that risk is traditionally underestimated with short records.

Practice is to design drainage structures to pass the flood which will be exceeded on an average of once in a specified number of years. Thus mean storm return periods of 2, 5, 10, 20, 50 or 100 years, depending on the severity of exceedance, are often used as the basis for the design flood. In general, areas sensitive to more extreme events plan for more remote possibilities. Whereas such rules of thumb are useful design guides, it is better to select a probability of exceedance based on hydro-economic risk analysis.

13.1.1 Probability distributions

Although rainfall and runoff are to some degree random, they have limits imposed by the climate and environment and even follow some trend. Rainfall or storm intensity generally follows a distribution pattern with a mean and variation. Total storm precipitation depth for any selected storm duration can be correlated with probability as in Figure 13.1 (Gringorten, 1963).

Rainfall follows a more distinctive probability distribution than runoff. Runoff or stream flow can be described with a distribution curve but the parameters are more complex than for storm distribution. No mathematical expression can be fitted to runoff distribution or even the relationship between runoff and rainfall, on account of such factors as antecedent ground water conditions, alternative sources of flow (groundwater, surface runoff, unnatural discharges), changing land use, storage-discharge relationships for the catchment, and complex topography, all of which result in a non-linear rainfall-runoff relationship (Chow et al, 1988).

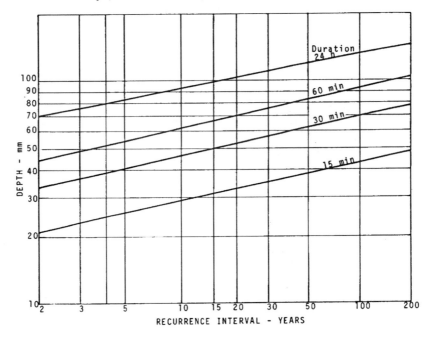

Figure 13.1. Rainfall depth – Duration – Frequency relationships

Various mathematical approximations have been attempted to fit the flood discharge spectrum (and, not of interest in this context, drought flows). Peak runoff has some frequency distribution (e.g. Fig. 13.2) which may or may not coincide with a mathematical function. Parameters which describe mathematical or other distributions are indicated below. The storm drain designer is primarily interested in the upper extreme values of flow.

In the expressions, the following terms are employed:

Arithmetic mean: The centre of gravity of the distribution. The population mean is designated μ, and the sample mean $\bar{x}$.

Population mean
$$\mu = \int x\, dp \qquad (13.1)$$

Sample mean
$$\bar{x} = \frac{\sum x}{N} \qquad (13.2)$$

where x is the variate and N the number of observations and p is the expectance of x.

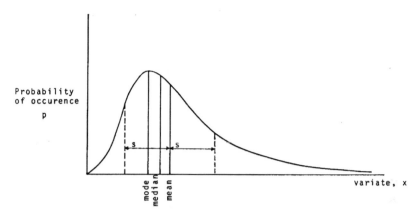

Figure 13.2. A probability distribution

Median: Middle value of variate, such that it divides the frequency distribution into two equal portions.

Mode: Value of variate which occurs most frequently.

Standard deviation: A measure of variability. Variance is the square of the standard deviation.

Population standard deviation
$$\sigma = \sqrt{\frac{\sum (x - \mu)^2}{N}}$$
13.3)

Sample estimate
$$s = \sqrt{\frac{\sum (x - \bar{x})^2}{N - 1}}$$
(13.4)

Skewness:

Population skewness
$$\alpha = \frac{1}{N} \sum (x - \mu)^3$$
(13.5)

Sample estimate
$$a = \frac{N}{(N - 1)(N - 2)} \sum (x - \bar{x})^3$$
(13.6)

The distributions in Figure 13.3 are often used in hydrological analysis (Haan, 1997; Yevjevich, 1972a; Bury, 1975).

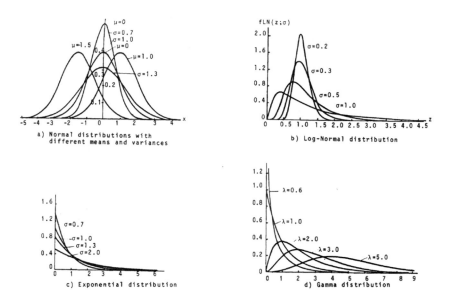

Figure 13.3. Shapes of various mathematical probability distributions

The *Normal distribution* represents the distribution of a completely random number about a mean. It is a symmetrical bell-shaped distribution with the area under the curve equal to unity.

The *Log Normal distribution* represents a similar distribution of the log of the variate.

The *Gamma distribution* has a skewness and passes through zero. It is used in drought flow analysis in particular for estimating reservoir.

Pearson distributions (Types I, II and III) were devised by Pearson (1930) to fit virtually any distribution and are of an exponential type. Log Pearson distributions are also used.

Extreme value distributions (Types I, II and III). Fisher and Tippet (1928) found that the extreme values of many distributions approached a limited exponential form as the number of points in the sample increased. They fitted equations to the upper extremes. Gumbel (1941) first applied the Type I extreme value theory to floods. His work is now accepted for the analysis of hydrological extremes (1958).

The distribution is of the form

$$p = 1 - e^{-e^{-y}} \qquad\qquad (13.7)$$

where: p is the probability of the flood being equalled or exceeded.
          e is the base of Naperian logarithms, and
          y is a mathematical function of probability.

The equation for the extreme value of the variate as a function of recurrence interval T resulting from Gumbel's theory is:

$$x = \bar{x} - \sqrt{\frac{6}{\pi}}s\{0.5772 + \ln[-\ln(1 - 1/T)]\} \qquad (13.8)$$

Gumbel prepared graph paper which causes the variate (flood peaks), to plot as a straight line against probability or its inverse, return period (e.g. Fig. 13.4). Alternatively the variate may be plotted to a log scale (Fig. 13.1). The distribution, together with the log Pearson type III is often applied in flood hydrology.

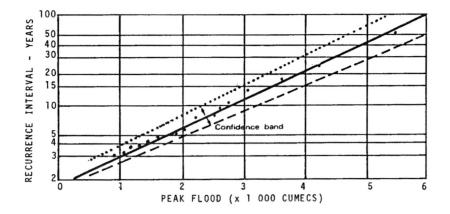

Figure 13.4. Frequency distribution of flood peaks

13.1.2 Analysis of records

Presuming that flow records are available or can be synthesized (e.g. Yevjevich, 1972b; Fiering, 1967), then the extreme flow distributions can be analysed by means of the following procedure. The record is usually produced in chronological order as a complete duration series with peak flows for each day or month or year identified. The record should be divided into years (with the beginning of the hydrological year preferably at the start of the wettest season).

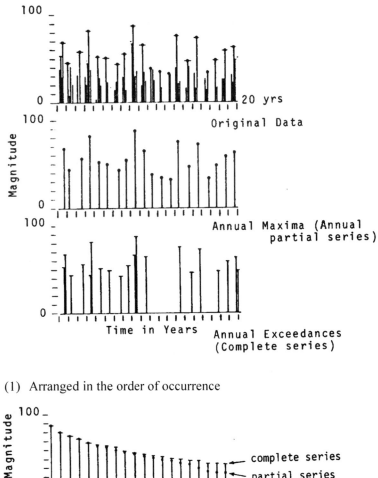

(1) Arranged in the order of occurrence

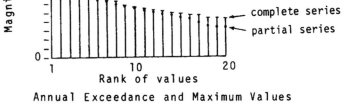

(2) Arranged in the order of magnitude

Figure 13.5. Hydrological data series

The annual maxima should be selected and ranked, taking only the maximum flood in any one year. If the annual flood peaks are arranged in order of magnitude (Fig. 13.5b), we have an extreme value exceedance distribution. One thus has what is terms an annual partial series. If every flood on record was included, we would have a complete series. The difference is only of interest for low recurrence intervals giving a lower recurrence intervals over 10 years both series yield practically the same results.

The probability of a flow being equalled or exceeded in any hydrological year is the inverse of the frequency of it occurring. Thus a flood which is equalled or exceeded on an average once every 50 years has a two percent probability. It is common to use the frequency, or recurrence interval or return period in hydrological analysis, in preference to probability. Thus

$$P(P \geq x) = \frac{1}{T}$$
(13.9)

So the probability that the annual maximum flood X is less than x is

$$P(X < x) = 1 - \frac{1}{T}$$
(13.10)

where T is the recurrence interval of the flood of magnitude x.

The estimation of T from the ranked sample has been done in different ways. Once a sample has been ranked in descending order of magnitude, the recurrence interval T may be estimated from the Weibull formula

$$T = \frac{N+1}{m}$$
(13.11)

where N is the number of events (or years or record) and m is the rank proceeding from 1 for the highest value. Some other formulae for estimating recurrence interval and the corresponding origin are indicated in Table 13.1.

The resulting recurrence intervals or so-called plotting positions may be plotted on a suitable graph such as in Figure 13.4.

Table 13.1. Formulae for recurrence interval T

Formula		T	Distribution	T for N=50, m=1
California	(1923)	$\dfrac{N}{m}$		50
Haxen	(1930)	$\dfrac{2N}{2m-1}$		100
Weibell	(1939)	$\dfrac{N+1}{m}$	Normal and Pearson III	51
Blom	(1958)	$\dfrac{N=0.25}{m-0.375}$	Normal	80.4
Beard	(1962)	$\dfrac{N+0.4}{m-0.3}$	Pearson III	72
Gringorten	(1963)	$\dfrac{N+0.12}{m-0.44}$	Exponential, Extreme Value I	89.5

The constants in Gringorten's equation actually vary slightly depending on length of record

13.1.3 Confidence bands

An infinite length of record will yield a true distribution with a very low minimum and an extremely high maximum. Any record of finite duration will have a more limited range and can only approximate the true distribution. The shorter the record, the less representative of the true distribution it is likely to be. In fact, the variation of a number of sample means about the true mean will be distributed as a normal distribution with a mean μ and variance σ^2/N.

The possible extent of the true distribution each side of the available data can be described in terms of confidence limits. There will be a confidence band above and below the plotted line on an extreme value plot such as Figure 13.4. The width of the band will depend on the degree of confidence accepted and on the scatter and sparseness of the data.

Confidence limits can be estimated in the case of a normal distribution from the data set (Haan, 1977). Thus there is a 68% probability that a sample mean will be within $\pm \sigma = \sqrt{N}$ of the true population mean. The probability of lying within a certain band about the mean, $F(c)$, for different c is given in Table 13.2.

For other plotting positions, the band width increases in accordance with a factor G. The factor is dependent on the population distribution, length of record and recurrence interval. It may be deduced that G is dependent primarily on recurrence interval T, and the values in Table 13.2, apply for over 20 years of record.

Table 13.2. Confidence bands

Degree of confidence, c (%)	95	90	80	68
$F(c)$ (about mean)	2.0	1.7	1.3	1.0
Recurrence interval, T (years)	2	10	100	1 000
$G(T)$	1.0	1.5	2.2	2.7

Figures for the 90% confidence band are presented by Viessman et al (1977) and Beard (1978). Yen (1970) presented a chart from which it is possible to read the probability of an event of rank m = 1, 2 or 3 in n years of record corresponding to an event of average return period T.

13.1.4 Design discharge

The construction of a culvert, bridge waterway or even drain to pass a flood of a selected recurrence interval involves some risk. The probability of the discharge capacity being exceeded at least once during the operation of the structure is greater than $1/T$ where T is the recurrence interval of the design flood.

Although a minor overtoppping may result in only inconvenience, a severe overflow of an embankment could cause scour and wash-away. This would cause economic damage as well as risk to traffic and life. A probabilistic approach to the selection of design capacity is therefore desirable. Young et al (1974) applied risk analyses to the design of highway culverts and indicated damage costs. Allowance for cost uncertainty and interaction between culvert capacity and flood magnitude was studied in detail by Mays (1979).

13.1.5 Spread risk

The simplest approach to the computation of design flood in the case of a structure which is to function over a long time or indefinitely is on an annual basis. The costs of the structures which can discharge different flood magnitudes are added to the corresponding probable cost of damage. All cost figures are converted to a common time basis, e.g. annual costs. The method is explained with an example below.

13.1.5.1 Example

A highway authority is to construct a low bridge over a river. The flood frequency distribution is indicated in Figure 13.4. The capital cost of the bridge is a function of the discharge capacity of the waterway beneath. The capital cost corresponding to various design discharge rates is converted to an annual amount representing interest and redemption on the loan to meet the cost, and added to annual maintenance cost. Annual interest plus redemption figures can be obtained from interest tables or from a formula (Institution of Civil Engineers, 1969). The capital cost is multiplied by the factor

$$\frac{r(1+r)^n}{(1+r)^n - 1} \tag{13.13}$$

where r is the interest rate (a fraction) on the loan, assumed here equal to the interest rate on the redemption or sinking fund, and n is the loan period in years. To get a true picture, the loan should be renewed for as long as the economic life of the structure.

The resulting annual cost, as a function of waterway capacity. is plotted in Figure 13.6 (curve A). Now corresponding to each of the trial waterway design capacities is a probability of exceedance in any year, as indicated in Figure 13.4. There is a different probability corresponding to either the best estimate or the upper confidence limit of the flood frequency curve.

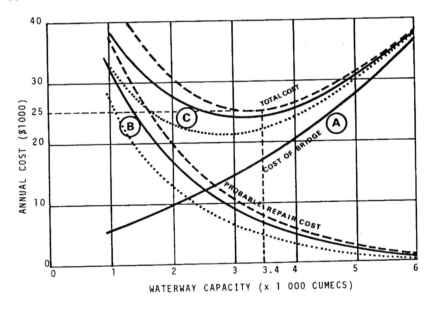

Figure 13.6. Overall costs for spread risk example

The probabilities of each flow being exceeded are multiplied by the cost of a flooding. This may include repair costs to the bridge. damage to surroundings, and interference with transport. Assuming in this case that the cost of exceeding the design capacity is \$100 000, the probability cost is this figure multiplied by the probability of exceedance of the discharge. The resulting annual costs are plotted in Figure 13.6 as curve B, for both best estimate and upper confidence limit.

The total average annual cost is the sum of A and B and this is indicated as curve C. The minimum cost corresponds to a waterway capacity of 3 400 m^3/s, provided the upper flood curve is chosen. This is the usual procedure as it provides a margin for uncertainty in flood estimates. It will be observed however that the best estimate curve will result in a slightly smaller optimum waterway capacity. Thus the effect of uncertainty in hydrological data is to increase expenditure.

In real situations, the computations will be more complicated. There is always the risk to human life which is difficult to evaluate. The damage costs may very well increase with increasing flood magnitude above the design capacity. In this case, the probable economic lost must be evaluated. It is the integral of the probability of various extreme floods occurring multiplied by the corresponding cost, i.e.

Probable annual cost = ΣPC (13.14)

where P is the probability of the flood being in a certain range and C is the corresponding damage cost.

The aspect of cost escalation also arises. If damage costs are likely to increase over the years, they should be evaluated separately for successive years, and discounted to a present value. Thus each annual cost is multiplied by

$$\frac{1}{(1+f)^n}$$ (13.15)

where f is the inflation rate as a fraction. The total of the present values for each year is obtained and this may be added to the capital cost of the bridge. Again the capacity with least total cost is selected.

13.2 RELATIONSHIP BETWEEN PROBABILITY, RISK AND HAZARD

13.2.1 Definitions

Frequency is the number of occurrences and could be a number, a percent or a fraction.

Recurrence interval is the average number of years from one event to another.

Probability is a statistical measure of frequency or probability. Various events will have different probabilities of occurrence. The total of probabilities equals one.

Risk is the likelihood of a specific undesirable event, e.g. shortfall of specific magnitude. The events could be floods, pollution, cost.

Reliability is the opposite of risk.

Hazard is the level of danger, e.g. deep flood, high velocity, poisoning.

Vulnerability is the exposure to a hazard, where the hazard be purposely or accidentally created.

In other words, probability and recurrence interval have to do with how frequently an event occurs. Hazard is what damage the event can cause, and risk is the probable damage.

13.3   SPILLWAY DESIGN FLOOD

Dams are designed to discharge their design flood. Overtopping of the spillway could be dangerous as it could damage the dame and lead to failure. Dam failure can be catastrophic as a high flood wave proceeding down the river could damage property and endanger lives. The correct flood to design for has been considered by various national committees of ICOLD (International Committee on Large Dams). Generally, the dam, depending on the importance, should be designed for a selected design flood, where factors of safety should be 1.3 – 1.5 or more, but should also be checked for a safety evaluation flood when safety factor should be above 1.

Table 13.3 Dam specifications

Dam classification

Wall height, m	Size class
5 – 12	Small
13 – 30	Medium
30 +	Large

Hazard rating

Loss of life	Economic loss	Hazard rating
None	Minimal	Low
≤ 10	Significant	Significant
> 10	Great	High

Recommended design discharge recurrence interval (RDDRI)

Dam size	Hazard Rating		
	Low	Significant	High
Small	$Q_{20} - Q_{50}$	Q_{100}	Q_{100}
Medium	Q_{100}	Q_{100}	Q_{200}
Large	Q_{200}	Q_{200}	Q_{200}

Q = flood, subscript = recurrence interval
Add to freeboard for waves, setup, etc.

Safety Evaluation Flood (SED)

Dam size	Hazard Rating		
	Low	Significant	High
Small	RMF –	RMF –	RMF
Medium	RMF –	RMF	RMF +
Large	RMF	RMF +	RMF +

RMF = Regional Maximum Flood, taken from regional Francou-Rodier type envelopes (Rodier, 1984)
RMF + often results in the PMF outflow (Probably Maximum Flood)

13.4   RISK FACTOR IN WATER SUPPLY

It is often not recognized how important risk management is in planning and operation of water services. Engineers are accustomed to plan using deterministic projections and explicit project formulations. They also assume certain financial constraints and at the most perform sensitivity studies on some of these assumptions. However, the practice of probability planning and even recognition of some of the risks may change future plans.

The various uncertainties involved in servicing communities with water and drainage can be divided into technology-related risk, financially-related risk and external factors.

13.4.1 Technological risks

- The design by a consulting engineer may not be the most appropriate.
- The construction cost is not known until the contract is awarded and even completed.
- The durability and life of the structure is not known with certainty.
- Life-cycle costing may not have been incorporated in awarding tenders or in the design.
- Operating costs and maintenance costs may not be properly considered at the design stage.
- Losses of water are not known and are related to maintenance level which is also not known.
- The structural factor of safety and loading on systems can affect damage costs.
- The dependency of one item on another can lead to chain reactions.

13.4.2 Financial risk

- The interest rate payable on a loan to finance a project is not known until the loan is taken out, and renewed loans will have an even greater uncertainty with regard to their interest rate.
- The exchange rate is very influential on the cost of the project if the money is borrowed internationally.
- The recovery of revenue from consumers is often uncertain and time-delays and legal costs subtract from that revenue.
- The interest during construction and during low demand periods increases costs.
- Low operating factors during initial years after installation of a large project and under estimates in future demand increase unit costs of water supplied or services provided.

13.4.3 Natural and external factors

- Droughts can result in shortage in water and the frequency and extent of droughts cannot easily be predicted.
- Floods may damage services and increase costs due to repairs.

- Earthquakes, landslides and other geotechnical problems may require reconstruction and result in periods without services and also dangers to health.
- Losses of water due to theft and other unaccounted-for losses increase the amount of money which has to be collected from paying customers.
- Wars, political and civil commotion can damage installations and cause under utilization.
- Legislative framework and legal requirements may change over time.
- Ineptitude and fraud will increase costs.
- Changes in government and government policy can result in under or over utilization of services.
- The vulnerability of services to sabotage and accidents requires costing and planning.
- Climate change could reduce availability of water.

13.5   PLANNING TO MINIMIZE EFFECTS OF RISK

Engineers should be well aware of the consequences of adopting standard designs and safety factors on the cost and even long-term viability of projects that are constructed.

As an example, there is an acceptable level of leakage and bursts in pipelines which can be tolerated within a water service company's framework. There are regular teams in operation to maintain urban water infrastructures, but the existence of long transmission lines is a separate problem. Whereas a 1 km pipeline may be regarded as acceptable if the number of leaks is of the order of one a year, if the pipeline is considerably longer, for example 100 km, then based on the same standard one could expect 100 leaks a year. The latter is probably not acceptable as the number and duration of outages will be excessive. A way of reducing the risk could b to construct intermediate or end storage.

Thus a series of structures or actions in chain formation depends on the weakest link. Such a system in series should be designed to a higher standard than a number of facilities in parallel. For example, a number of sources of water in parallel, either as alternatives or feeding together simultaneously reduces the damage due to shortfall if one or some of the systems fail. Therefore, a different, i.e. lower, factor of safety or lower standard of design may be tolerable in the latter circumstances.

At the planning stage, the engineer and economists making the decisions as to which projects to construct, need to consider the cost of outages or other failures. Whereas the financial cost may be readily derived, the economic cost of a number and lengthy outages can be high but is not often included in costing. That is, shortage of water to a city will not only inhibit production, it will also discourage developers from establishing industries in the town subject to these water shortages. This is an economic cost, as opposed to a financial cost such as interest rates.

The planner needs to consider the possibilities of lower as well as higher demands than his projections based on deterministic or statistical models. Each possibility needs costing and the consequences need building into the model to produce the final recommended answer. Thus, a shortfall in demand will severely affect the financial viability of a project based on a full demand. Not only initial build-up years but also future consumption patterns need to be considered to ensure adequate payback. This requires a simulation of the demand pattern and costing methods using tabular and other financial models.

Developing areas in particular are being ravaged by factors which are likely to affect future demands for water. These include the effect of AIDS on population numbers and advancement. Countering this is the rapid growth in the standard of living of many of the population so that the upper and lower extremes in demands are wider than are accustomed to in developed countries. A decrease in immigration to urban areas may come about due to changing values and government incentives to stay on the land. The urban immigration has been large over the last decade internationally and this has increased growth in service requirements in urban areas and peri-urban development such as squatter camps has posed a great problem in many developing countries.

The international economy is at a level of growth lower than has been experienced for over half a century and the future growth rates and growth pattern are highly unknown. The world has reached its highest standard of living, particularly in developed countries, and whether the standard of living will continue to increase by consumers spending more or whether there will be a return of the pendulum or even pressure from environmentalists to become more self-sufficient in the present unsustainable world, has yet to be seen. Environmental pressures are particularly high in arid and semi-arid countries and even greater conservation practices may have to be sought.

Thus the cost of lack of knowledge in correct planning can be large. Not only will it increase cost of services to future generations, it will adversely affect the credibility of borrowing nations if they cannot pay back loans, as well as the value of the currency which manifests in exchange rates.

13.5.1 Economic risk

Risk costs money and money also has to be spent to minimize its effects. For example, Figure 13.7 shows that the higher the variability in river flow the greater the storage needed. Conversely, if consumers are prepared to accept risk, their costs can be reduced. Figure 13.8 illustrates the overall cost of water increases with risk, whereas economic sense is to minimize expenditure in the face of risk.

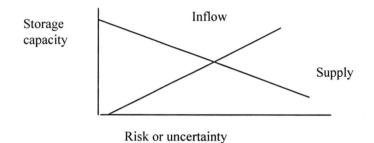

Figure 13.7. Uncertainty and storage

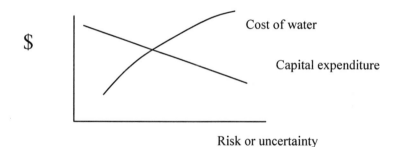

Figure 13.8. Uncertainty and cost

The consequences of over- or under- design are described below and a probabilistic method for determining the optimum design capacity is advocated.

Although river flows are caused by natural meteorological and geophysical processes, we are not able to predict the flow easily and tend to regard the flow as a random variable superimposed on known trends. River flows have a random variation from year to year and month to month. However well we can predict climate they are still innumerable unknowns which makes the runoff and river flow even more of a random variable. We therefore often design storage reservoirs to meet a certain probability of drought associated with any level of assuredness as a cost (Fig 13.9).

Cost per unit of water

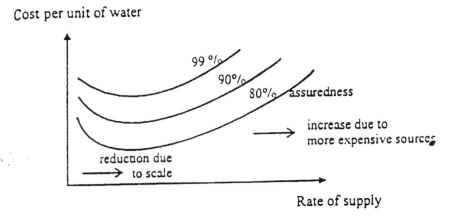

Rate of supply

Figure 13.9. Effect of assuredness on cost of water

On a normal design basis when the drought of the selected risk does occur, then the reservoir stored volume would drop to zero. Similarly, if the reservoir were oversized the spill would be reduced and the evaporation increased so that there is also a reduced yield from the system.

Other sources of water such as ground water have similar unknown probabilities of being recharged. Whether it be harvesting of rain water or cropping of rainfed farms, there is always a degree of uncertainty in the amount of water available.

One way of minimizing the risk of failing to supply is to reduce the release rate from the reservoir as the volume of water stored decreases. Such operating rules can be derived by trial and error or the releases at different storage levels can be optimised using economic principles (Stevens et al, 1998).

Variable draft operating rules are derived by considering both the probability of different inflows occurring and the economic value of water at difference rates of supply. Thus, if the demand has some elasticity, then higher rates of supply would be of lower unit value whereas the last amount available is of very high value to consumers. If the economic value at different supply rates and the hardship to consumers at different supply rates can be put in numbers the system can be optimised to minimise the probable economic loss or maximize the probable economic benefit.

Operating rules can be derived for conjunctive use systems where risks are considerably diminished by looking at the probability of independent events occurring simultaneously (see Fig. 13.10).

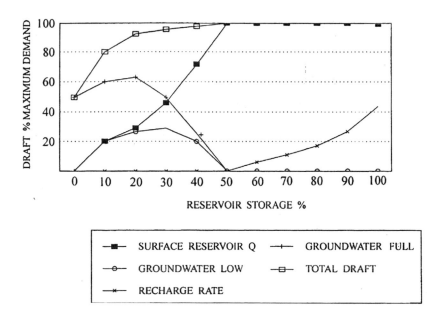

Figure 13.10. Operating rules to minimise economic losses

According to the Berthoux (1971) if the amount of uncertainty is high it pays to decrease the Investment in projects. However, his theory was intended primarily for the demand side and in the case of supply the effect of increase risk of availability of water could be to increase the capital size of the works.

However if a higher risk can be taken in having to drop the supply rate then the size of the reservoir can be reduced, i.e. cost savings can be achieved if the operating rule could be optimised allowing for the probability of having to decrease the supply rate.

Similarly when the drawoff is maximised the unit cost of water is reduced because the capital investment is divided by a larger figure. In the end the classical supply and demand relationship is complicated by this non linearity.

13.5.2 Effect of uncertainty in demand estimates

The greatest unknown in planning of a water project is frequently the demands in the future. Where public works are being planned then the rate of increase in demand in the supply needs to be estimated and this may not have a finite limit as would be the case for designated irrigation areas. Both the time horizon needs to be decided as well as the rate of growth up to that time.

However, the old fashion method of projecting on a logarithmic scale the demand using historic growth rate and selecting an arbitrary time horizon can be very expensive. Figure 13.11 illustrates what would happen if the demand were overestimated. This has occurred in South Africa after the planning of the Lesotho Highlands project, costing approximately US$1 billion to date. The expected demand has not materialized with the result that an expensive capital project has to be funded by dividing by a lower supply rate. The unit cost of the water increases even more with the result that the sharp price increase reduces demand even more (see Dandy and Connerty, 1994).

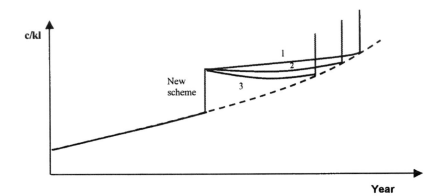

Figure 13.11. Effects of tariffs on demand (demand 1 is low, demand 3 is high)

One of the solutions of this dilemma is to plan in the face of high uncertainty with high operating, low capital cost projects. Whereas with a high certainty of demand, the higher capital cost project is more viable particularly if the capacity factor is high and the load factor is high.

The capacity factor is the range of average water consumption to the capacity of the project. The load factor is the ratio of the average water consumption at a particular time to the peak consumption at that time. The peak could be less than the capacity of the project for a number of years as the demand increases.

Thus, the effect of uncertainty is to reduce capital expenditure as indicated by Berthoux (1971). He also suggested another way of allowing for uncertainty. That is to artificially plan with a larger discount rate than would be used for a deterministic system. A third way is to do a probability study whereby a probable minimum cost project is selected (see Fig. 13.12).

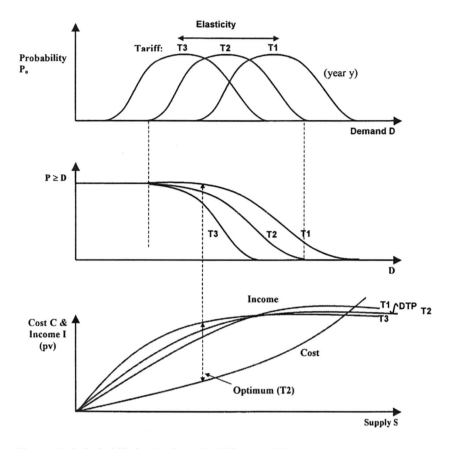

Figure 13.12. Probabilistic planning with different tariffs

The effect of under-designing, i.e. underestimating the demand, would also be to increase the cost. This is because a smaller project than necessary would be constructed. The demand would therefore reach the supply capacity of the system sooner than expected and another project would have to be planned. This may be a more expensive project increasing the marginal cost of water or it could be a parallel project. The cost of two pipelines to take a certain flow is considerably more expensive than one larger pipeline, i.e. the advantage of scale in hydraulic engineering is to decrease the marginal cost of water (Stephenson, 1998).

Even thus there are fluctuations in the water demand over time. Apart from the compound increase in water demand due to increasing population and increasing standards of living, there is also the variations in consumption due to seasonal and climatic factors.

13.5.3 South African case study

It has been described above how overplanning of water projects could increase the cost of water unnecessarily and reduce consumption. This vicious circle is detrimental to the water supplier and to the consumer because a lower water consumption would imply sacrificing amenities, whether they be luxuries or health related.

There have been a number of other factors which have decreased the water consumption growth rate in South Africa. The AIDS epidemic is of particular concern. Estimates have indicated 10-25% of the population have HIV or AIDS and as a result the mortality rate is expected to increase in the future and the population growth rate will decrease. The economic effects of AIDS were compounded by the international economic slump being experienced. The high cost of AIDS is due to wasted cost on training, early retirement and deaths in the workplace, lack of skilled people, aging of the workforce and high medical care costs.

The demographic movements due to economic and social pressures must also be added to the equation. Some of the businesses have become concerned by the extent of trained personnel emigrating. Their places are taken by less skilled people and the economic productivity of city dwellers is greatly decreasing. There is also a peri-urban migration from rural areas. These people can frequently not afford the urban system and informal housing and servicing develops around the cities. The inability to pay for services and the pressure of the richer people to minimize health risk drains economic resources. There is cross-subsidization resulting in discontent amongst some of the richer sectors.

In attempts to decide the growth rate in the future, demographers, politicians, health scientists and engineers were consulted. A wide range of future scenarios emerged. The end result is not only a decrease in growth rate but also a possible decrease in water consumption in a decade or so (Fig 13.13). This is compounded by a great uncertainty in the future demand. Both factors therefore point to a lower planning horizon. Alternatives to large capital cost projects must therefore be sought. These can range from improved loss-control to more efficient water use through better plumbing and greater awareness amongst consumers. Demineralization and recycling are also relatively low capital but high operating cost sources.

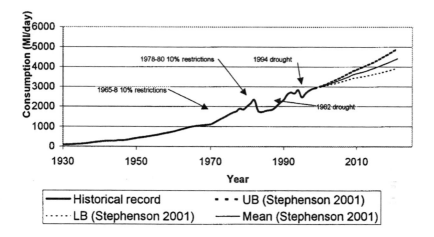

Figure 13.13. Changes in demand growth pattern

The tightening of the reins can only be done so much and then an expansion must occur again. In the medium term they solve some of the problems, but the poor economic conditions will remain for many decades despite job creation policies and higher educational standards.

It is thus apparent that there are many uncertainties in planning for water supplies:

- Uncertainty in the river flow rate in the case of surface water resources.
- Unknowns in aquifer and catchment boundaries and runoff.
- Uncertainty in water demand in the future due to demographic and economic factors.
- Economic uncertainty (international and local growth) affecting industrial demand.

The economic relationship between value of water and supply rate is non-linear, so hedging against risk at an early stage of drought could reduce the effects of deficits if the reservoir must dry, i.e. a hedging rule is advisable.

In the face of high uncertainty it is wise to avoid high capital cost expansions (e.g. dams, long transmission conduits) and instead revert to operating intensive schemes (e.g. demineralization, recycling, pumping). This can be affected by using high discount rates. Underdesign is advisable in the face of uncertainty.

Underdesigning of reservoirs saves cost, and unexpected demands can be met by rationing.

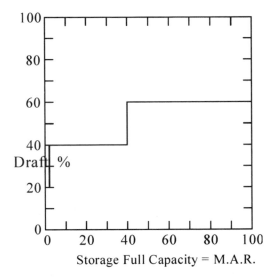

Figure 13.14. Variable Draft Operating Rule to hedge against severe drought risk

13.6   RELIABILITY OF WATER SUPPLY NETWORKS

The reliability of a water supply is defined as

$$R = 1 - P \qquad (13.16)$$

where P is the probability of a risk as a fraction.

The risk could be one of flow deficit Q_d, or pressure head deficit H_d or exceedance of concentration pollutant C_d. The risk could also be defined as a weighted combination of different deficits, e.g. 0.4 x flow deficit ratio plus 0.3 x head deficit ratio plus 0.3 x quality deficit ratio, where deficit ratio is the shortfall divided by target figure.

If a demand pattern over a period is known, as well as a relationship between flow and head and water quality, a distribution system can be simulated by computer. A network analysis could be done for selected flow patterns, or a continuous simulation performed over a period of time, to obtain the frequency or probability of the deficits. Such studies are the subject of a number of current research projects (e.g. Tanyimboh, 2003).

13.7   VULNERABILITY

Since the attacks on the World Trade Centre, increasing diligence has become the norm particularly in public services. Water services are of particular concern because poison dosing could reach millions of people in a short time. Attacks to structures and infrastructure are of less concern, but the vulnerability of computer systems could be a problem. There has long been a danger of attack to water infrastructure in some politically unstable countries and we could learn much from countries already in a state of preparedness, e.g. Israel. It is a fact that security costs money, so the risk needs to be assessed, i.e. probability as well as the hazard.

The following programmes need considering or implementing:
- Water quality monitoring
- Water pressure monitoring
- Visual and audio surveillance
- Vulnerability assessment
- Likelihood assessment
- Hazard assessment
- Crisis management
- Security officer training
- Public awareness campaigning
- Source and reservoir protection
- Distribution network and fittings protection
- Review hydraulic, pneumatic, electrical, mechanical, thermal, chemical, nuclear systems
- Check sources of services, water and discharges of wastes
- Stresses, structures, links, materials, and time dependent strength
- Natural hazards, flood, earthquake, geological, wind, fire
- Security hardware and access control installation
- Protective clothing, shields, nose, mouth, eye and ear protection
- Neutralizing and washing facilities
- Isolation rooms
- Staff and visitor identification and screening at different levels]
- Communications protection
- Duplication of vital systems
- Data backup
- Computer and internet systems
- Vehicle protection
- Threat identification

The identification and prioritisation of vulnerability in water services could be guided by studies done in the nuclear industry, for computer network firewalls, and by security planning done by water authorities in the past. The implications of damage, interference, accident or probability can be assessed from economic and danger points of view. Loss of facilities or data and danger to humans and the environment should be considered. Differentiation is needed between installations for water supply or sewerage and operation. I.e. if vulnerability is considered at planning stage a different design may result, but there will always be the necessity for alertness during operation. Risks due to intentional or malicious damage cannot be easily quantified, but accidents and incorrect estimates can be attached probabilities and costs. Differentiation should be made between probability and damage, and hazard-risk indices are suggested (Stephenson, 2002). The same approach can be used for financial risk, IT risks, natural hazards (e.g. drought, earthquake) and economic risks (e.g. changes in water demand, AIDS, interest rates). Vulnerability can be reduced by preparedness and another index will account for this. Margins of safety such as in structural design have cost implications and should be evaluated. Such studies have already been done. There has not been much published on vulnerability of water services, but a number of books exist on safety and security in general (e.g. Grimaldi and Simonds, 1984). Industrial safety has long been a concern and safety programmes and insurance schemes and legislation has been developed for this.

Service authorities from large Water Boards to small local municipalities should be informed for introducing safety programmes. The cooperation of security officers and local disaster management organizations should be solicited. The experiences of security organizations may also prove useful in finalizing guidelines. The appropriate level of security may vary with the risks, community and affordability. However, all avenues of attack or danger need to be sought out.

The consequences of attack are expensive, often far more than the cost of protection, but the probability needs evaluating before embarking on expensive protection systems. The following may result from attack:

- Structural damage
- Increased danger
- Human or animal death
- Injury or disability
- Illness
- Trauma (this may happen even due to stress of anticipation)
- Inconvenience and cost of disruptions
- Loss of information and know how
- Breakdown of computer, electrical or mechanical systems

The evaluation of water services is an important step in deciding the measures which should be taken to protect the works against damage. The value of water infrastructure, i.e. water supply and sewerage systems for urban and rural domestic and industrial use, is hundreds of billions of dollars. More particularly, the population and industry are at the mercy of single water suppliers should any disaster occur. Although asset registers are incomplete and of varied format, available data will form the basis of extrapolations for selected areas.

The steps in a vulnerability study would be:

(1) To quantify the extent of domestic and industrial water service infrastructure. Where available, lengths of pipes, distribution networks, reservoir capacities and treatment works will be documented. Extrapolations will be done for unavailable data.

(2) To assess the vulnerability of water services to damage and risk. This will include indices categorizing possibilities and consequences of damage.

(3) To test the methodology on pilot institutions

(4) To develop strategies to minimize risk and damage to the infrastructure

(5) To inform water services providers of measures which should be undertaken to avoid damage. This will include guidelines and training packages.

Step 1 of the above objectives will be the most onerous, that is to quantify and estimate the extent of water services infrastructure. Registers of departmental installations and water plans are submitted by service providers, but in general soliciting data from local and rural authorities can only be done by sampling and extrapolating. This approach was done for the first phase of the OFWAT Asset Registers in the UK. The resources to go to the second phase of surveying all infrastructure will be assessed. This will also guide national government water departments and water authorities on efficient data collection. Standardized data format and database packages will be compared and developed for this.

The identification and prioritisation of vulnerability should be guided by studies done in the nuclear industry, for computer network firewalls, and by security planning done by water authorities in the past. The implications of damage, interference, accident or probability can be assessed from economic and danger points of view. Loss of facilities or data and danger to humans and the environment will be considered. Differentiation is needed between installations for water supply or sewerage and operation. I.e. if vulnerability is considered at a planning stage, a different design may result, but there will always be the necessity for alertness during operation. Risks due to international or malicious damage cannot easily be quantified, but accidents and costs can be attached probabilities and costs. Differentiation will be made between probability and damage and hazard-risk indices will be suggested (Stephenson, 2002). The same approach can be used for financial risks, IT risks, natural hazards (e.g. drought, earthquake) and economic risks (e.g. changes in water demands, AIDS, interest rates). Vulnerability can be reduced by preparedness and another index will account for this. Margins of safety such as in structural design have cost implications and should be evaluated. There has not been much published on vulnerability of water services, except AWWA, but a number of books exist on safety and security in general.

The cooperation of security officers and local disaster management organizations will be essential. The experiences of security organizations may also prove useful in finalizing guidelines. The appropriate level of security may vary with the risks, communities and affordability. However, all avenues of attack or danger need to be sought out.

13.8   REFERENCES

Beard, L.R. (1978) Impact of hydrological uncertainties on flood insurance. *Proc. ASCE,* **104**(HY11), 14168, p1473-1484, Nov.

Berthoux P.M. (1971) Accommodating uncertainty in forecasts. *J. Am. Water Works Assoc.,* **66**(1), 14

Bury, K.V. (1975) *Statistical Models in Applied Science.* John Wiley & Sons, N.Y.

Chow, V.T., Maidment, D.R. and Mays, L.W. (1980) *Applied Hydrology.* McGraw Hill

Dandy, G.C. and Connerty, M.C. (1994) Interactions between water pricing, demand managing and sequencing of water projects. *Proc. Water Down Under,* Adelaide, Austrln. Inst. Eng. 219-224

Fiering M.B. (1967) *Streamflow Synthesis.* Harvard Univ. Press, Cambridge, Mass., 139pp.

Fisher, R.A. and Tippett, L.H.C. (1928) Limiting forms of the frequency distribution of the smallest and largest members of a sample. *Proc. Cambridge Phil. Soc.,* **24**, p180-190.

Grimaldi, J.V. and Simonds, R.H. (1984) *Safety Management,* 4th Ed., Richard Irwin Inc., Illinois.

Gringorten, I.I. (1963) A plotting rule for extreme probability paper. *J. Geophys. Res.,* **68**(3), p813-814, Feb.

Gumbel, E.J. (1941) The return period of flood flows. *Ann. Math. Statist.,* **XII** (2), p163-190, June.

Gumbel, E.J. (1958) Statistical theory of floods and drought. *J. Inst. Water Engrs.,* **12**(3), p157-184, May.

Haan, C.T. (1977) *Statistical Methods in Hydrology.* Iowa State Univ. Press, Iowa, 378pp.

Institution of Civil Engineers (1969) *An Introduction to Engineering Economics,* London, 182pp.

Mays, L.W. (1979) Optimal design of culverts under uncertainties. *Proc. ASCE,* **105**(HY5), 14572, p443-460, May.

Pearson, K. (1930) *Tables for Statisticians and Biometricians,* 3rd Ed., Part I. The Biometric Lab., Univ. College, Cambridge Univ. Press, London.

Rodier, J.A. de Roche M (1984) *World Catalogue of Maximum Observed Floods,* IAHS, Pub. 143, Wallingford.

Stephenson, D. (1998) *Water Supply Management,* Kluwer, 308pp.

Stephenson, D. (2002) Hazard risk indices for flooding. *Urban Water,* Sept.

Stevens, E., Stephenson, D., Chu, S. and Huang, W-L (1998) Management of a reservoir for drought. *Water SA,* **24**(4), 287-292

Tanyimboh, T. (2003). Reliability analysis of water distribution systems. *Proc XXX Congress, IAHR,* Thessaloniki, Aug.

Yen, B.C. (1970) Risk in hydrologic design of engineering projects. *Proc. ASCE,* **96**(HY4), 7229, p959-966, April.

Yevjevich, V. (1972a) *Probability and Statistics in Hydrology.* Water Resources Publications, Fort Collins, 302pp.

Yevjevich, V. (1972b) *Stochastic Processes in Hydrology.* Water Resources Publications, Fort Collins, 276pp.

Young, G.K., Childrey, M.R. and Trent, R.E. (1974) Optimal design for highway drainage culverts. *Proc. ASCE,* **100**(HY7), 10676, p971-993, July.

14

Economics and financing water services

14.1   SOURCES OF FINANCE

It used to be the local authority or a water board which was responsible for installing the infrastructure for supplying water and removing waste water from towns. These were the most logical water service providers as they had access to finance and were familiar with revenue collection and appointment of consultants and contractors to execute works. At a higher level, i.e. provincial and national government, the ability to control the services is generally limited by remoteness and experience. On the other hand, they are more able to finance larger projects and to bridge risk and other operational problems.

There has been some clamour to rearrange water services on a physical catchment basis rather than on a municipal or borough basis. The catchment is the natural watershed which could be the source of water for a consumer such as a town, as well as the drainage conduit for the town. Water quality managed by the same service provider would be logical as a catchment authority would be more inclined to enforce effluent standards than isolated effluent treaters.

The problem with private investment in water services is that the investment is large, the revenue is not always controlled by the company operating the service and the risks are not only large, they are often unknown. Hence, private companies are reluctant to take over completely the water services or even catchment resources for supplying towns.

Whereas this is a problem for private investors, there is another problem of private control which the authorities have to contend with. That is, that water services are inevitably monopolistic in nature, i.e. there are no competing service providers in any single town or zone. It is expensive and impractical to install duplicate water supply pipes and more particularly drainage pipes. This means the operator of the systems has a monopoly which could be manipulated to financial advantage unless there were controls in place.

Therefore an independent regulator is desirable for control of privatized water services providers. The regulator in turn should be independent, i.e. neither a contractor nor the owner of resources, such as the government. Indeed, the regulator has to adjudicate between the two and his biggest dilemma is to select standards of service without unduly taxing the consumers or users. The method of control of income and profit of a private water company has to have some flexibility to enable the operator to make a reasonable profit, but at the same time not to overly tax consumers and to provide a reasonable level of service. This will require a maintenance and even a rehabilitation schedule to ensure assets are maintained in a correctly acceptable condition at all stages. It may be somewhat arbitrary as to the level of losses which the regulator will permit so that he will employ benchmarking or open meetings to make decisions regarding levels of maintenance and service.

The French companies involved on an international basis in private operation of water companies have extensive experience in obtaining the correct level of service while maintaining a satisfactory profit level.

Developing countries are particularly in need of know-how and finance in providing services to their communities. This means that they not only have no experience, they also do not have the institutional capacity to handle such service provision in many cases. The necessity for consultants is therefore apparent. Whereas there have been some public-public partnerships to assist such developing countries, most clamour and marketing is done by the private companies wanting to assist the governments in establishing and operating services.

However, in parallel with the lack of technical knowledge concerning the installation and operation of water services in developing countries, there may also be a lack of appreciation of the skills and ability to negotiate with companies will to provide the services. Particular care may be required where there is a lack of competition and where the client is also dependent on private consultants for obtaining financing for the installation of water services.

It is envisaged that an alternative form of public-private partnership may emerge for developing countries. This may entail heavier community involvement, particularly to avoid dishonest practices in the negotiation of the contract. The co-option of community representatives could not only monitor the award of contracts, it could also hold the interests of the community at heart. There are many other requirements in developing communities which are not considered in the western world. These include particular technical training and institutional capacity building in order to manage such services. This should not only be confined to courses and field training, but should also be accompanied by long-term follow on from an independent trainer and mentor as the learning process cannot be speeded up particularly in a fast developing country. It is in this role that public-public partnerships and even academic-public partnerships may stabilize the decision making.

By involving the community at different levels, not only will greater transparency be necessary but there will be a considerable management burden and timing will not be as fast as for a sophisticated society. The community would and should benefit by inputting labour, i.e. embarking on a learning process as well as contributing in kind to the project. The economics are therefore more complicated that a simple commercial deal. In fact, the economics becomes considerably more complicated. Many of the benefits will be economic not financial, so that the private operator may not be able to access the benefit without surcharging the other profits.

Due to unstable currencies in many cases and even unstable governance, short-term contracts may be more attractive to private operators. Hence, long-term financing, for example via an international aid organization plus technical input on a consulting basis may be an alternative until private companies can formulate groups capable of handling the ways of the developing country.

Another problem then arises, as the smaller the subsidiary or company operating the service is, the more liable it is to be damaged by risk. Collapse or withdrawal of a company could affect the maintenance and operation of the service and therefore a wedge-type, local operation is seen as a solution.

Indeed the risks are much higher for developing countries. The collection of revenue, for instance, is more erratic, and law enforcement and policy decisions frequently need to be made. Some form of guarantee by the government would therefore be required by the private operator.

However, the contact with private industry, whether it be in the form of banks or technical advisors, is unavoidable. Access to financial markets, and therefore public relations and transparent governance are necessary to obtain the capital to finance a service in a poor country.

One of the advantages of privatization is quoted as relieving the public authority of its role in service provision. This enables the public authority to pursue its policy decision making process or even regulatory ability in an unbiased way and uncluttered with technical staff. International experience can also be brought in by shopping on the open market.

The outright purchase of a water company, such as by Lyonnaise d'Eaux of the Northumbria Water, is no longer likely. This is particularly the case in developing countries where devaluation is prevalent and therefore renders the profitability unattractive to private international companies.

Nevertheless, between 1990 and 1995, private sector infrastructure financing in developing countries soared from US$ 2.7 billion to US$ 37 billion. In 1994, thirty developing countries collectively awarded seventy-five private companies a total of US$ 10 billion in service provision.

14.2   BENEFITS OF RURAL WATER SUPPLIES

The high per capita cost of water supply to many rural areas is an inhibiting factor for development. The benefits of water supply are largely long-term and therefore direct funding is a problem. Benefits include better health, irrigation opportunity and more time for employment or education. Studies conducted in villages in South Africa attempted to evaluate these benefits. Situations before and after, and villages with and without water, were compared, but differences were difficult to distinguish. Household surveys and discussions proved the most fruitful method for obtaining data. The data are used to rank and implement projects as funds become available.

Many areas in Africa, and in particular rural areas, do not have piped potable water supplies. The traditional method of collecting water in vessels is becoming more difficult because sources are being exhausted or are located further away from the water source. Lack of river storage in rural areas means that communities are more susceptible to drought conditions because the base flow is drawn on more extensively as the demand increases.

As population densities increase, so pollution of natural sources also increases. Major environmental concerns are the over-exploitation of natural resources and the lack of adequate potable water from the health point-of-view. Water-borne diseases are endemic in Africa and a modern water supply system will alleviate much of the problem by purifying and disinfecting the water. The problem is not always with the water, but also with the hygiene habits of the community.

Collection of water in rural areas is done by three groups of people; adult women, youths and children over the age of five. People usually travel in groups of two or three when collecting water when it is located far from the village. If the water scheme brings the water source closer to the village, this pattern might change.

Before water supply schemes were constructed in the study area, Eastern Cape in South Africa, the number of water sources, which was mainly groundwater, was limited to one or two per village. The actual collection of water is done only within certain hours during the day. This pattern is very much governed by the school time, time of the year, and other responsibilities of the ones involves in fetching, such as dipping cattle or driving the cattle to the grazing camps.

Experience has shown that these hours of collection do not change even when the scheme is constructed, unless taps are provided at each household. The collection hours are mainly two hours in the morning, two hours in the afternoon/evening and about 1.5 hours during lunch break/noon.

Water is mostly carried in containers on women's heads or in wheelbarrows, terrain permitting. The volume of these containers does not usually exceed 20 ℓ, as this is a convenient plastic container size. The water consumption of the people is governed by the number of people in a kraal who collect water and how often they go. The volume each person can carry per trip will not exceed the original volume (20 ℓ) even after the scheme is constructed, since the weight of water beyond 20 ℓ becomes unbearable.

The volume of water used ranges from a daily consumption per person, in tropical developing countries, of a little over a litre, to about 25 ℓ for consumers with tap connections or standpipes. For village dwellers who use public standpipes, the daily use averages 10 to 15 ℓ per person.

In a field survey conducted by the Department of Agriculture and Forestry in 1987 in Sterkspruit in Transkei, it was estimated that although a project had a capacity for providing 25 ℓ per capita per day, the people could only collect about 12 ℓ per capita per day because of the aforesaid limitations. In other words, unless water is provided right at the household, there is a practical limit (approximately 12 ℓ per person per day) to the amount of water people can or will collect regardless of the design capacity. There are exceptions, such as when the community has a vehicle, then the per capita consumption can be higher and often washing takes place at the source.

In a study of fourteen communities ranging in size from 800 to 5 000 inhabitants, with newly installed piped supplies and without alternative supplies in north-east Thailand, Frankel and Shonanavirakel (1973) found the range of consumptive usage adopted for design purposes to be 50–80 ℓ per capita per day. Actual use they found to vary from 9.6–36.8 ℓ per capita per day at standpipes, and 24.4–65.2 ℓ per capital per day for house connections. This led them to believe that a design criterion of 25 ℓ per capita per day for house connections would be more realistic than 50–80.

In Tanzania, where Warner (1973) made studies before and a year after the installation of water supplies which reduced distances from water source for most households, the quantity used increased, but not by more than a few litres per capita per day. The lower the initial use had been, the greater the increase was following installation of supply.

The lack of water can affect the economy and economic development of communities. Apart from poorer crops and consequential malnutrition, the physical process of collecting water manually affects the social habits of rural communities. The time taken, up to four hours a day, for one to three members of each family means that time is diminished for maternal health care, housekeeping, farming or other economically oriented activities, or for assisting with the education of the children.

The time spent, particularly by women, means that they are unable to spend much, if any, time in improving their education or training and, therefore, they are less productive than in the western system. Similarly, the education of children suffers because it is customary for children to collect the water in the morning, resulting in them often being late for school. Poorer pass rates or lack of learning may result, having a long-term effect on the education and training of the community and therefore the economy of the country. Oakley and Marsden (1984) discuss some of the ways of improving such situations.

This study seeks economic feedback which may justify provision of piped potable water to rural communities. It attempts to evaluate the benefits of such water supply. It stops short of discussing financing of schemes though.

14.2.1 Justification for rural water supplies

The per capita cost of a basic water supply in a rural setting can be less than that in an urban context, notwithstanding the fact that the scale of the project is considerably smaller. The savings are brought about largely because of the meager consumption in rural areas (less than 30 ℓ per capita per day even with piped water, as opposed to urban consumption of up to 300 ℓ per capita per day). In addition, pipe networks in rural areas can be minimized by installing standpipes to serve a number of dwellings at strategic points. This saves internal plumbing which can cost two to three thousand Rand (one Rand is approximately 12 US cents, 2003) for a house, as indicated by builders in South Africa.

Another factor which can reduce cost in rural areas is the reduction in storage capacity, even though there may be an added risk of failure. The alternative in times of storage depletion, such as importing water, is relatively less costly than for larger urban communities. The low-cost materials used and the cheaper or free labour input by the community can further reduce rural costs.

14.2.2 Evaluation methods

Data can be gathered on a broad scale by means of surveys or calculated indices. An ideal index would be a measure of economic growth or gross domestic product (see, for example, Fig 14.1). These data include school pass rates over the period of installing water supplies. Alternatively, increased farming activity could be monitored by means of an economic product index. The treatment or mortality rate logged in clinics or health departments could be plotted. The above methods would be applied to conditions before and after installation of the water supply facility. Alternatively, demographically similar villages or regions, with and without water, cold be compared.

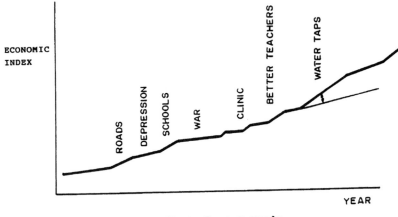

Figure 14.1. Economy index to illustrate various effects on economy

The evaluation of the economic benefits of better education is more difficult because they are long-term and many other factors mask the benefits solely due to water. Therefore assumptions have to be made to calculate improvements in economic well-being. It is indicated below that even with conservative assumptions, the economic improvement of communities justifies the early installation of water systems.

The macro indices described above are subject to many side effects and these cannot always be isolated. Therefore, the results have to be viewed with some circumspection.

Sociologists and economists prefer a micro-type survey with intensive data collected from interviews with individuals representing the community. House-to-house surveys can collect facts on the changes in household incomes and habits or monitor changes before and after a project. They can also assess the time taken by households to collect water and use it. The personal approach can bring to the fore many unexpected and unknown factors for assisting in planning future water supplies (see Oakley and Marsden, 1984).

Many economic surveys have been conducted in the former Transkei, South Africa, over the last few decades. Such economic surveys have been conducted by the Human Sciences Research Council (HSRC, 1993) for enumerating and assisting with income estimates and other economic evaluations. The World Bank (1980) has sponsored economic studies throughout South Africa and even more recently, the Department of Agriculture and Forestry, Engineering Branch in the former Transkei (1995), has initiated surveys. The results of these surveys have been analysed and some of the results are indicated and evaluated below.

14.2.3 Study of water collecting time

A project initiated by the Department of Agriculture and Forestry in Transkei measured the time taken by household members to collect water, prior to constructing water systems with pipes and taps. The Department is constructing water supply systems in fifty of its rural villages.

These projects are being implemented based on community negotiation and involvement. Contributions are made by both the Government and the receiving villages. The levels of these contributions are arrived at through numerous negotiations. The pilot project was run for seven days in each of six villages. Observers were chosen from the residents.

14.3   RESULTS OF FIELD SURVEY

The following is a summary of the analysis of the findings of the survey:
  (1) The majority of the water carriers were female below the age of 19.
  (2) Almost all the collections are done between 05h00-08h00, 13h00-15h00 and 16h00-17h00. This confirms the findings of international surveys done in this field.
  (3) A lot of time is spent on socializing on the way to the water source. The time spent per family per day on water collection differs considerably

from village to village, the minimum being around 70 minutes and the maximum being around 350 minutes.

(4) The number of times each carrier collects water varies between two and seven a day.

14.3.1 Health benefits of water supply

The health and therefore economic and social wellbeing of most Africans is limited by lack of clean water (WHO, 1982; Water Supply and Sanitation Services in SA, 1994). The problem is to evaluate it. To calculate the intrinsic value or shadow price for supplying domestic water to people (see Stephenson and Petersen, 1991), the following information was collected from the Development Bank of Southern Africa report (1987) on Transkei.

- 95% of the population can be classified as rural
- the average household contains about 8 persons
- approximately 41% of the population is economically active

Bembridge (1984) estimated the average per capita income to be US$ 194, of which only 10% consisted of cash. He further found that cattle, sheep and goats are the largest resource of the population. It is significant that the majority of the household tasks are carried out by women, so it is their time that will be most affected by the provision of domestic water.

Many of the diseases that affect rural populations result from poor nutrition, poor sanitation and unhygienic living conditions. It was also found by Stone (1984) that 90% of the households drew water from sources open to contamination by stock. Provision of clean, potable water will therefore have a direct effect on the health of the population. Average use of water is 12 ℓ per person per day. With unlimited supplies of water, a maximum of 25 ℓ per person per day would be used. This figure is low compared to world standards (20 ℓ per person per day being considered the minimum), but can be expected to rise rapidly as the standard of living increases.

Estimates of the benefits derived from the supply of clean water are sadly lacking. The World Bank report (1980) put the cost of health care for the average family in a low density country at 5-10% of their income. It was further stated that 90% of diseases are due to contaminated water. Figure 14.2 shows an estimated shadow cost of not supplying clean water.

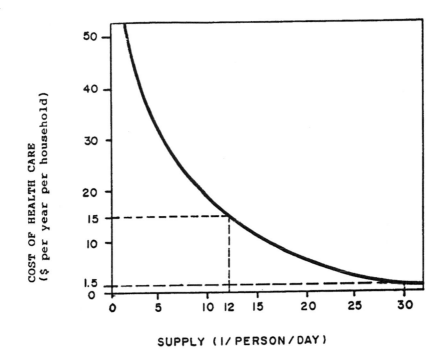

SUPPLY (l/ PERSON / DAY)

Figure 14.2. Health costs resulting from insufficient water supply

14.3.2 Educational benefits

14.3.2.1 Analysis of school results

School marks obtained from the Department of Education were analyzed but did not show a clear trend of improved marks immediately after installation of water schemes. Some schools even show a decrease in marks on the period surveyed (1985-1992). On average, there is an increase in marks and pass rate, but further work needs to be done to distinguish the improvement due to water supplies from improvement due to better teachers, political change, better facilities or standard of living.

There are common trends in many schools, e.g. drops in marks in 1989 or 1990, then increases in 1991 or 1992. Some results only go up to 1986, so trends could be misleading.

14.3.2.2 Comparison of school pass rate – 2 districts with and without water supply

The comparison of school pass rates for two districts, one without water supply and one with piped water, was made. The Libode district which obtained piped water from standpipes, received a water supply in 1991-2. The control district without water was Ngqeleni to the south of Libode and a similar distance from the main road. Topography and other factors were similar.

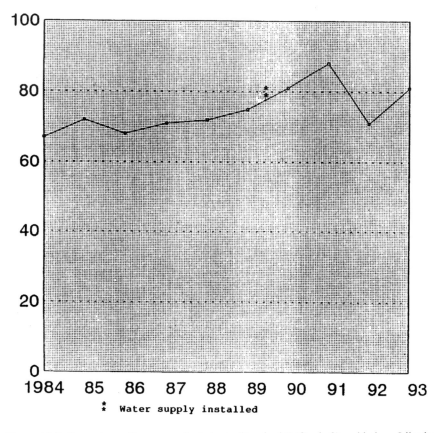

Figure 14.3. Percentage of passes – all students, Standard 7 (Grade 9) and below, Libode

Comparison of the average Standard 7 (Grade 9) school pass rates in two schools were made for the period 1984 to 1993. Before 1989, Ngqeleni had a 10% higher pass rate, but there appeared little difference between the pass rate since 1989. It is possible that water supply in Libode installed after 1990 brought the pass rate up.

Table 14.1. Summary of Libode district water situation

	Adequate	Needy	Critical	Desperate
Water available ℓ/cap/day	> 20	10 - 20	- 10	- 10/polluted
Distance m	- 750	750 - 1000	> 1000	> 1000
Gradient %	< 6	6 - 12	> 12	> 12
Total no. of precincts	27	27	16	41
% of total	24.55	24.55	13.64	37.27
Total no. of people	15 387	15 728	8 362	29 664
% of total	22.25	22.75	12.09	42.91

14.3.2.3 Educational benefit

The survey data indicated water supplies reduce the time children miss school. A daily time of 2 hours missed by 50% of school children was quoted. From this, a hypothetical improvement in household income is calculated below.

Time spent collecting water = 2 hr/d x 50% of children. 50% of children with 20% increased pass rate, results in 10% additional matriculants, i.e. 10% more are better qualified for jobs. A reasonable guess is that half of these obtain better jobs at an average salary of R20 000 per year. ½ x 10% x 4 children per house @ R20 000 = R4 000 per year increased income per household in 10 years time, the time from primary school to employable age. (R1 = US$ 0.22, 1997).

R4 000/year over 10 years @ 5% annual discount rate has a present value of R4 000 x 7.72 = R30 800. Assuming 50% of income returned to home for 10 years, this is equal to %15 440. The present value, if the benefit started ten years from now, would be: R15 440 / $(1.05)^{10}$ = R9 480. This household benefit far exceeds the cost of water supply of R100-R1 000 per household, even if the computed benefit was considerably less.

A significant remaining problem is financing the installations. Someone (the government or a 'donor' or lender) must invest for ten or more years until payback is possible. If it is the government, who theoretically have the money, that amount of money must be deducted from alternative investments which could limit growth.

14.4 STUDY OF HOUSEHOLD ECONOMY

In an attempt to correlate income with water supply, the following surveys were conducted. The results of the Human Sciences Research Council (HSRC, 1990) for Transkei were to be used as a base. A subsequent survey was set in progress to try to detect effects of water supplies to the economy. The regions studied were Libode, Lusikisiki, Lady Frere and Herschel in Transkei, South Africa. At this stage, a questionnaire had been drawn up with input from the HSRC. The questionnaire is a simplified version of a 1990 questionnaire. About sixty houses (thirty close to the water supply standpipes) in Libode were surveys as a pilot study. The results (incomes) were compared with the previous results. Again, villages without water distribution were used as a control. Limited change could be expected from the results, as the time since water was supplied is short in economic terms But the path is set for continued monitoring.

14.4.1 Average household monthly income

Average household monthly income figures were listed for Libode, Lusikisiki, Cacadu and Herschel in South Africa in 1993 values. The data derive from a baseline household survey of some 9 000 household throughout the length and breadth of South Africa. The study was performed by the World Bank, working in conjunction with the South African Labour and Development Research Unit (SALDRU, 1993). The results were made available in August 1994 on tape and in a manual.

The research also accessed Bureau of Market Research data and calculated "personal disposable income by population group and magisterial district/region, for the period 1970-1985". The data have been disaggregated by dividing these totals by de facto population figures, provided by the DBSA's regional profile of the Southern African population and its urban and non-urban distribution.

The data suggest that, with the exception of Libode, income levels decreased between 1980 and 1993. A general decline in income could be associated with political unrest and sanctions. Libode, which was the recipient of water supply, resisted this general trend, but it is not possible to evaluate the benefit owing to the general decline in income.

14.4.2 Micro survey of household economies

A technician conducted house to house surveys with the aid of a questionnaire. He selected thirty households at random in the Libode district, half with and half without piped water. In all cases with water, the water was supplied at a standpipe. In some of these cases, there was a shortage of standpipes or the standpipe was still some distance away. This was also evident in the returns (see Table 14.2) on collecting time – still a large portion of the day. In fact, the amount of water used does not appear to increase significantly for households with water standpipes compared to those without. Neither was there any evidence of improved health. The major proportion (about 75%) of water is used for washing, which is not as sensitive to purity as drinking water. The small change in lifestyle of those with standpipes confirms the figures of the HSRC survey.

One notable village was Gunyeni. There they have adequate water and standpipes conveniently placed. The respondents were ambitious and suggested training programs for improvement in living standards and hygiene. Deductions on the economic benefits of water are interesting. No-one thought of any economic use for water, but all respondents wanted more water for crop gardening. Food was often quoted as a problem and the thought of self-sufficiency was a high priority.

Excuses for no economic progress were generally of two types: the infrastructure ("The Government") was conceived as an obstacle to economic improvement. With regard to more time for children's studies, many respondents said children's school attendance did not noticeably improve because they had other duties and they still played hooky.

It appears a major stumbling block to self-improvement is negative attitudes. This may be improved by better communication with national, socio-economic programs. In fact, some respondents requested talks and training in hygiene and household management.

It is thus apparent that to evaluate the benefits of water supply is fraught with difficulty. Engineers like to imagine their water supply schemes improve life, but the scale and random effects make evaluation not as easy as design.

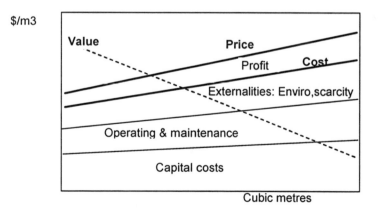

Figure 14.4. Relationship between cost, price and value of water

Table 14.2. Summary of household survey

Village	Water source	No. in home	ℓ/day	Mins/day	Health	Education/Aspiration	Complaints
Gunyeni	SP	12	80	60			Gardening
Corana	SP	19	80	?	River bad	Pilot	Gardening, still time lost
Rhini	SP	6	100	40	" "	Manager	None for garden
Khuleka	-	9	80	40			
Misty Mount	-	6	100	5	-	Pilot	None for garden
Khuleka	-	7	80	40	-		
Khuleka	-	7	80	45			
Misty Mount	-	6	100	45			None for garden
Mdlankomo	-	11	80	40			
Mdlankomo	-	8	80	40	-	Business	Gardening, still miss school
Rhini	SP	5	60	30	-		None for garden, still takes time
Gunyeni	SP	10	60	20	-	Ranger	None for garden
Gunyeni	SP	13	60	20	-	Doctor	Garden, wants training
Magcakini	SP	15	100	40	No problems	Teacher	None for garden
Gunyeni	SP	5	80	20		Teacher	Garden, Govt.
Zithathele	SP	13	80			Police	None for garden
Rhini	-	10	100		Broken pipes	Nurse	None for garden
Misty Mount		10	80	60		Soldier	None for garden, Govt. complaint
Magcakini	SP	7	60	40		Doctor	None for garden, Govt. complaint
Mahankato	-	12	100	60			
Masememi	-	8	100	90			
Mkhavkoto	-	11	100	60	Diarrhoea		
Corana	SP/R	5	100	60			Time saved, business
Corana	SP/R	19	80		Clean water	Pilot	None for garden
Corana	SP/R	7	120	60	Hygienic		Car washing
Michonaxato	-	6	100	90	Diarrhoea	More time available	Need more water
Ngqeleni	R	9	100	90	Diarrhoea	More time available	Need more water
Corana	SP/R		Short				
Rhini	SP	7	100	40		Manager	None for garden
Magcakini	SP	6	100	45		Inspector	None for garden, Govt.

Legend: SP = Standpipe;　R = River

Before– and after–piped water supply situations were compared, but it will take many years to observe significant change in the economy as a result of water supply, if indeed the effect can be isolated from other developments. The effects on health and economy are also difficult to evaluate.

Collecting time of water reduces with installation of piped water. This can beneficially affect available time for attending school but affects social patterns. Only in one of the before– and after–water supply cases could an improvement in pass rate be deduced.

14.5   REFERENCES

Bembridge, T.J. (1984) Aspects of Agriculture and Rural Poverty in Transkei. *Second Carnegie inquiry into poverty and development in South Africa.* Cape Town.

Development Bank of Southern Africa (DBSA) (1987) *Transkei Development and Information.* Midrand.

Frankel, A. and Shonvanavirakel, B. (1973) *Report to World Bank on water consumption in Thailand.*

HSRC (Human Sciences Research Council) (1990) *Survey of economy of Transkei.*

HSRC (Human Sciences Research Council) (1993) *Water supply economic survey of Libode district, Transkei.* Pretoria.

Oakley, P. and Marsden, D. (1984) *Approaches to participation in rural development.* I.L.O.

SALDRU (South African Labour and Development Research Unit) (1993) (+ World Bank)

Stephenson, D. and Petersen, M.S. (1991) *Water Resources Development in Developing Countries.* Elsevier, Amsterdam, 289pp.

Stone, A. (1984) *A case study of water resources and water quality of Chalumna/Hamburg area of Ciskei*, CCP 148.

Water Supply and Sanitation Services in SA (1994) *South Africans Rich and Poor: Baseline Household Statistics*, Univ. of Cape Town.

Warner (1973) *Tanzania – Case study.* Private report.

World Bank (1980) *Poverty and Human Development.* Oxford Univ. Press, N.Y.

W.H.O. (1982) *Activities of the World Health Organization in Promoting Community Involvement for Health Development.* Geneva.

15

Development issues

15.1   BACKGROUND

Underexpenditure in the field of water services is most excruciating in developing countries. Although the developed world has political problems in allocating sufficient funds for maintenance and improvement of the water services, the developing world lacks the knowledge and funds to put in even the simplest services.

The problems of the developing world are compounded by rapid population growth, migration to peri-urban areas, financial disadvantage, extreme geographic conditions and lack of knowledge concerning water health.

The amounts invested by donor agencies and countries are insufficient to keep up with the population growth in the developing world and serves largely as token money. The broader solution will have to be achieved by these countries themselves.

15.1.1 Water supply problems

Despite 100 billion US dollars' worth of capita investment in the 1990's, the water services sector still faces some major obstacles (Retali and Étienne, 2000):

- 25% of the world's population lacks access to safe drinking water. This figure could rise as high as 65% by the year 2025, but statements emanating from the World Water Forum (2003) aim to reduce it to nil in that time.
- 50% of the world's population does not have access to satisfactory sanitation.
- Global water consumption is increasing at an annual rate of 2.5%, or twice as fast as population growth.
- 27 countries are currently classified as water-stressed (less than 500 m^3 per capita per year).
- 25 million people die or contract serious illnesses due to contaminated water each year.

Under such conditions, a rational approach towards water resource management has become increasingly important in many countries, given the significance of resource deficits in constraining economic development opportunities.

It must be recalled however, that the water supply sector is characterized by a set of specific constraints, the most critical of which are highlighted below:

15.1.1.1 Water has often been considered as a free resource, or a "gift from heaven"

For example, the policy of low priced water practiced in the past in many countries has led to a major resource wastage; on a number of occasions, public organizations, lacking the necessary funding, have been unable to raise the requisite capital investment. Such policies often penalize the poorest segments of society which, by being denied access to the supply network, are required to pay up to 100 times more and face tremendous health risks.

Many organizations, including the World Bank, have encouraged political leaders to bill water at its actual price, which would not exclude introducing certain selective subsidy programs. This type of approach enables avoiding wastage while generating for the water supply service the necessary financial resources, provided it is being well run. At what stage in the developing economy water can be charged at its full value is debatable. Also at what level to service needs consideration.

15.1.1.2 Water is often considered as part of the national heritage

Control of water resources by a public authority is compatible with the delegated management of water and sanitation services. Moreover, in many forms of delegated management (operation and maintenance contracts, leasing or concession), the public authority retains ownership of all facilities and only engages in service delegation by granting the right to use facilities to a private firm for a designated period of time. The national government may have to subsidize water until it is affordable by all.

15.1.1.3 Water represents a public health issue

Delegated management contracts stipulate precise quality standards, along with a series of regular controls. European Community regulation serves as an example by specifying standards for over sixty distinct parameters. Consequently, public health is often ensured to a greater extent in systems run by private firms.

15.1.1.4 Water represents an important social concern

This precept is not incompatible with private service management. In particular, the public authority may elect, provided the contract's overall economic equilibrium remains intact, to grant preferential service rates to the least well-off population segments.

15.1.2 Population of developing countries

The end of the 20th century was one of relatively high population growth. Of the world's population of 6 billion, 75% (4.5 billion) can be classified as poor; they earn less than 20% of the world's income. Of these, 1.5 billion live in abject poverty (insufficient means to maintain nutrition and health). The majority live in Asia, but within 25 years, the majority (1000 million) will be in sub-Saharan Africa (see Stephenson, 2001).

Also 1.5 billion people do not yet have access to safe drinking water and 3 billion do not have adequate sanitation facilities. Over 54 million people die each year from water related diseases. UNCED (1993) indicated that about $20 billion should be spent a year on water supply and sanitation in developing countries, which is only $5 per capita, so is probably too low to achieve a significant improvement in standards. The 1980's was seen as a decade of hope. 1.6 billion more people received water. The number of urban people with access to water increased to 80% and with sanitation 50%. But still the number of urban people without sanitation increased by 70 million (see Black, 1994).

Bhatia and Falkenmark (1993) suggest $11-14 billion is required for water supply and sanitation infrastructure in urban areas in developing countries. Forty-five percent of the world's population lived in cities in 1990 and over 50% in the year 2000, thus compounding the population growth rate. The majority of the urban growth will in the figure be in developing countries. There are now 320 cities with over 1 million inhabitants, and this could double to 600 by the year 2025.

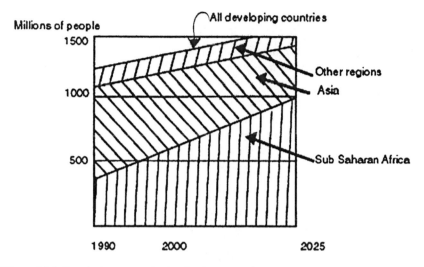

Figure 15.1. People in poverty in developing countries

Poorer countries have doubtlessly often been affected adversely by their historical relations with developed nations. Some of the reasons for this may be, to a lesser or greater degree, dependent on the country:
- Colonization, raising expectations and leading to breaks in social customs.
- Foreign aid packages.
- First world dealers placing their goods by unfair practices.
- Inability to create institutions capable of handling aid efficiently.
- Modern medicine resulting in longer lives and thereby higher populations (even though now the raging AIDS epidemic has brought in a new dimension, especially in Africa.

15.1.3 Financial limitations

The debt of developing countries stood at about US$1.7 trillion in year 1995 and was increasing at 5% p.a. with little sign of improvement in payback ability. In fact, conditions are deteriorating at an increasing rate. Until the 1980's, aid by the developed countries was largely in the form of capital grants to the poorer nations. The British Commonwealth and French colonies and earlier the Spanish and Portuguese settled territories, received preference by their respective mother countries during the colonial period. Money was used to build infrastructure for use largely by the expatriates or to benefit the colonial power. When colonization stopped, so did much of the capital flow. Further aid was largely token aid. War risks in certain areas have made economic investment even less attractive. Figure 15.2 shows the relative distribution of world income.

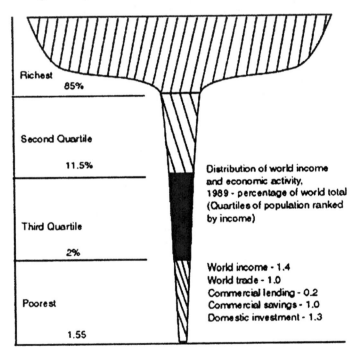

Figure 15.2. Global income and economic disparities (UNDP, 1992)

Now the pattern has changed even more drastically. The new emerging countries of Eastern Europe and Asia will absorb a lot of the capital from Western Europe and the US. The strategic importance of Africa as a destination of massive investment has been lost with reduced values of resources. The Far East is showing signs of economic development and is a vast market so is also attracting capital. The economic incentive to invest in Africa, the rest of Asia and South America has, in general, been decreased. The poorer countries are now left to service debts far greater than their gross domestic production, in many cases. The World Bank indicated a cash flow of $157 billion to developing countries in 1992, of which 60% is now from private organizations as opposed to public aid money.

The trade barriers imposed by developed countries are widening the poverty gap even more. Developing countries cannot sell resources or crops unless their currencies are heavily devalued. Yet they have to purchase pumps, computers and technology from the developed countries to be 'westernized'. This colonial legacy is a problem of democracy and the anger built up is liable to result in instability and retardation of 'development'.

The total input is still falling behind the debt, which increased at about 4% p.a. from $137 in 1986 to $179 billion in 1992 from the poorest countries. The debt service charge for Sudan and Somalia is 2000 times the value of their exports; it looks like they can never catch up on repayments. In addition, aid is decreasing to Africa at about $2 billion per year. Of Africa's 51 countries, half have debts in excess of their GNP. Moçambique's is 400 times their GNP. However, the economies of these countries are very low. The debts in Latin America are far higher in absolute terms, e.g. $616 billion in 1992.

The new policy of structural adjustments to obtain IMF funds is causing confusion in developing countries and favouring developed countries who still subsidize crops.

The changing focus of aid from capital to institutional projects is also a source of greater corruption and less money may get through to the workers in the short term. It may be a recipe for degeneration of morality in much of Africa, for corruption spins off down the line. The elite in these countries remain relatively untouched by removal of subsidies on transport, food and welfare, as they hardly use them.

The World Bank (1988) found that the rate of failure of its projects in sub-Saharan Africa was twice as large (20%) as the average. The average rate of return was half that of Asian project. Institutional projects were even less successful (28% success). Evaluations are generally made at the end of the project duration. But after 5 to 10 years, all African projects dropped in productivity (from 10% to 3% average annual return). Livestock and plantation projects showed no success at all, or negative returns.

In the early 20th century, water-funding agencies concentrated on irrigation and hydro power. During the 1970's. emphasis swung to water supply. This was spurred by the UN Water Decade in the 1980's, the goal being to supply everyone with at least 20 ℓ/d of water. A mass balance was not carried out on finances available and finances required, with the result that installation fell short of expectations.

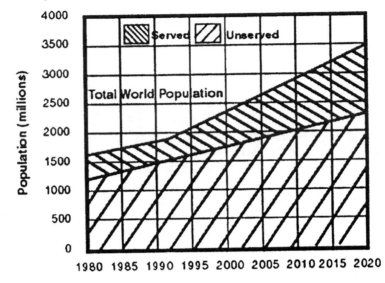

Figure 15.3. Urban population and water supply coverage in developing countries (Bhatia and Falkenmark, 1993).

Money is not a limited resource like water or coal. Although finance may be limited in the short term, economies can grow and multiply. By circulating a finite amount of money, it can be made to work twice, three times and more. And economies of relatively poor countries can go further than the same money in a rich country, because expectations are less and costs can be less. Some things can be done without money, e.g. labour or management by the community in return for services. So it is not necessary or even desirable to run a service in a developing country on the same lines as a developed country.

15.1.4 Institutional problems

Problems due to poor institutional involvement can be:
• Reluctance to accede to community wants, as it appears to diminish officer's standing.

- Perception that alternative technologies are better, notwithstanding public preferences.
- Under-recovering of costs reduces financial sustainability.
- Subsidised costs annoy the subsidisers.
- Cross-subsidies result in decreased usage by the high payer, and over-usage by the low payer.
- In the limit, the utility may provide free services to avoid costs of collection.
- Low contributions of poor users means suppliers do not feel obliged to extend of maintain services.
- The priorities of public servants will put richer users and public salaries before poorer users.
- Cuts in services to non-payers wastes the capital installation cost and would result in users finding alternatives, possibly not as hygienic.

15.2   WATER QUALITY AND HEALTH

More people die of poor water quality than due to lack of water. Yet people will sacrifice more to get water than to purify it or otherwise ensure its safety. The sewage problem still remains more serious in developing countries than water supply, from the health point of view (Wright, 1997; Varley, 1995).

A minimum of 2-3 litres of water is required per person per day to live. Additional water would be required for washing people and more for utensils. As the community becomes more sophisticated, it will require more cleansing and even luxury water, for example for gardens, washing motor cars and topping up swimming pools. The demand of sophisticated communities can exceed 300 litres per person per day.

The probability of contracting a water-related disease is statistically related to the concentration of pathogens or viruses. Improved laboratory techniques have resulted in the identification of harmful inorganic matter. These may occur in water sources naturally but also from industry and agriculture. Of concern to health also are heavy metals such as mercury, cadmium and lead, as well as pesticides such as DDT and carcinogenic compounds.

Pathogenic bacterial-related diseases include cholera, typhoid and dysentery, whereas viruses, which are smaller, require a host and include hepatitis. It is difficult to detect bacteria in water, so evidence of excreta is sought and indices such as *E. coli* are used.

15.2.1 Water-related diseases

Water-related diseases may be grouped as follows:

15.2.1.1 Waterborne diseases

Waterborne diseases are those which are mainly spread through contaminated drinking water. The main infecting organisms are bacterial (*Vibrio cholerae, Salmonella typhi, Shigella*), viruses (herpatitis A, orgiviruses, rotaviruses and enteroviruses) and protozoa (*Giardia lamblia, Histolytica*). Contamination of the water occurs with faecal matter entering the water.

15.2.1.2 Water-washed diseases

These diseases are mainly infections that can be significantly reduced by an improvement in domestic and personal hygiene. They depend on the quantity of water that is available, rather than on the quality. All diseases with faecal-oral transmission fall into this category – such as typhoid and cholera. Others are skin and eye diseases, such as skin sepsis and trachoma, and infections carried by parasites on the skin surface, such as lice. Most of the intestinal worms also belong in this group, including roundworm, threadworm, whipworm and pinworm.

15.2.1.3 Water-based diseases

In the case of water-based diseases, the pathogen has to spend part of its life cycle in the water. The best known of these is *Schistosomiasis* (Bilharzia). It is a water-contact disease that has infected many millions of people in the tropics. It is spread through schistosome eggs in human excreta, which hatch on reaching water. The resultant larvae invade suitable snail hosts and multiply.

15.2.1.4 Water-related insect vectors

Water provides the environment necessary for the development of many insects that transmit diseases. Malaria is a water-habitat, vector-borne disease, certain mosquitoes being the host. Other such diseases are filiarasis and elephantiasis (also transmitted by mosquito), and onchocerciasis (transmitted by the black fly).

15.2.1.5 Shortage of water

Lack of water for drinking results in dehydration and susceptibility to disease, by weakening. Also the stress of collecting water reduces quality of life.

The transmission of disease is a complex process and hence no direct relationship between water supply or sanitation improvement and the occurrence of disease can be formulated. However, the following steps in combination with education may cause a marked reduction in the occurrence of water-based diseases:

- Disinfection of domestic water supplies
- Provision of well-designed and constructed latrines
- Increased quantity of water for domestic use
- Provision of laundry facilities, thus reducing contact with open water bodies
- Provision of adequate drainage and disposal of waste water
- Management of open water surfaces, e.g. level variation, or spraying

Water quality refers to substances or living organisms dissolved or suspended in water. Water used for domestic purposes need not be completely pure but could have limited dissolved salts for taste and to minimize the corrosive potential of the water. Furthermore, it has been estimated that only one out of every 20 000 strains of bacteria is pathogenic, and the mere presence of bacteria in drinking water is not necessarily cause of concern.

The approach to water quality control in water supply projects should therefore include the following steps:

- Protection of all components (including the source, storage units and pipelines) against possible contamination by pathogenic organisms
- Removal of basic pollutants, .e.g. suspended solids
- Improvement of the existing water quality to ensure aesthetic acceptability (turbidity, odour and taste)
- Neutralize possible irritants, e.g. pH, temperature, hardness
- Educate consumers as to basic precautions for the collection, storage and use of water.

Apart from diseases, many other factors in water can render it unsafe. High dissolved salts can lead to high blood pressure. Alkaline water can affect the arteries and cause skin disorders. Heavy metals and some organic pollutants concentrate in the life chain and can reach lethal proportions. The factors causing cancer are only beginning to be understood.

Then there are physically dangerous factors: flooding, drowning, land slides due to groundwater pressures, pipe bursts, dam break, wave action, temperature and drought.

15.2.2 Water quality standards

Water quality alone is not the only determinant in the good health of a community. Evidence is that the availability of water, sanitation and hygiene are also significant factors.

These factors in a developed society assume a lesser significance because of the institutional infrastructure. Table 15.1 (Briscoe et al., 1986) demonstrates their relative importance.

Table 15.1. Importance of water related activities to health

Infections	Water quality	Water availability	Dispose excreta	Excrete treatment	Hygiene cleanliness
Agents:					
Viral	2	3	2	2	3
Bacterial	3	3	2	2	3
Protozoal	1	3	2	2	3
Poliomyelitis, Hepatitis	1	3	2	2	3
Worms:					
Ascaris	1	1	3	3	1
Hookworms	1	1	3	3	1
Enterobius	1	3	2	2	3
Tapeworms	0	1	3	3	1
Worms aquatic:					
Schistosomiasis	1	1	3	2	1
Guinea	3	0	0	0	0
With 2 stages	0	0	2	2	0
Skin, eye, hair (lice)	0	3	0	0	3
TOTAL	14	22	24	23	22

15.3   LEVEL OF SERVICE

Depending on the investment and requirements of the community, different levels of supply could be planned. Figure 15.2 shows a progressive range of options. They could be one of the following:

   (1) *Simple source and disinfection.* Protection of sources, particularly springs or flowing streams, against contamination is perhaps the most basic form of protection of the water system. Transport of the water would still be in containers.

Washing and drawing water from stream or pond

Borehole, windmill & container

Communal reservoir

Slow sand filtration

Township standpipes

Purification &disinfection

Single domestic connections

Multiple house connections, hot and cold and sanitations

Figure 15.3. Evolution in water supply standards

(2) *Pump and supply line to a reservoir or standpipe.* This would shorten the distance to travel to collect water but still provides a communal collection point. The number of standpipes should be sufficient to ensure there is not a long queue or else an open reservoir for dipping containers in, in which case disinfection is important.

(3) *Individual standpipes or connections.* A standpipe for each house, or else each pair of houses back to back, further improves the availability. Going a step further, it could be provided in a basic lean-to laundry-cum-kitchen structure. At this stage, it may also be possible to consider water-borne sewage for a communal toilet.

(4) *Individual household supplies.* At this stage, the householder may have multiple connections to kitchens, ablution facilities and toilets. He may also consider a roof tank for storage if the supply rate or pressure are inadequate for meeting peaks. Generally at this stage, there is also communal storage for balancing demand against the supply rate which would be at a uniform rate in order to optimally use pipes.

The minimum water requirement is 5-10 litres per day if access is difficult. With piped water supplies, the consumption can increase to between 10-20 litres per day in the case of standpipes, and 30-50 litres per day in the case of individual household connections with minimal ablution facilities. The per capital consumption will only increase above this with communities earning above the breadline salaries. Provision should also be made for community centres and commercial development. If the water is economic, it may be made available for limited agricultural purposes, but generally piped and purified water cannot be economically used for irrigating.

Losses can also be considerable in substandard systems in particular. Leakage and illegal connections can account for up to 50% of water use, where a realistic target is less than 10%.

The City of Durban (Macleod, 1997) embarked on a water supply program to supply some 200 000 families within 20 years. The problem of installing standpipes was that residents refused to pay if they had to walk over 200m to collect water. They adopted a 3-option policy:

(1) *Water tank system*: Small bore pipes feed 200 litre tanks next to houses. Water is paid for when the tank is filled.

(2) *Semi-pressure system*: Water is fed through a restriction to roof tanks in each house so that peaks are reduced.

(3) *Full pressure system* to cope with multiple connections under normal supply conditions.

The installation program is accompanied by extensive consultation and an information program.

Non-payment of water accounts was initially a problem, but defaulters' water was then restricted to 200 ℓ/day with a restrictive washer, resulting in a lowering of bad debts to 0.1%.

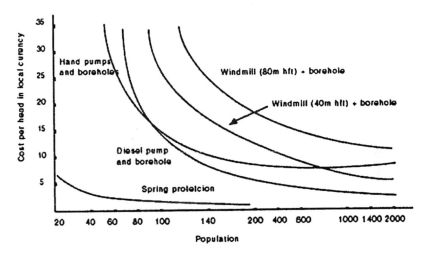

Figure 15.4. Cost graph for comparison of various types of water supply against population (Cairncross et al., 1980)

15.4   ALTERNATIVE SYSTEMS

To reduce costs and make water supply systems more flexible, the following are possible (see also Chapter 6).

Use pipes for grade B water (partially treated wastewater) for some washing operations and gardening. Pipes should not go into houses to reduce the possibility of drinking the water. Potable water could be provided by tanker, container or bottle, or separate supply pipe network. But considering the amount of pure water actually drunk without boiling, this requirement is most likely to be met with containers. But bottled water is notoriously expensive in the western world. This is largely due to bottling, transport and marketing. Subsidization of containerisation of water will in many cases be less than purifying all water to potable standards and reticulating it. Grade B water could be coloured or otherwise marked by taste, or odour.

Small bore water supply piping can be a means of supply control, and ultimately even population density control. Demand control by financial means such as tariffs (Stephenson, 1999) is not feasible in sub-economic communities, as they do not pay, or subsidized anyway. Small bore systems (sometimes with distribution storage) can be extended eventually from standpipe distributions to household distribution if roof tanks are used. Then the pipes supply over 24 hours at an average rate and peaks are met from the tanks. This is one way of minimizing costs and phasing development. It also reduces pipe pressure classes as hydraulic gradients are high. Distribution by container is one way of inducing payment and therefore responsibility into communities (Watson, 1995).

Conjunctive use of alternative sources can reduce costs. For example, when a local spring dries up, there may be a further reservoir source. Or rainwater could be collected for peak supplies, if the piped supply was inadequate.

15.5   PROBLEMS IN SUPPLY

According to Lyonnaise des Eaux (1998), in view of the complexity and diversity of situations encountered, one is obliged to simplify matters considerably and only take two groups into account:

(1) Poor neighbourhoods which are to be equipped
These areas may be in cities or in marginal urban areas. They are equipped with a minimal street system and sometimes networks. The residents' situation is either in line with the law or is due to be regularized by the authorities. This category includes the "barrios humildes" of Argentina, most of El Alto (Bolivia), etc.

(2) Areas of spontaneous improvised and temporary housing
These areas may also be found in cities or in marginal urban areas; they are characterized by illegal occupation of land and lack of public services.

They are not definitely established from a strictly legal point of view, even if they have existing for a number of years, until the situation is regularized.

The latter group is extremely heterogeneous and can be broken down into many subdivisions, taking account of the social organization of the community especially. It may seem absurd, for example to group together the highly structured "zonas invadidas', where street systems are planned, homes are "solid" but land occupation is (for the moment) illegal, and the shanty towns of Port-au-Prince, where buildings are impermanent and services non-existent.

Private concessionaires cannot take over the authorities to "officialise" areas of spontaneous housing, nor can he replace the town planning service. They can, however, "support" the town planning services, making it easier to integrate these communities into an official framework by providing them with adequate services.

Generally speaking, two main difficulties can be highlighted for poor neighbourhoods:

(1) Connection costs are too high

For example, in Buenos Aires expansion area, users have to pay infrastructure charges which are set by the regulator and are intended to cover expansion costs amounting to around $500 for a water connection and $1 000 for a sewer connection. In certain areas, the average monthly income stands at $240.

(2) Customer management costs are too high

There are various reasons for this, and they are not always specific to poor areas. Related problems are:
- A high percentage of unpaid bills
- A high rate of unbilled or fraudulent consumption
- Low consumption resulting in relatively high collection costs
- High network maintenance costs

(3)   Consumers' water procurement strategies

Each city should be considered as a specific case in a given socio-economic and cultural context. Table 15.2 illustrates a situation which is frequently observed in large urban centres in emerging countries.

As regards water supply, populations may be:
- Supplied by the network
- Supply by other method
- Standpipes, wells, with or without intermediaries
- Water deliverers
- Not supplied, resorting to uncontrolled sources (ponds, rivers) or wells.

15.5.1 Payment

The willingness to pay for sewers is generally less than for water supply. Wright (1997) quotes where several sewer lines were installed only about 10% of the design number of households connected up their houses. Sanitation is not seen as a life necessity as it has a longer term effect on health than lack of water. On the other hand, subsidization of poor people's sanitation by richer city dwellers is probably more pressing, so sanitation may be more likely to become an institutional project than water supply.

Some utilities add a surcharge on water bills to cover sewerage. In fact, the cost of sewerage generally exceeds water supply and under-recovery of costs is common, resulting in disintegration of the administration.

Table 15.2. Water procurement strategies (Lyonnaise des Eaux, 1998)

Supply method	Conditions for producing water	Use	Consumption (ℓ/c/d)	Water quality	Payment method	Cost of water per m^3
Individual service line	Private tap in the home or close by (yard)	All use	40 to 250	Generally good	Monthly billing	p
Public fountain	Carrying, waiting at water supply point	Food	10 to 50	Possible con-tamination at water supply point	Payment in cash	10 * p
Delivery	Delivered to the home	Food	50	Variable	Payment in cash	30 * p
Private well	Access close to dwelling	Toilet, washing, etc.	50	Variable (user is not always aware of risks)	Individual mainte-nance	"Free" (except electricity)
Uncon-trolled sources (pond, river)	Access close to dwelling or carrying	Toilet, washing, etc.	50	Bad (con-tamination)		"Free"

15.6   COMMUNITY PARTICIPATION

The concept of community participation in development started on a small scale in the 1950's. The concept then broadened and grew during the 1970's, stimulated, in part, by the failure of many of the large development projects of the previous decade; today it has become a significant factor in development planning. The early models of development, in the immediate post-colonial era, were purely economic and were based upon the "trickle down" hypothesis, whereby (in simplistic terms) economic growth or benefits in one sector (e.g. urban industrial development) would eventually spread to other sectors, such as the rural economy. This model has proved to be invalid and, if anything, the reverse has happened, in that while a small elite has prospered, the large majority of the rural population has become poorer.

There are exceptions to the above of course, and undoubtedly there have been cases where large scale investment in specific geographical areas (generally those with good agricultural potential) have benefited the majority in the area. Thailand is one such case. However, the Thai experience also showed that there was no spread of benefit outside the particular geographic area being assisted and that the net result was in fact an overall increase in the gap between rich and poor nationwide.

Public involvement programs usually have four primary general objectives:

(1) Identifying local needs, preferences and priorities
(2) Identifying all impacts of alternative and recommended programs
(3) Promoting understanding and support for the identified objectives and solutions proposed
(4) Providing opportunity for those affected by water resource development to influence decisions regarding development.

Public views are probably both more important and more difficult to obtain in rural areas in developing countries. The difficulties stemming from a lack of communication are primarily due to:

- The dichotomy between national and local priorities
- Much of the water resource planning affecting rural areas is conducted by urban personnel of the national government
- Much of the water resources planning is conducted by or jointly with foreign agencies and foreign private consultants.

15.6.1 Affordability

Although it has been generally assumed that rural populations in developing countries are too poor to pay for adequate water supplies and sanitation, Churchill (1987) indicates otherwise. The ability to pay and the desire for good services are underestimated. People in some situations will pay up to 10% of their income. The number of villages that are on or below subsistence level is diminishing and there are signs of wealth practically everywhere. This may appear in the form of portable radios or bicycles and clothing.

In many situations, a scheme has failed because they have only provided communal standpipes at some distance from individual dwellings, whereas the people would have been prepared to pay for connections in their own houses. A careful examination of the wants and means of the population is therefore warranted.

Appropriateness of the technology of a water supply scheme is important to ensure its continuity,. High-tech systems including pumps, windmills and even hand pumps will deteriorate and unless there is a source of spares and a means of obtaining them, the entire project could be rendered useless in a few years. Therefore gravity systems and those that will not exhibit less efficiency if left unattended need to be earmarked. Alternatively, arrangements need to be made for maintenance on a regular and ad hoc basis.

15.7   POLICY

The focus must thus be on developing institutional arrangements that provide services at least at cost and in a way that is responsive and accountable to beneficiaries. Where public finance is scarce, significant additional resources can often be mobilized within local communities. Attention should be paid to ascertaining from the poor what level of service it is that they want, discuss what is feasible and then to provide a range of affordable levels of service.

There needs to be a change of emphasis on the part of governments worldwide (not only in developing countries) from single focus projects (e.g. a focus on technology, or on preventative health, or on hygiene education), to a more integrated approach to the provision of infrastructure and the improvement of health.

Table 15.3. Summary of problems for developing countries

Problems	Solutions	
	Short-term	Long-term
Finance	Loans	Charging
Equipment	Donation	Self-made
Exchange rate	Soft loans	Self-sufficiency
Technology	Expatriates	Training
Administration	Consultants	Capacity building
Government	Deposed	Democracy
Water-related health	Medicine	Sanitation
Water quality	Disinfection	Treatment
Water supply	Sources	Household taps

This change of emphasis needs to be accompanied by other shifts in emphasis, particularly:

- From top-down approaches, to approaches that are a judicious mixture of top-down and bottom-up
- From focus on construction costs of facilities, to lifetime costs of facilities (i.e. including operation and maintenance), and
- Recognition of the need to ensure the financial and environmental sustainability of projects.

15.8   DEVELOPING PEOPLE

Not only is the majority of the world population underdeveloped and under-skilled, there is an increasing technology gap as the first world advances into new management styles. There is a switch from hard technology to people orientated skills and business awareness. The developing world has not even caught up with the hard technology.

Due perhaps to pressures from government and public servants, expenditure on capital intensive works is being minimized and emphasis is shifting to passive maintenance and consumer alerting. This may eventually result in a stalemate or hopefully reversal in the trend, but the point is that a short-term approach is being adopted more and more as technically ignorant people emerge as the leaders of water organizations.

The managers could perhaps be directed by the politicians who are appointed on a shorter term basis than engineers and operators, because the latter base their skills on experience as well as training. On the other hand, public servants and, more particularly, politicians are appointed for their communication skills.

Different skills are required in different ways depending on the profession.

Table 15.4. Training for various skills in the short and long term

Who	Short-term training	Long-term training
Public	Press	School
Administrators	Hands on	Courses
Engineers	Site training	University
Developing countries	Mentoring	Experience
Public servants	Checking	Transparency

Skills of technology and the newer skills of business acumen are held by the developed countries and they tend to offer their services to the developing countries. Rather than sitting back and thinking or adopting a longer term strategy, the developing countries appoint these skilled people even on a short-term basis, but nevertheless extracting from these countries resources due to payments and reduced local experience.

Training of local people both in the technology and in the attitudes to handle the development of water services are required. This will require salary adjustments to attract the correct calibre of people, establishment of academic training institutes and the installation of long-term mentoring systems to develop to a high standard of service provision.

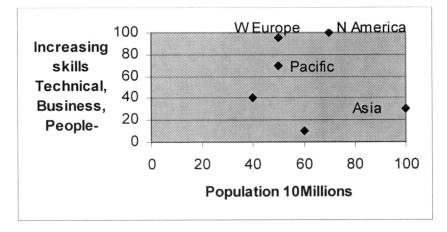

Figure 15.5. Graph showing population skills throughout the world

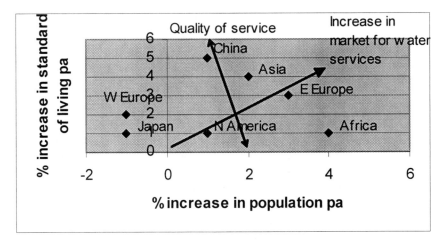

Figure 15.6. Increasing population numbers and standards of living resulting in greater water services requirements.

New methods may have to be developed by the developing world to cope with financing and the correct technology may not be the same as used in the western world. Fortunately, improved communications technology means that remote training and mentoring is becoming a reality. Internet contact can be direct and personal with mentors thousands of kilometres from the site of a waterworks. Up-to-date and interesting teachers can be switched on and the only remaining obstacle is to facilitate the connection of the trainer to the worker by means of a reputable institution. That means that reputable institutions may have to offer their services to poorer organizations or countries in order to speed up the technology transfer.

15.9   THE FUTURE

The waiting time to catch up by the developing countries in technology and service provision may be leapfrogged by changing technology. For example, the switching from hard engineering to soft engineering means controls and management can be programmed by computer. However, the basic purification works, pumping and pipelines must still be provided. These may be on a different scale as demand management and alternative systems materialize. New pipe materials, new types of pumps and new electrical machinery and controls will no doubt reduce costs and resource requirements.

With reducing availability of potable water sources, the emphasis on demineralization and recycling will become greater This will mean larger and more economic purification processes being developed.

In the field of natural resources, the ability to accommodate risk and the long-term prediction of recharge rates will enable capital costs to be minimized.

Management styles will change and non-technical managers will emerge to achieve different balances between consumers and suppliers. Consumers will become more aware of the limitations and change their demands to suit their resources. The development of community projects on a small scale may obviate the necessity for high technology engineering and enable stage development matching available financial resources to be obtained. Consumer awareness will improve health in the home, and on site disposal is likely to become more popular, especially in rural areas. Costs are likely to increase due to poorer technical input, but this may be blamed on increasing scarcity of resources.

Consumers' priorities are changing. In the developed world, we have seen an increasing environmental awareness and conservation consciousness. This will disseminate to developing countries, but owing to different pressures may take longer to evolve. Nevertheless, greater resources awareness will result in better consumer management of their services and resource usage. This picture has started to emerge in non-United States countries in the energy sector, and the green awareness also receives greater acceptability outside the United States of America. Thus while the developing world's wheels are slowly turning and improving the quality of life, the wheels of the developed world are also turning and may result in a less ambitious quality of life from the materialistic point of view and greater appreciation of the resources of the world and the importance of sustainability.

15.10 REFERENCES

Bhatia, R. and Falkenmark, M. (1993) Water Resources Policies and the Urban Poor: Innovative Approaches and Policy Imperatives. UNDP-World Bank, Water and Sanit. Prog.

Black, M. (1994) Mega-slums: The coming sanitary crisis. London, *Water Aid.*

Briscoe, J., Feacham, R.G. and Ahaman, M.M. (1986) *Evaluating health impact.* Water Supply, Sanitation and Health Education, UNICEF-ICDDR-IDRC report.

Briscoe, J. (1997) Managing water as an economic good: rules for reformers. *Water Supply.* **15**(4), 153-172, Blackwell Science Ltd.

Cairncross, S., Carruthers, I., Curtis, D., Feacham, R., Bradley, D. and Baldwin, G. (1980) *Evaluation for Village Water Supply Planning.* John Wiley and Sons, N.Y., 179pp.

Churchill, A.A. (1987) Rural water supply and sanitation. World Bank discussion paper, Washington D.C.

Lyonnaise des Eaux (1998) *Alternative Solutions for Water Supply and Sanitation in Areas with Limited Financial Resources.* Paris.

Macleod, N. (1997) Tariffs and payment for services – The Durban Story, *Civil Engg., SAICE,* June.

Retali, D. and Étienne, J.M. (2000) Water and Sanitation Services, in *Financing of Major Infrastructure and Public Service Projects*, Eds Perrot, J-Y and Chatelus, G., Pont et Chaussées, Paris.

Stephenson, D. (2001) Problems of Developing Countries, in *Frontiers in Urban Water Management*, Eds. Maksimovic, C. and Tejado-Guibert, J.A., IWA Press.

Varley, R.C.G. (1995) *Financial services and environmental health; Household credit for water and sanitation.* Applied Study No. 2, Environmental Health Project, Arlington, VA.

Watson, G. (1995) *Good sewers cheap. Agency-customer interactions in low cost urban sanitation in Brazil.* Water and Sanitation, The World Bank, Washington DC.

Wright. A.M. (1997) Towards a Strategic Sanitation Approach: Improving the Sustainability of Urban Sanitation in Developing Countries. *UNDP-World Bank Water & Sanit. Prog.*

16

The energy factor

16.1 PUMPING ENERGY

The cost of energy in providing water services is a considerable proportion of the total costs (see Figs. 16.1 and 16.2).

Energy is used for pumping in the primary instance but in fact the cost of many commodities is linked to energy costs. The production of many materials such as plastics is a direct by-product of the energy industry. Energy is used for mining and all forms of manufacture. Even manpower can be measured in terms of input in the form of energy. Thus apart from charges for abstraction of natural resources, the cost of commodities and their final sale value is directly proportional to the energy put in at different stages. For this reason, total cost remains the best measure of value, but this section is concerned particularly with energy management and the direct use of energy in the form of electricity or thermal fuel in providing water services.

Energy is used for pumping, primarily. This affects water supply schemes where water is abstracted from natural resources which are generally lower than the consumer. On the other hand, drainage solutions are usually gravitational because of the complexity of lifting sewage and stormwater to a higher elevation. There is also a failsafe mechanism in gravitational systems in that flow will be naturally downhill if operational systems fail.

©2005 IWA Publishing. *Water Services Management* by David Stephenson.
ISBN: 1843390809. Published by IWA Publishing, London, UK.

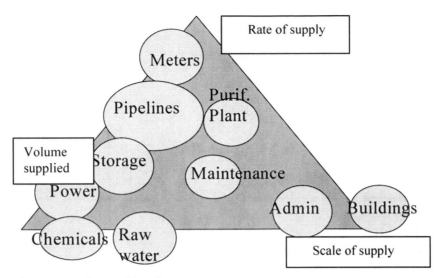

Figure 16.1. Influence of three factors on water cost

The cost of friction in pumping is the only component which can really be really be managed. Static heads or lifts to get potable water to consumers is a function of the topography and town layout and cannot be changed, but the judicial positioning and operation of reservoirs can minimize this cost.

Instead of pumping all water to the highest reservoir in a city, it is often more economical to have a number of smaller reservoirs and to pump into them individually or at different times. Therefore, correct control of the system is economic in minimizing total energy consumption. The same applies in pressure zoning. Pressure zones are used in water supply to reduce leakage, i.e. by minimizing pressure on leaky systems. However, they are also of use in minimizing costs of plumbing installations by pressure control and direct routing.

Pumping energy can be minimized using the following techniques:

- Provision of reservoir storage to enable pumping to be carried out over 24 hours a day. Storage will meet the peaks in demands.
- Minimizing friction costs by pumping at an average flow rate instead of short periods at a higher rate, since the energy for friction is proportional to the flow rate squared.
- Scheduling energy use in times of off peak energy costs.
- Minimizing pressures to minimize losses from the system.
- Minimizing stopping and starting to minimize water hammer damage to the system.

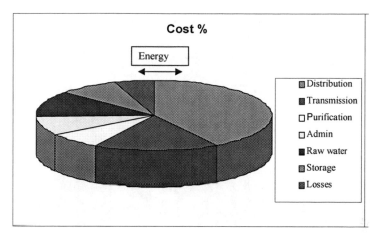

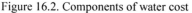

Figure 16.2. Components of water cost

The following example demonstrates how spreadsheets can be used to minimize the cost of pumping where electricity is charged on a time of use basis.

16.1.1 Programming pumping systems

PLC's (Program Logic Controls) are used on most electrical panels for pumping systems nowadays. They enable decisions to be made automatically depending on flow requirements and pressures in a system. They may also be linked to SCADA systems (Supervisory Control and Data Acquisition systems) so that data can be logged and reviewed to ensure the system is operating most efficiently. Thus, a heuristic system could be built up but optimization methods can also be used to minimize the sequence of pumping.

Many studies have been done in the field of water pipe network optimization but the problem of optimizing the pumping rate during the day in order to minimize operating costs and keep reservoirs full, is a separate, more time-dependent problem. This section (Ilemobade, 2003) presents a methodology for optimizing the pumping policy for a pump and reservoir system using historic and predicted demands. The pumping policy fits pumping rates into a minimum cost cycle over a given period based on reservoir volumes. This requires knowledge of reservoir storage capacities, demands, pumping capacities and electricity tariffs, which are often a function f time of use. Cost penalties for violating reservoir limits, switching pumps on, and storing water in the reservoir are used as constraints. The Libanon system on Gauteng's West Rand, South Africa, is used as a case study. The model is developed using the Downhill

Simplex method and it achieved cost savings of about 2% compared with conventional human intervention operation.

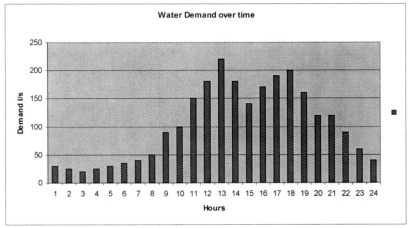

(a)   Water demand over time

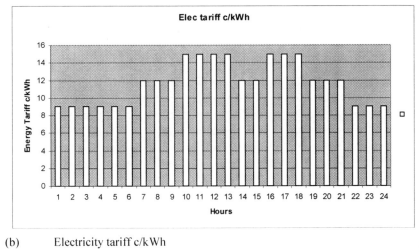

(b)   Electricity tariff c/kWh

Figure 16.3 (a) – (b). Energy and water costs

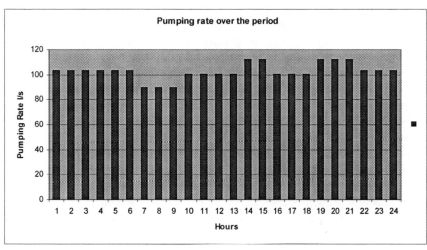

(c)   Optimum pumping rate over the period

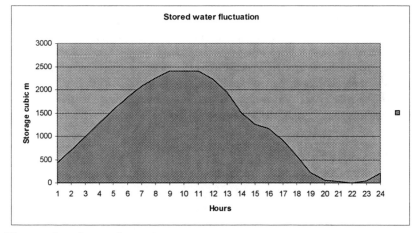

(d)   Stored water fluctuation

Figure 16.3 (c) – (d). Energy and water costs

   A number of past studies have concentrated on the optimization of pipe networks and reservoir systems to minimize costs at design stage (Lansey and Mays, 1989). A separate problem is that of the operator who needs to make regular decision s as to which pumps should operate at any given time and the rate of pumping in order to meet varying demands and ensure the reservoirs do not drain (Ormsbee and Lansey, 1994). To optimize system operation, i.e. to ensure the above situation while minimizing operating costs, is a more complex problem and is often not attempted.

The solution tackles the operation problem from the point of view of determining an optimum pumping policy for a system based on demands and reservoir volumes (historic or predicted) over a given period. In determining a pumping policy, electricity tariffs, pumps switching, storage and storage penalty costs are considered. A pumping policy refers to the set of rules scheduling pump operations at different reservoir levels that will result in the minimum operating costs for a given period and operating conditions.

The system looked at is a pumping main from a source of water with an unlimited capacity, multiple pumps, and a supply into storage reservoirs at the head of the demand system (Fig. 16.4). The limitation of the pumps, i.e. head versus flow and the relationship for various combinations of pumps, is accounted for.

If pumping rates are too low, the reservoirs may drain during peak demand seasons, and if pumping rates are too high, the reservoirs may overflow but more importantly, it will result in wasted electricity costs. A balance is required. The lower the load factor of the pumping system, the higher the energy loss due to friction. This results from the non-linearity in head loss versus flow through the pipelines. Higher load factors and reduced energy loss during pumping is therefore desirable. Another problem is the variation in electricity tariffs from hour to hour in an attempt to shed load or to distribute the load more evenly in order to enable the power stations to operate at as high a load factor as possible. This in turn, results in a minimum electricity cost to the supplier. The user of energy is therefore encouraged to use off-peak energy with a preferential tariff system. Optimal system operation also requires the consideration of this factor.

16.1.2 Model formulation

The Downhill Simplex technique, which is a multi-variable, constrained, non-linear optimization tool, was employed to determine an optimum pumping policy. It is confirmed using dynamic programming which is a simpler method confined to optimizing the overall pump flow rate on sequential time periods.

The Downhill Simplex method, also known as 'Amoeba' (Nelder and Mead, 1965; Press et al., 1992), finds the minimum of a function that has more than one independent variable. It only requires function evaluations and does not require any derivatives. It can be very efficient for constrained coefficient optimization and especially for problems that have a small computational burden.

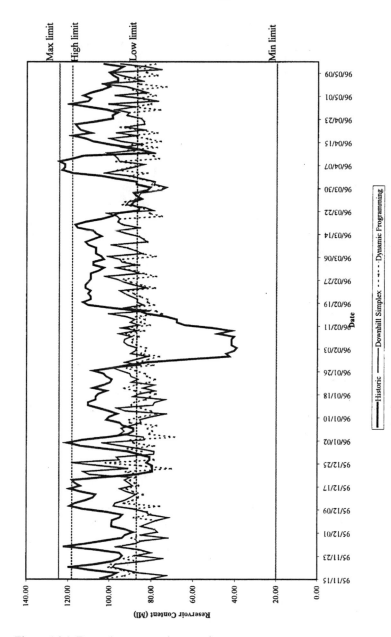

Figure 16.4. Example pump and reservoir system

The Downhill Simplex method starts from $N + 1$ points, defining the initial simplex. During optimization, the general idea is to keep the minimum solution within the simplex while at the same time decreasing the volume of the simplex. Many of the steps involve moving the point of the highest function value through the opposite face of the simplex to the point for which the function is expected to have a lower value (reflection). Such a step preserves the volume of the simplex and its non-degeneracy. The simplex may expand in order to speed up its movement in a given direction. When the simplex reaches a minimum, it contracts itself in the transverse direction and makes its move 'downhill'. It may also contract in all directions and pull itself through the best point (the least value of the function), just like an amoeba would. A limitation is that termination criteria are not easy to specify – hence the use of constraints.

A pumping policy decision is determined based on the reservoir volume at the start of a time step, the demands expected for the time step, and the pump settings during the previous time step. Pump settings refer to various combinations of pumps and thus, pump flow rates. A reservoir change level is assigned to each pump setting.

The optimization is initially given a starting guess, i.e. a simplex with $N + 1$ reservoir change levels. It then finds the optimum pumping policy (i.e. each reservoir change level and reservoir don't change level that minimizes operating costs over a given period and system description).

Variable speed pumps may be modeled as a number of discrete settings. Each reservoir change level is the maximum predicted reservoir volume at which a pump setting may be used, e.g. pump setting 1 should be used if the predicted reservoir volume is between reservoir change levels Y and Z. x_i, x_j, y_i, y_j, z_i and z_j represent reservoir levels constraints. Incorporating these in the example above, pump setting 1 is recommended between y_i and z_j. Any pair of X:Y:Z values will therefore represent a starting point on the 3-dimensional plane. The optimum pumping policy will then find the values of x_i, X, x_j, y_i, Y, y_j, z_i, Z and z_j that will result in the least system operating costs.

By employing the Darcy-Weisbach equation, total system head H_{System}, which is a function of flow rate Q, may be mathematically represented as:

$$H_{System} = H_{Elevation} - H_{Reservoir} + \Delta H_f + \Delta H_m \qquad (16.1)$$

where: $H_{Elevation}$ is the height of reservoir base above pump discharge
$H_{Reservoir}$ is height of water in reservoir
ΔH_f is head loss due to pipe friction
ΔH_m is secondary head loss.

$$\Delta H_f + \Delta H_m = \frac{8\lambda L Q^2}{\pi^2 g D^5} + k_L \frac{8Q^2}{\pi^2 g D^4} \qquad (16.2)$$

$$\Delta H_f + \Delta H_m \cong KQ^2 \qquad (16.3)$$

where: λ is the Darcy friction coefficient
L is pipe length
D is pipe diameter
g is acceleration due to gravity
K is approximately constant for a particular system. (Accumulated pumping data may likely show that the value of K for a particular system varies negligibly.)

For each pump setting, a curve relating pump head H_{pump} to flow rate Q may be expressed as:

$$H_{Pump} = f(Q) \qquad (16.4)$$

At the operating point for the pump and system head versus flow curves:

$$H_{Pump} = H_{System} \qquad (16.5)$$

By substituting equations 16.1 and 16.4 into 16.5 and re-arranging:

$$Q = f(V) \qquad (16.6)$$

For each pump setting therefore, there will be an equation that relates pump flow rate Q to reservoir volume V.

16.1.3 Operating cost components

Two major constituents of pumping costs are the peak demand charge (per kW peak) and the active energy charge (cents/kW/hr). The Electricity Supply Company of South Africa defines the peak demand charge as "the charge for the highest power delivered during peak periods". The switching on of pumps during peak demand periods increases this charge. It makes economic sense therefore to regulate the switching on and off of pumps in addition to solving the trade-offs between incurring the other operating cost components during especially peak periods. The Downhill Simplex method uses the reservoir 'don't change level variables' to minimize pump switching and therefore indirectly, the peak demand charge.

For each pump setting, pumping costs may be calculated as the energy consumed multiplied by the applicable tariff:

$$C_{Pumping}(P_{s,t}) = \frac{\rho g Q H_{Pump}}{1000\eta} x E_{Cost} \qquad (16.7)$$

where: ρ represents density of water (kg/m^3)
         η is pump efficiency (%)
         E_{Cost} is the energy tariff (cents/KW/hr)

Alternatively, energy costs may be determined from energy versus pump flow rate relationships generated using historic pumping data.

16.1.4 Predicting consumer demands

For a pumping policy to be optimal over a given period in the present and future, historic and predicted demands should be considered. For prediction purposes, historic demands must be modeled, analyzed and disaggregated into their various components:

$$D_{f,t} = D_{s,t} + D_{p,t} + \sum_{i-1}^{q} \phi_{i,q} D'_{r,t-i} + \mu_{D_n} \qquad (16.8)$$

The daily demands were predicted over six months – the same period as the historic set. Since pumping data and costs were generated for the historic data, this section aims to validate the methodology presented. The negative slope in the linear secular trend is caused by the gradual decrease in demands typically experienced between late spring and late autumn (the seasons over which the data were generated). This slope would have been eliminated if at least a full year's worth of historic demands were used in the analysis. To aid comparison, the predicted demand set is displaced on the ordinate axis by 100 Mℓ. Both sets exhibit similar trends and statistical properties.

Figure 16.4 presents the comparison between historical and optimized pumping policy operation. From the values of the pumping costs, the Downhill Simplex method may be seen to be capable of determining pumping policies that are optimum, practical and cost effective. Despite the marginal savings in pumping costs in this case study, the Downhill Simplex method presents the added advantage of recommending formal pumping policies that (a) do not require the lengthy training of human operators to achieve, and (b) may be used to successfully operate a human or semi-automated system.

The higher the penalty costs in relation to the higher operating cost components, the less likely the pumping policy will violate them. The graphs clearly show the sensitivity of the Downhill Simple and dynamic programming procedures to minimizing these penalty costs on a daily basis. The graphs also present a comparison between historic and optimized pump operation.

It is expected that the policy would optimize pumping around hourly tariffs by pumping less during peak periods and further minimizing operating costs. In order to obtain realistic operating costs, storage, pump switching and penalty costs are needed. From the results, the methodology presented has been shown to be effective, achieving about 2% savings in pumping costs alone as compared with human intervention operation. Length of base data has been shown to be of negligible effect in the generation of an optimum policy and the methods are sensitive to violating reservoir limits.

16.2   ENERGY INTENSIVE VERSUS CAPITAL INTENSIVE PROJECTS

Energy costs can be minimized on an operational basis but capital costs require a longer term planning process and optimization. Not only is there thus a difference in the planning approach but they can also be offset against each other to minimize total costs. For example, where there is high uncertainty in future demands for the service, it may be prudent to minimize capital expenditure. That is, instead of providing large surplus capacity during initial years one can construct new supplies in phases. Alternatively, one can rely on a higher operating cost system. For example, a high capital gravity system would be more prone to price sensitivity if demands varied from those assumed at planning stage, than an energy intensive system. If small scale projects with high pumping costs were installed then the energy could be tapered to meet the demand without incurring excessive costs which would otherwise have to be met in the form of capital repayments.

Thus, greater attention may have to be paid to pumping from groundwater resources or demineralization of wastewaters as these are largely energy related. This is also in line with reduced size of dams and major water works which are becoming more of an environmental and social issue.

16.3   PUMPING SEWAGE

Pumping systems for sewage are not favoured, not only for their impracticality but also because of operating problems. The high gas content in sewage is often in a corrosive environment but with clever management can be put to advantage. Gases from digesters at wastewater treatment works can be used for heating and driving small generators. Alternatively, they can be used for direct energy production or even incineration on site.

The future changes in values may dictate which way to go. That is, whereas in previous chapters the possibility of onsite disposal of sewage was considered, centralization would be required in order to generate viable proportions of gases such as methane for direct utilization. Thus the food crisis versus the energy crisis will have to be solved and changes in values may dictate which was to go.

16.4   HYDROPOWER

Although hydroelectric power is a major subject not normally treated as part of water services for an urban community, the increasing attention to renewable and hybrid power systems means that attention to self-contained systems may be the order of the day in the future. Storage of water at a high elevation may end up a more economical form of energy storing than batteries or thermal systems. Therefore, household scale or micro-scale power generation systems may use hydropower primarily for storage, and solar power and wind power for primary power generation at times of availability.

The idea could be taken further and use of household waste for biogas generation which could also be used for pumping water to high head and generation locally during peak power demand areas.

16.5   SOLAR-HYDRO HYBRID SYSTEM FOR RAPS

Often remote areas do not receive power supply with direct connections to the main grid. RAPS is remote area power supply. Although the theoretical cost of energy supplied by the grid is low, the connection to a distant net would increase total costs, so that the option is not economical. As an economic alternative, an off grid power supply scheme is proposed.

The solar-hydro hybrid system scheme for power generation would consist of a solar panel, a storage water reservoir and a system of pump/turbine connected to a motor/generator. For the operation, the solar panel will generate the energy necessary to the daily consume and the energy necessary to pump water into the storage reservoir. During the night, the operation of the turbine with the water stored in the reservoir will produce the necessary energy. If a stream is present in the vicinity of the house, its water can be used to produce energy by operating a small turbine. Depending on the amount of energy required, the system can be operated also in parallel with the solar panel. In case of drought, the panel will supply all the energy during the day.

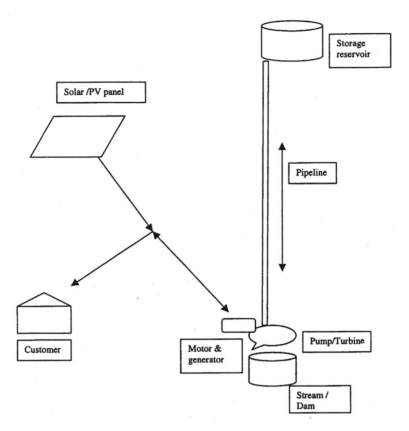

Figure 16.5. Proposed solar-hydro hybrid system for RAPS (Remote Area Power Supply) in South Africa

16.5.1 Hybrid energy systems for rural communities

It will be desirable to provide renewable energy to communities who at present do not have such facilities and as a result are using natural resources in an unsustainable manner. Many small communities in Africa and Asia are using wood fuel and other biomasses including dung for heating and cooking. Owing to the perspective of these people, there is little consideration for sustainability of these sources or of the environment. The result may be a depletion of the resources and an irreversible degradation of the ability of the land to sustain biomass and, as a consequence, fauna.

It would be desirable to provide communities with more sustainable forms of energy and additional energy to enable them to improve their life styles. Access to electricity would provide lighting which could be used for reading and educational purposes. The switching on and off of electricity enables energy to be conserved when it is no longer required. Power for communications, i.e. radios, television and telephones, will further improve the life style and lead to an educational type of improvement in family life. It is in fact this improvement which eventually will take the communities out of the subsistence living level into development which will input to economies instead of draining national and foreign resources merely to subsist.

It is not practical to provide grid electricity to small communities owing to the high cost per unit of suspending transmission lines over long distances for small power delivery rates. Not only the cost of such power per unit but also the inaccessibility and the ability to maintain small supply rates over long distances are doubtful. Therefore, off-grid electricity is more desirable.

The availability of alternative forms of energy in rural areas may be high relative to the energy requirements. For example, river flows may be greater than needed to provide power. Large solar radiation rates are wasted and more wind energy may be available. However, not all these forms of energy may be available at any one time. There are different periods during the year and during the day that solar power is available, and/or wind or hydropower (on a micro scale) so that alternatives would be a renewable natural energy resource supplemented by a backup system.

The backup system could be in the form of batteries, pumped storage into water reservoirs, burning of fossil fuels, for example coal, diesel, petrol or biomass. The use of biomass including domestic sewage should be evaluated from the calorific point-of-view and the ability to efficiently convert it to mechanical or electrical energy. Mass balances need to be done for all these sources of energy in order to determine the carrying capacity of them in various areas.

Alternative energy for hydraulic pumping could also be obtained from a thermal expansion of bladders or containers which in turn are encapsulated or connected to water containers. A series of one-way valves could draw water in when the bladder contracts during cold periods and be forced up to tank during hot periods.

The sustainability of the mechanism for generating electricity needs also to be investigated. This will probably require pilot installations but it is essential in order to test whether the system would be suitable for poorer communities. The communities invariably have little technical knowledge and fewer morals regarding maintenance and community ownership. The result is often due to poor institutional capacity, subject to corruption and even theft of parts. This system needs to be considered in total if any such energy source is to be tapped.

16.5.2 Development of alternative energy sources

Some sustainable energy sources have been investigated in previous studies. These include a preliminary assessment of micro hydroelectric power potential in South Africa (Balance, Stephenson and Chapman, 2000). Combined energy sources such as wind and micro hydropower generation were also investigated by the EU in conjunction with the University of the Witwatersrand in South Africa.

However, other sources have not been investigated and in particular the economics of alternatives, and the optimization of the system by using hybrid schemes needed to be investigated.

- Micro hydro using run of the river for generating. This relies on the fact that flow rates are often excessive but it is uneconomical to construct dams for micro scale hydroelectric power. Semi-submerged water wheels which rely on flow more than head have been investigated but need to be tested and evaluated. The advantage of such systems is that they can be put on floating rafts in order to maintain a satisfactory submergence and to be lowered as the water level is lowered to actually provide an additional head loss. When rivers flow in flood, the raft will rise but the higher flow velocity will maintain a satisfactory generation rate. Such systems are very inefficient but it is only the capital cost which is really of concern for such systems. In a similar way, centrifugal pumps which are available economically can be used by discharging through them in reverse.

- The use of thermal expansion of air containers to force water through pipes. Research will depend on the outcome of the solar radiation studies, i.e. the energy available for expanding the bladder. The system may be competitive with hydraulic ram type pumps, particularly where the flow of water is limited. Hydraulic ram pumps are well known for pumping to high heads without energy except the water energy of the stream. That is, a large flow rate is used to isolate the water in the container and generate water hammer heads to force the water column uphill.

- Domestic biomass combustion. The volume of biological waste from homes is relatively large, which can be seen by looking at the input to the household. Energy balances should be done. But above all, calorific values of domestic wastes should be investigated. A big energy waste is drying of the wastes and solar methods may be suitable for this. Another impediment is the large storage needed during the stabilization process and the containment to prevent obnoxious smells and also to catch gasses released during the stabilization process.

It is possible that the thermal content of biomasses will be insufficient for energy requirements for a complete household, but the optimum conjunctive use of alternative energy sources needs to be investigated.

Finally, the life cycle of the alternative energy sources and the optimum utilization of it must be made. This must extend beyond the mechanical life and include the cost of deteriorating environments and loss of biomass and natural habitats using indirect costing procedures.

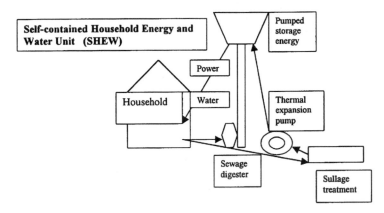

Figure 16.6. Self-contained Household Energy and Water unit (SHEW)

16.6   REFERENCES

Ballance, A., Stephenson, D., Chapman, R.A. and Muller, J. (2000) A geographic information systems analysis of hydro power potential in South Africa. *Jnl. Hydroinformatics*, **02**.4, IWA, 247-254

Ilemobade, A.A. (2003) *Decision Support System for Rural and Peri-urban Water Distribution System Design and Operation.* PhD thesis, Univ. Witwatersrand, Johannesburg.

Lansey, E.K. and Mays, L.W. (1989) Optimisation models for design of water distribution systems. *Reliablity Analysis of Water Distribution Systems.* Ed. Mays, L.W., ASCE, 37-84.

Nelder, J.A. and Mead, R. (1965) Simplex method for function minimization. *Computer Journal,* **7**, 308-313.

Ormsbee, L.E. and Lansey, K.E. (1994) Optimal control of water supply systems. *Jnl. Water Resources Planning and Management,* **120**(2), 237-252.

Press W.H., Flannery B.P., Teukolsky, S.A. and Vellerling, W.T. (1992) *Numerical Recipes in Fortran, The Art of Scientific Computing.* Cambridge Univ. Press, 402

17

Project management

17.1 INTRODUCTION

Planning and management are needed throughout the life of a public works to ensure a good service and cost effectiveness. The activities which require good management include:

- Planning and design
- Construction of new works
- Operation, of pumpstations, treatment works
- Loss control – water, stock, money
- Maintenance and refurbishment
- Revenue collection
- Assets
- Environmental protection

The project management techniques described here were developed primarily to improve construction efficiency but are equally applicable for planned maintenance and general corporate functioning.

17.2   CONTRACT PROCEDURES AND DOCUMENTS

The common method of organizing a construction contract is to invite bids from contractors. A design with drawings and specifications is issued to tenderers and a contractor appointed after considering a priced schedule of quantities and the ability of each of the tenderers. Any subsequent variations to the design are paid at scheduled rates, or on a dayworks basis, or re-negotiated. Disputes are handled by mediation or arbitration (the latter involving a court-like argument). Guarantees and insurances are required from the contractor and retention money is paid after a specified maintenance period.

Contract documents rely a lot on standard specifications prepared by engineering organizations. These may refer to national or international standards for materials and workmanship. In addition, the following documents form part of the contract:

- Invitation to tender
- General and special conditions of contract
- Description of works
- Laws of contract
- Qualifications by tenderer
- Technical specifications
- Data sheets
- Schedule of quantities
- Completion time and other requirements
- Formula for price escalation, taxes, penalties, bonuses
- Lists of manpower, machinery and materials
- Engineering drawings

Open tenders are usually called for by public bodies. But tenderers may be pre-selected, and in the case of private enterprises, contracts may be negotiated with preferred contractors. Alternative designs may be submitted at tender stage.

Payment is made as work proceeds, based on establishment, work done and materials on site. 5 to 10% of the payment is withheld to ensure completion and part paid then and the balance after a specified maintenance period.

17.3   PROJECT STAGES

A civil engineering project such as a water pipe or sewer goes through a number of stages before it can serve its purpose, i.e.

- Needs identification
- Engineering planning
- Environmental impact assessment

- Financing
- Legal and regulations
- Community informing and feedback
- Route selection
- Surveying
- Design
- Contract preparation and award
- Construction
- Training and commissioning
- Maintenance
- Decommissioning

It is the design to construction stages which are regarded as the project which may require project management (e.g. Bennett, 1985; CDM, 1994). Whereas the civil engineers appointed to do the design may manage the project through construction, this is not always the case. On larger projects,, the engineers may be selected by bidding on a quality and cost basis, and different engineers may supervise construction. Alternatively, the project may be done on a design and construct basis. Or the project may be done at the risk of the contractor or on a BOT (build, operate and transfer) basis.

The method of paying for the construction costs is usually at tendered rates to the selected contractor. This is terms a cost reimbursible contract. Alternatively, the contractor may do the job for a lump sum. And there may be incentives for doing a job under the target figure and penalties for exceeding the target price. But the risk of variations in design and conditions needs to be tied up first. There may also be penalties and bonuses for finishing the project later or early respectively.

Incentives built into pricing may be for encouraging local content, labour intensive work or meeting targets in price and time.

17.4   MANAGEMENT TECHNIQUES

Systematic ways of managing complex projects have become common with the use of computers. However, many of the older techniques remain in use, .e.g bar charts. The following methods (see Neale and Neale, 1989) are available:

Bar charts: Each major activity is plotted as a bar over its duration. The resulting chart is easy to read and modify. It is not so easy to indicate slacks, dependencies or to make changes. The bars can be linked to show succession. They are usually drawn from the earliest possible start.

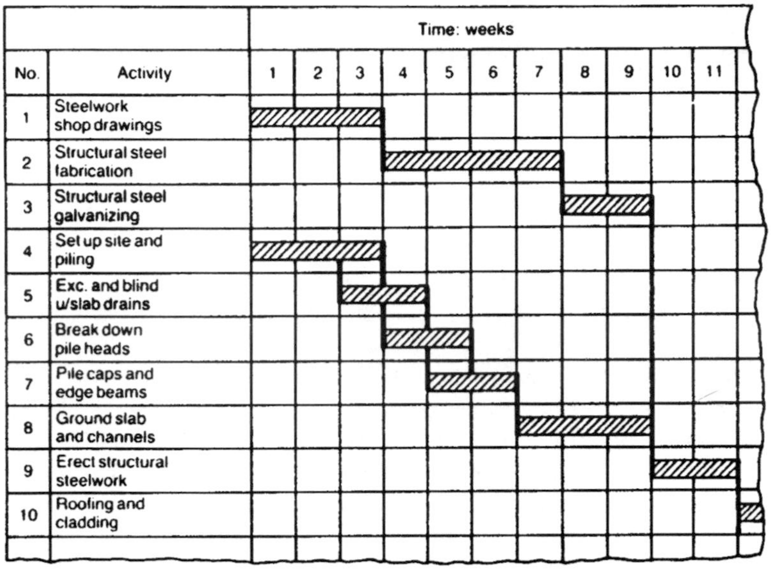

No.	Activity	Time: weeks										
		1	2	3	4	5	6	7	8	9	10	11
1	Steelwork shop drawings	▨	▨									
2	Structural steel fabrication				▨	▨	▨	▨				
3	Structural steel galvanizing								▨			
4	Set up site and piling	▨	▨	▨								
5	Exc. and blind u/slab drains				▨							
6	Break down pile heads					▨						
7	Pile caps and edge beams						▨					
8	Ground slab and channels							▨	▨			
9	Erect structural steelwork										▨	
10	Roofing and cladding											▨

Figure 17.1. Linked bar chart for a factory extension (Neale and Neale, 1989)

Logic charts: A flow diagram us created by drawing arrows from one activity to a successive one.

Gantt chart: Gantt developed the bar chart into something more useful by adding slack lines after the activity bar. Precedence arrows may also be added.

Line of balance: Each line represents a common type of work or trade and the vertical axis is project number (e.g. a lot of houses) and horizontal axis is time. So the lines are diagonal across the diagram as the tradesmen proceed from one house to another.

Linear programme: This is used for long projects, e.g. railways, roads, canals, or pipelines. The horizontal axis is still time but the vertical axis is now chainage (distance along the works). And different diagonal lines are drawn on the chart for each activity.

Critical path method (CPM): Arrows representing each activity are drawn on a chart of activity on the vertical axis against time on the horizontal axis. The duration and slack are indicated in a box for each activity. Activities with no slack are said to be on the critical path.

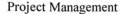

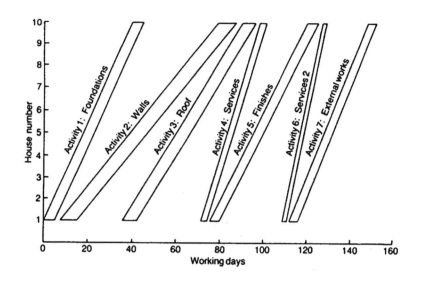

Figure 17.2. Line-of-balance for ten-house project (Neale and Neale, 1989)

Project evaluation and review technique (PERT): A similar plot to CPM also indicating the probability of finishing by a specific date.

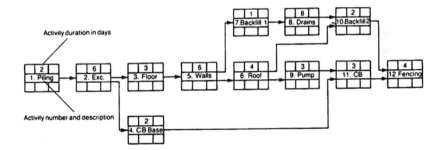

Figure 17.3. Precedence network for land-drainage chamber (Neale and Neale, 1989)

The latter methods are called Precedence Network Methods and are particularly amenable to computer methods. Microsoft's Project is used by many disciplines. Regular updating is possible and distribution to all parties by e-mail is quick.

Beneath charts are often lists of resources and updating notes.

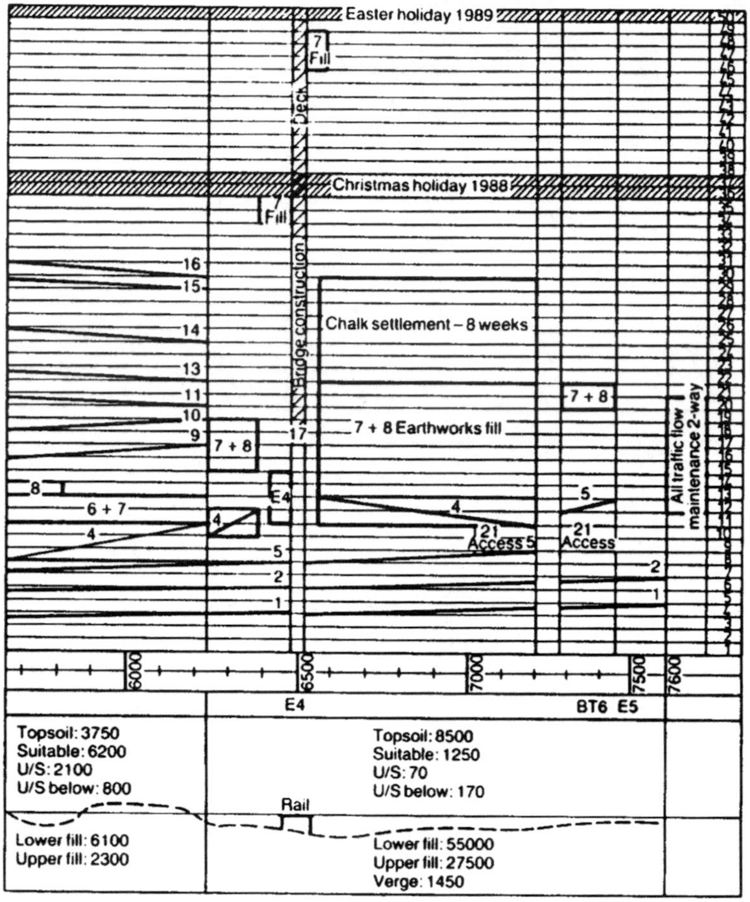

Figure 17.4. Part of a linear programme for a trunk road scheme showing activities from chainage 5500 to 7600 from start up to week 50 (Neale and Neale, 1989)

1. Setting out
2. Site clearance
3. Fencing
4. Pre-earth work drainage
5. Topsoil strip
6. Bulk earthworks

7. Lower fill
8. Upper fill
9. Drainage
10. Capping
11. Sub-base
12/. Kerbs and gullies

13. In situ channels
14. Road bases leanmix
15. Road base: flexible
16. Surfacing
17. Verge earthworks
18. Accommodation works

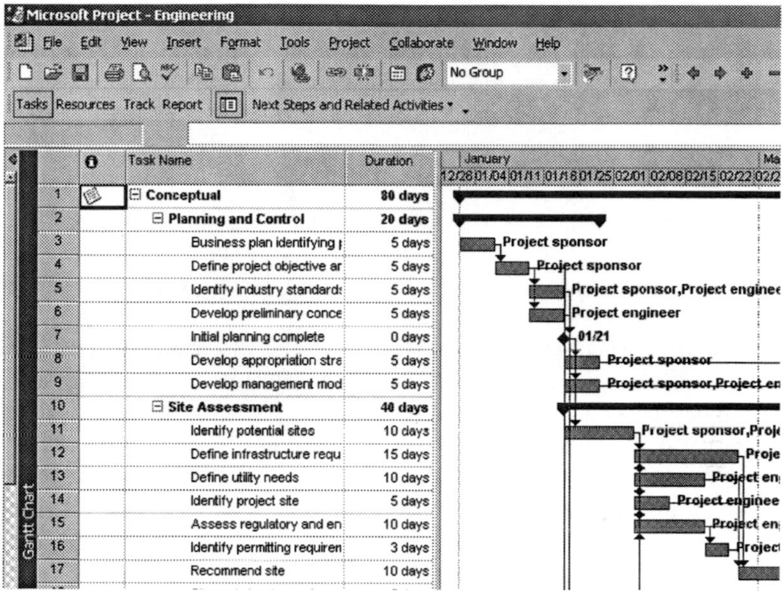

Figure 17.5. Microsoft Project network

Microsoft Project. Figure 17.5 is a screen from Microsoft Project, a widely used program. The process is generic to many planning processes.

Project management involves organizing tasks and resources to meet an objective in the best possible way. It organizes time and resources to meet the objectives. It involves plan building, tracking, managing and completing the project.

Time is measured by schedules, and resources (people, materials and equipment) are measured in terms of cost. The scope of a project is defined in terms of goals and tasks. The trio of time, money and scope define the project.

The advantages of standardized packages such as Microsoft Project include speed of preparation and interpretation. However, over-use for purpose of report padding is a danger to trainees. Human intuition may be superior for small projects.

Software like Microsoft Project maintains a database of schedules and resources to assist the project manager. Information is kept in fields such as tasks or durations. The following information is entered into the database:

- Tasks
- Durations
- Dependencies
- Resources
- Costs

Charts and tables are given to assist the manager. These can be arranged in groups or with filters to facilitate envisaging the resulting information. Reports can be printed, e-mailed or published on a web.

17.5   RESOURCE MANAGEMENT

A contractor needs certain resources to complete a project. The limiting resource can affect the completion time. But each resource represents a value or asset and a balance must be made between investment and returns. The following resources are part of a project:

- *Manpower* – Includes management, skilled workers and labourers. Sometimes the lesser skilled workers are outsourced, or employed casually. Labour laws and overheads will influence the decision.
- *Plant* – Construction equipment (cranes, excavators), or transport vehicles and services (compressors, generators) represent an expensive investment. Some plant may be hired for a job.
- *Materials* – Consumables such as cement, aggregates, pipes, fittings, are purchases as the project proceeds. On the other hand, the owner of the final product may want a stockpile of spares for maintenance purposes. Contractors may also keep stocks of spares for plant and equipment at remote sites, depending on accessibility to suppliers.
- *Finance* – Cash is needed to pay workers until the client pays. In addition, there are other expenses, including dividends to shareholders, running costs, tendering, payments for equipment purchases and offices.
- *Subcontractors* – A contractor will tend to specialize in a certain type of work, and will outsource some specialized work. A network of such sub-contractors may be via a formal company link or informally.

Detailed resource aggregation and allocation is required for a company engaged in many contracts. Resource smoothing is possible with some network planning techniques, i.e. the slacks are used to reduce demand peaks on resources. Some resource research will be required at the time of tendering to estimate a contract completion time.

Monitoring of resources is needed throughout a project. This is for the following purposes:

- Resource allocation
- Cost allocation
- Movement from site to site
- Risk management (accidents, cost overruns, new contracts).

17.6   COMMUNICATION

Engineers communicate primarily with engineering drawings. Works are located, set out and built from drawings. Drawings are now frequently prepared by computer, using packages such as AUTOCAD. Locational and regional maps may be prepared in GIS format (Geographic Information Systems). Data is fed in from survey equipment, including GPS (Geodetic Positioning Systems).

Drawings are supplemented by Specifications. These are frequently standardized for different types of work and materials. Building codes of practice and safety regulations may supplement this data.

At inception and throughout a contract, regular meetings should be held between contractor, engineer and client. Problems and changes are presented at these meetings and decisions should be minuted. Variation orders and dayworks orders should be made out on official paper forms.

Computers are used more and more to prepare drawings, or for correspondence and to transmit these to the relevant parties. E-mail is used for directed transmission and internet for more general searches for information. Backups and paper copies are advisable as computer data can be more readily lost or corrupted than hard copies.

Informal or formal contracts between engineer and contractor at all stages is advisable. It can short circuit problems, and of concern, disputes at a later stage. Legal intervention is much more expense than on-site solutions, so compromises should be sought at an early stage.

Civil engineering contracts are often not won with a large profit margin. So the contractor cannot afford to yield much in disputes. On the other hand, the engineer may lose face with the client if he seeks more money for unexpected problems, or grants the contractor more money. The client therefore needs to be in the picture at the investigation stage (when money could be saved on subsurface investigations, for example) and to realize the consequences when difficult or unexpected terrain is encountered. A contractor will try to optimize any new or unexpected work to increase his profits, so the tender schedule of quantities should be as complete as possible.

17.7   QUALITY CONTROL

Quality control has become a formal process. ISO 9000 has been developed to ensure high standards in all processes from design to construction. This involves:
- Check systems
- Documentation
- Security
- Benchmarking

382 Water Services Management

BS 5750 is the British version of ISO 9000 and European Standard EN 29000 series.

Quality assurance is defined in BS 4778 as 'All those planned and systematic actions necessary to provide adequate confidence that a product or service will satisfy given requirements for quality.' BS 4778 is the same as ISO 8402.

Apart from quality, a purchaser is interested in cost and timing (CIRIA, 1992). In a free market, price can influence the other two, or vice versa in the case of the supplier. In the case of public commodities, the quality may be influenced by a regulator, and in some cases purchases have constitutional or legal rights to quality.

Many processes have inherent checks, e.g. word processing software now checks spelling, computer files cross check for duplication, deletion and viruses.

In design, the older engineer may have an intuition as to the correctness of calculations, but this is no longer sufficient. A check system has to be demonstrated.

Many firms have opted for ISO 9000 accreditation and obtain certificates. The same is possible for environmental protection in accordance with ISO 14000. These certificates try to ensure correctness for the client and public. Audits at accreditation and subsequent design audits are thorough, but there is always a chance of some new process slipping in. The extra cost of the check systems is said to be worth it to the client, but the cost effectiveness may depend on the importance of the final product.

Construction sites usually have laboratories for standard tests, e.g. soil density, moisture content, concrete mix control and compression tests. More specialized tests can be sent to commercial laboratories, but materials with standards marks take a lot of the onus off the contractor. Records of all tests and even witnessing of tests should be seen by the engineer as well as the contractor. Deviations may require demolition or re-supply and construction.

In the case of sewers and water pipes, some form of pressure testing is normally required on completion. High pressure pumps will be used for water pipes, but sewers can be tested with gas.

17.8 HAZARDS RELEVANT TO WATER SERVICES

Construction in the water industry is particularly dangerous owing to depths of excavation, high pressures and health hazard of dirty water (see CIRIA, 1993, 1997). The following dangers may arise:
• High pressures in pipes
• Air under pressure and explosive release

- Tunnelling under high external pressures
- Collapse of trenches not shored
- Structural collapses
- Stacked pipes
- Heavy loads from cranes
- Floods
- Objects falling into excavations
- Gases from the ground or groundwater
- Sewage gases, methane, explosive or toxic
- Contagious or dangerous diseases from sewage, rats, insects, snakes
- Distance from access manhole to working place
- Slippery surfaces
- Construction equipment – heavy excavators, trucks
- Electricity and conduction in water
- Deep pressures and surfacing of divers
- Disposal of waste which may not have been properly treated
- Loud compressors
- Explosive gases
- Inhalation of dust, asbestos
- Damage of underground services – gas, electricity, sewage, pressurized pipes
- Traffic
- Weather
- Drowning

Special protective clothing, gas masks, pressure equipment, lights, eye protection, gloves and ear muffs are required. Notices and guard rails should be prominent. Whereas construction workers should be trained in safety and first aid, casual visitors, e.g. designers, inspectors, may be unaware of dangers. People with disabilities, e.g. smell, colour blindness, vertigo, claustrophobia, need to be identified before entering sites. Safety officers should be appointed (apart from security officers who guard equipment and stores, and for malicious damage or loss). A permit and colour coded access permit system may be used. Washing facilities, first aid kits, gas detectors, radio contact, CCTV's, safety ropes and winch, life jackets, are special equipment not necessarily found on surface construction sites.

Water services therefore demand a higher level of design, maintenance and vigilance. Because water and drainage are essential life services, they warrant a skilled technical management, and their management is likely to absorb an increasing proportion of household budgets in the new millennium.

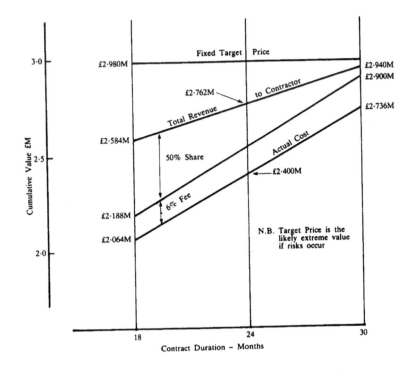

Figure 17.6. Target contract fixed target (Perry et al., 1982)

Table 17.1. Some hazardous substances in use in the water industry (CIRIA, 1997)

Substance	Natural state	Supplied state	Corrosive	Poisonous (1)	Flammable/ Explosive
Sulphuric acid (and other acids)	L	L(96%)	X	X	
Sodium hydroxide (caustic soda)	L	L(47%)	X (2)	X	
Calcium hydroxide (hydrate lime)	S	S(Powder)	X	X	
Potassium permanganate	S	S		X	
Aluminium sulphate	S	S or L		X	
Ferric aluminium sulphate	S	S or L		X	
Ferric chloride	S	L	X	X	
Polyelectrolytes	S or L	S or L		X	
Chlorine	G	L	X	X (3)	
Sodium hypochlorite	S	L		X	
Sulphur dioxide	G	L	X	X (3)	
Hexa fluorosilic acid	L	L	X	X	
Sodium silicate	S	L		X	
Phosphoric acid	L	L	X	X	
Carbon dioxide	G	L		X (4)	
Oxygen	G	L			X (5)
Ozone	G	(6)	X	X	X (5)
Ammonia	G	L	X	X (3)	
Petroleum	L	L		X	X
Solvents	L	L		X	X

Key:

L = Liquid form

G = Gaseous form

S = Solid form (powders, etc.)

Notes:

(1) If ingested in the form as supplied

(2) Corrosive to skin

(3) Asphyxiating gas

(4) Displaces oxygen in confined spaces – causes oxygen deficiency

(5) Promotes combustion

(6) Usually generated in situ

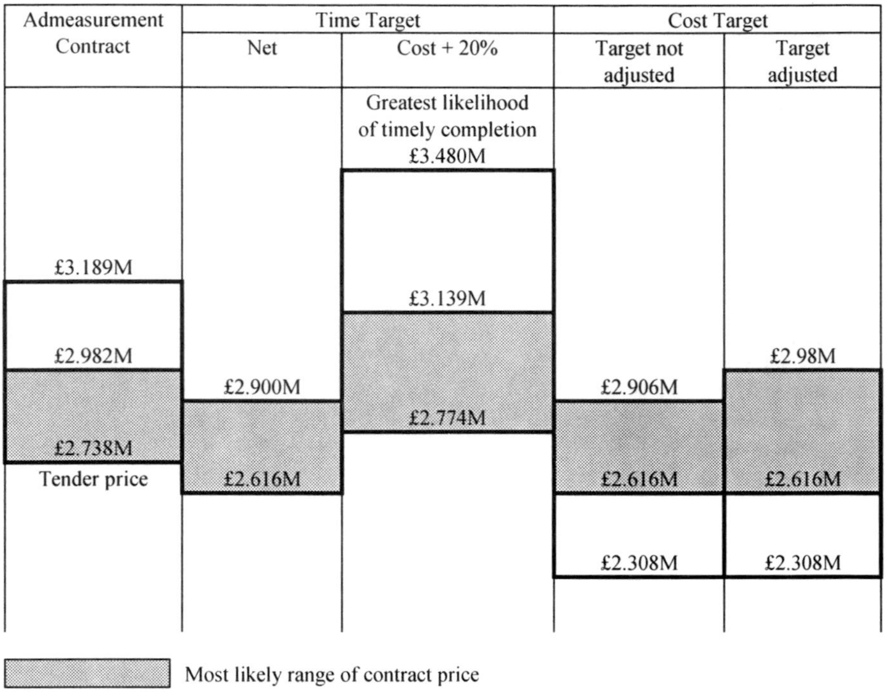

Admeasurement Contract	Time Target		Cost Target	
	Net	Cost + 20%	Target not adjusted	Target adjusted
		Greatest likelihood of timely completion £3.480M		
£3.189M		£3.139M		
£2.982M				£2.98M
	£2.900M		£2.906M	
		£2.774M		
£2.738M				
Tender price	£2.616M		£2.616M	£2.616M
			£2.308M	£2.308M

▒▒▒▒▒ Most likely range of contract price

Figure 17.7. Summary of alternative contractual mechanisms (Perry et al., 1982)

17.9 REFERENCES

Bennett, J. (1985) *Construction Project Management*. Butterworths, London.
CDM (Construction Design and Management) (1994) *Regulations*. London
CIRIA (Construction Industry Research and Information Association) (1992) *Quality Management in Construction*. Publication 88, London.
CIRIA (Construction Industry Research and Information Association) (1993) *Trenching Practice*. R97.
CIRIA (Construction Industry Research and Information Association) (1997) *Site Safety for the Water Industry*. Special Publication 137, London.
HSE (Health and Safety Executive) (1995) *Working with Sewage – the Health Hazards*. IND(G) 197, London.
Lewis, V.L. (2003) *Project Management*, 2nd Ed., Kogan, London.
Neale, R.H. and Neale, D.E. (1989) *Construction Planning*. Thomas Telford, London.
Perry, J.G., Thompson, P.A. and Wright, M. (1982) *Target and Cost – Reimbursable Construction Contracts*. CIRIA Report 85, London.

18

Organisation and operation

Generally water resources, in rivers, lakes or ground, are owned and/or controlled by government. This may be for an environmental, conservation or nation-wide reason. In some countries, the dams and abstraction works are built and owned by the water company and in some by a government department, e.g. Department of Water Affairs, Department of Environment, or by a provincial or catchment organization.

From the water resources management point of view, it would be most convenient to have all water controlled by one organization. If there could be interbasin transfers, that authority may have to be national, but otherwise river basins or catchments are the more logical natural boundary, with local exceptions if integral consumer bodies, e.g. towns, span watersheds.

The authority for water supply and wastewater control should technically be the same, as responsibility for water quality is otherwise compromised.

A large proportion of the cost of water can occur in the form of administration of the water company. High administrative costs may come about because of

- Deteriorating capital assets requiring more attention
- Greater necessity for public relations with customers and the public

- More professionals in the 'soft' professions as opposed to the technical professions resulting in more people-orientated activities
- Greater change and new technology in the supply of water

Even the process of managing water companies has changed radically of the last century. A greater emphasis is now paid to information technology (IT) and data management for managing systems. The computer based intelligence not only acts as a data gatherer and distributor but also makes decisions. These may range from modeling water distribution systems to management of accounts and procurement of resources. Whereas this helps to ensure a stable system and reduces the human element effects, it does incur a high initial cost in that higher tech people have to be paid to design the software systems and maintain them.

This may be contrasted with the older methods common in the smaller waterworks which sprung up over countries. The concept of water management companies is relatively new and it is only towards the end of the last century that the older engineers and managers of the water company who were there almost from the conception of much of the installation had retired. Their knowledge and experience was taken with them and it was now realized that hard copies of knowledge are desirable especially in view of the faster turn around in staff. Not only is the knowledge of systems to be converted to digital versions but also the skills in operating must be thought out by optimization programs.

This high tech approach may be contrasted with the other facet required of water companies and that is good public relations. One of the things that privatization has brought about is the realization that the payer should receive the service he wants. This may mean improvements to the system to suit consumers' requirements and the cycle is now turning towards good customer relations, but the cost implications may yet have to follow and there may be some streamlining of the system in years to come.

Advances in water supply management includes:
- Data management
- Leakage detection and management
- Analysis, design and rehabilitation of distribution networks
- Water quality management in distribution networks
- Network optimization / staged expansion / storage-capacity
- Network operations
- Water economics
- Water demand management (conservation and recycling)
- Reliability analysis

18.1   QUALITY CONTROL

The core business of a water company is to provide sufficient water at an acceptable quality, pressure and assurance of supply. To do this the water company needs adequate engineering skills but also a good level of administration to ensure the ideas are carried out. Therefore a management team has to be created to fit the asset installation and enhancement system into the overall picture of supply. In order to meet the objective, the water company needs to establish strategies and this is required for long-term, medium-term and short-term management. In the *long-term*, the following are required by the planners:

- Target setting in terms of standards such as leakage inspection, quality control
- Administration of the system to achieve the targets at minimum cost
- A balance between the multiple objectives of the company such as customer satisfaction and cost minimization
- Obtain and allocate funding
- Communication between the relevant stakeholders
- Timeous action

In the medium-term, the company should establish a programme to meet its targets and these in turn will include:

- Improvement in the quality of service such as attention to customer complaints
- Reduction of losses by leak detection
- Rehabilitation of older systems
- Cost minimization
- Monitoring of the various technical aspects of the system including water flows, pressures, quality and consumer reaction
- Training of employees

While all this is happening, there may be aberrations in the cycle, such as drought, changes in consumer requirements and staff disabilities as well as leaks, and other action is required. The management of the short-term objective is often achieved by outsourcing in order to optimize staff costs.

Monitoring systems are becoming more important as systems get older. Pressure reduction due to aging pipes, leaks and increases in demands on the system have to be continuously assessed and the necessary action taken. But monitoring of pressure, flow and water quality throughout the system should lead into decision-making programs such as reticulation network models and pump scheduling programs. However, it is not only the distribution system which needs monitoring, but also the resource systems such as the river flow or ground water levels and quality. These are required to ensure sustainability and reliability in supply.

Shorter term monitoring requirements may occur at pump stations where things such as pump efficiency, maintenance attention (example overheating bearings, and mechanical and electrical equipment) need to be continuously checked. Civil engineering works on the other hand require a different type of monitoring and this may not need to be on line but could involve scheduled inspections. Throughout the system, monitoring is needed for identifying and locating of leaks.

Consumer monitoring is done in the way of analysis of reports and complaints and checking the meter data. There it is useful to link the billing data to the technical side of the works so that they can pick up changes in consumption patterns and identify problem areas.

Computer systems are now available for doing much of this including SCADA (Supervisory Control And Data Acquisition Systems), GIS (Geographic Information Systems) and PLC (Program Logic Controls). Software for data management is probably more standard than user software, that is programs for modeling and real-time management of the hydraulic systems.

Water quality is becoming the greater component in the observation system because of regular media scares concerning dangerous water quality and new found pollutants, but more particularly because of increases of consumer interaction. Therefore control of disinfection systems, pH, taste and odour need observation at various points throughout the distribution system. On the resource side, the optimum level of treatment and dosage of chemicals can be obtained by monitoring the incoming water quality.

With single pipe systems, it is difficult to achieve an economic balance between quality of water to be purified and the water quality required by the consumers. Bear in mind that the majority of water is not used for drinking but for washing, but nevertheless strict quality control is required to ensure no health hazards.

18.2   CONTAMINATION

Even with adequate water treatment and monitoring at the source and water treatment works, there can be dangers of contamination further down the line. Sources of contamination can include:
- Cross connections of different systems or pipes
- Air valves which can draw air or polluted water in
- Corrosion of pipes and fittings
- Construction activities when the pipes are opened up
- Leaks and bursts
- Biological growth within the pipes and reservoirs
- Temperature changes
- Dead ends which collect bacteria and corroded material
- High velocity scouring of rust, etc., from the pipe walls

- Age of systems, particularly concrete or fibre cement
- Low levels in reservoirs or stagnant reservoirs
- Poor linings to pipes.

Particularly as systems get older, new sources of contamination may arise and this is not easily detectable by cursory inspection. It needs forethought and intrusive monitoring in many cases. This may require taking systems out of service for short periods and redirecting flows. But the looped networks are becoming more important than when the systems were new.

18.3   FINANCIAL MANAGEMENT

The cost of water is one of the more important factors to the consumer. Although the technical staff may optimize especially in the design and operation, there are a number of other factors which influence the cost of water. The component of cost imposed by the national government for resource conservation or environmental improvement component is only one example, but the most important factor is the financial charges. The interest and exchange rate payable on loans has an important, direct bearing on the cost of water. The financing of water companies can come from various sources:

- Revenue raised from the sale of water which is invested until it is required for new capital work
- Raised on the open stock market, usually in the way of guaranteed bonds with a fixed interest
- Capital introduced by private companies obtaining ownership or leases on the water company
- Money lent possibly as soft loans by aid organizations although this applies primarily to developing countries
- Loans arranged by banks who obtain private funding or government funding.

The interest payable is normally more attractive than by private borrowers because of the large and necessary nature of a water service. The exchange rate is however another thing and money borrowed by poorer companies overseas is often repaid with devalued local currency enforcing an increase in water tariffs. This is becoming even more of an issue with privatization and risk exists not only to the water company but also to the private company buying into the water company.

The financial advisors of a water company therefore need to find finances sufficiently far ahead of time, at the correct amount and at the best possible interest rate, exchange rate and loan period. The loan period is often less than the engineering life of the works it is to finance, so that the loans are often directly linked to one facet or component or the water works but are taken out when the float in the water company's bank drops below a target figure or when a large capital expansion is spent.

It is becoming more important for the planning engineer to realise that large capital loans can be taken in opportune times on the economic market and therefore some adjustments in the planning time horizons may be necessary to fit in with the financing limitations.

18.4HUMAN RESOURCE MANAGEMENT

It is becoming more important to maintain the technical competency of water works staff for various reasons. Not only is the technical know-how changing at a fast rate, particularly in the information technology sector, but also there is a higher turnover of staff necessitating training of new staff. Different types of training are therefore required, i.e. induction, ongoing or refreshing and advanced.

Although a basic education at university or technical college is becoming a more common sight in water company staff, the importance of on-site training, i.e. non-academic training, must not be overlooked. This is the type of training which is often lacking in developing countries because of the inexperience of staff.

The biggest expenditure by water services is likely to be on capital works. System planning and optimization forms a large part in providing a cost effective service. But it is the day to day operation which is most apparent to the public. And operational costs can approach and even exceed capital expenditure over time.

The efficient day to day operation of the works is important not only for minimizing expenditure but also for ensuring customer satisfaction and investor confidence. The latter functions were once confined to shops and banks respectively, but nowadays water service providers have to consider non-technical operational aspects as well.

Despite a rapidly changing business approach and fashions, knowledge and technical expertise will continue to be a valuable asset in a water company. Good management will use experience as well as new knowledge to provide the best service.

Interaction of water service providers by way of professional associations, publications and teaching, provides a rapid form of knowledge sharing. This establishes benchmarks for established organizations, and starting points for newer works and emerging countries. Organization such as IWA (International Water Association) and AWWA (American Water Works Association) are large, providing stages for numerous meetings, congresses and publications. Many smaller national organizations exist, and many more specialized organizations such as IAHR (International Association of Hydraulic Engineering and Research), IWRA (International Water Resources Association) and national engineering associations.

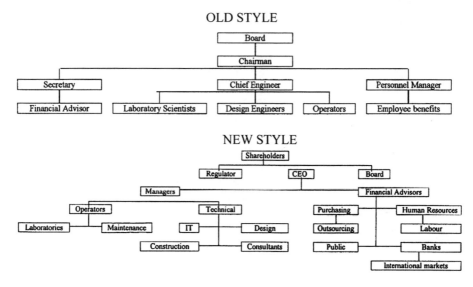

Figure 18.1. Management style

18.4.1 Human resources and training

The output of any organization is dependent not only on the number of staff, but also on their ability and efficiency. The days of rewarding loyalty and attendance are being replaced by dynamic management styles.

The new approach is characterized by:

- Lifetime jobs are less easy to get
- Continuing education is necessary to stay abreast of new technology
- Younger more mobile work force is evident in new technologies
- Pension possibilities are of less concern than tax efficient salary packages including travel allowance
- Work place equity is changing the old boy school nepotism

It is necessary for companies to expose and train older employees to keep up with the younger staff ability, or job reversal will occur, with disheartening consequence. But technical training is equally important to keep up with best technology. In the long term, University and College courses provide suitable backgrounds but short courses and workshops are also necessary at regular intervals. This is in addition to on-site training and learning.

In the case of water works in less developed countries, staff exchanges, visits and mentoring arrangement with more experiences organization s are also necessary.

18.5   MAINTENANCE SCHEDULING

Planned maintenance is important for maintaining serviceability and economic life of facilities. But even with passive maintenance, some benefit is obtained by documenting activities. Data lists of leaks and other problems can be assessed before deciding on replacements, rehabilitation or revised schedules. Data bases should contain location, description, method of solving, equipment, costs and cause of problems. These facts can then be perused regularly to identify problem areas.

A system of identifying features should be worked out. Apart from GIS maps, a numbering system can locate pipes and fixtures in the system, e.g. zone, section, pipe, fittings, as well as depth, condition, year of installation and value in the case of asset registers. The latter can be of more use than maintenance, it can enable decisions concerning value, planning and environmental impact.

Whereas water supply pipes do usually receive attention, the unblocking of sewers, stormwater drains, culvert and bridge openings, is probably not as accurately documented an account of the diverse nature of problems. Some problems may even be attended to by emergency service providers as opposed to water works operators. Feed back from fire fighting departments is also of use. They may draw heavily from hydrants, they may monitor pressures, and they should equally be made aware of water pipes out of service, floods or blockages.

Privatization of such services can restrict access to data so a public coordinator or IT department may be required.

Staffing of maintenance, operation and emergency services is not standard. Overtime, staffing during holidays and out of hours work make such jobs unpopular and that in turn reduces the image of the service provider. Publicity and rewards thus form an active part in public relations.

18.6   EMERGENCY SUPPLIES

There are many situations where fast connections of water to consumers is more important than standards. Community water supply (and other services) could be disrupted, so there is an existing demand. This is not the same as planning for expansion in demand due to natural growth. Emergency situations can arise due to:
- War
- Sabotage

- Refugee camps
- Natural catastrophes, e.g. earthquakes, volcanoes, landslides, floods
- Cutoff of existing source, due to drought, over-utilization, failure of supply system
- Pollution
- Drop in pressure.

It is not economical to have standby equipment in stock for these emergencies, but a detailed suppliers list should be prepared by purchasing in conjunction with technical staff. Compatibility with existing equipment, staff familiarity will operation and maintenance, reliability and speed of acquisition and installation are important. This is probably weighted more than cost, preferential suppliers, etc., used in normal sourcing. International aid is of value to countries affected by emergencies, but the value could be severely restricted unless the above factors are considered.

Alternative standard facilitating installation may include:
- Lower supply rates, limited by pressure, or time of supply
- Lower pressure
- Limited purification with disinfection
- Short specifications, design and constraints.

Supply may be laid on in phases, e.g.
- Containers
- Tankers
- Surface pipes
- Communal outlets, e.g. tanks, standpipes.

Use of private consultants and contractors is more common in emergency supplies, as their services may be temporary. But handover problems and long term implications may have to be discussed with communities and officials.

Water service providers should have some idea of national responsibility, not only to minimize monopolizing and profiteering, but also to act as a public service. National government may not have access to the necessary skills and international aid organizations will generally offer humanitarian aid but not technical or construction services.

In the case of emergencies, it is often necessary to compromise environmental and social requirements. There may be later legal implications but foresight could reduce the red tape.

18.7   LEGAL

18.7.1 Water law

Use and effects on water can be controlled at Government level, or provincial, or water authorities. The latter may have powers assigned to them by government.

A new water law (e.g. South Africa, 1998) was to establish the basis of water allocation, on the premise that water cannot be privately owned but belongs to all the people. It identifies past discriminatory laws and proposes the National Government now takes responsibility and exerts authority for allocation and use, transfers and international matters. The necessity for independent water courts will fall away.

The purpose of an Act is said to be to ensure water is conserved, managed, used and controlled to meet personal needs in a sustainable way, to facilitate national development, protect the aquatic related environment, to prevent pollution and to meet international obligations.

Licenses will be issued for water abstraction, whether surface or groundwater. Individuals may use water for domestic use or stock watering free of charge if they have lawful access as well as for recreation and emergency fire fighting.

The principal of integrated catchment management is recognized and to do this it is proposed:
- to identify management areas
- classify significant resources
- determine reserves
- identify quality objectives
- prepare allocation plans
- develop strategies for future development.

This requires the allocation of water by the relevant department, or at least to establish the principles of allocation. This could include inviting tenders after provision for licenses, reserve and international obligations. Management plans would recognize catchment characteristics, especially water, the environment, water uses, disaster effects, community involvement, economics, monitoring and responsibilities.

In establishing the plan, consideration is also given to enforcement, information dissemination, evaluation and time frames.

Water resources can be 'designated', thereby enabling them to have designated authorities allocating licenses for water use and abstraction. Licenses are also required for activities such as hydro-power generation, workings in the bed of the river or fish farming, and intercepting water.

There is also control of waterworks, including storage, abstractions, drainage, monitoring, treatment and stream diversion. These shall be registered and may be subject to approval, e.g. dam safety, environmental impact. Borehole drillers also have to be registered, as well as bodies discharging into water resources. Licenses may be withdrawn or amended. Access to land for monitoring or water works can be authorized by the Minister and he may expropriate property and set charges for water use or discharge.

A bill allows for a transition from riparian based apportionment to beneficial use in the public interest.

Included is the possibility of establishing a national water utility to draw together role players and advise the minister and manage water infrastructures.

An interesting feature is the possibility of trading entitlements to use or control water. Also covered by licenses is weather modification and importing water.

An Act should be broad in its wording and open to changing technology, interpretation and uses. Although much control is in the hands of the minister, this is intended to prevent private use or abuse not in the national interest and not to be construed as authoritarian.

There are many arguments for government intervention in water allocation (Dinar et al., 1997):

- Large capital requirements
- Economy of scale
- Joint (multipurpose) development
- Long time horizons
- Interdependence of resources, e.g. withdrawals upstream
- Risk management (floods, pollution) are public
- Strategic importance
- Sustainability of infrastructure
- Conservation.

18.7.2 Bylaws

Water use can be controlled by physical means (pressure control or obstruction or cutoff), appeals (newspapers, etc.), economic (tariffs), metering or by legislation. High tariffs on discharge to sewers can also improve water use efficiency. Water bylaws may be promulgated by national government, local government or water companies. These rules may require consumers to:

- Maintain fittings in reasonable repair to avoid leaks
- Prohibit garden watering, car washing, etc., on certain days
- Use standards of pipe and fittings approved

- Restrict overflows
- Restrict cistern capacity and certain types of flushing valves
- Restrict automatic sprinklers, flushers, etc.
- Limit use of water for secondary power, e.g. hydraulic actuators
- Control pollution discharges
- Prevent cross connecting to other water reticulation systems
- Ban wet cooling or air conditioning systems
- Limit pressure reduction, e.g. private booster pumps

Such regulations can be controlled by the water supply authority most easily. Fines or restrictions in supply can be used as penalties for transgressions, but basic health requirements should always be met. The monitoring by the water authority need not be confined to water fittings. Pollution of the catchment and export of water or wastewater from the catchment should also be performed by the water supply authority. A tariff based on water consumption volume, rate, use, pollution, loss and discharge would be a theoretical solution but the metering of all these parameters could be excessive. Hence spot checks and fines appear the most practical way of controlling water use and abuse.
An accountable water provider must comply with laws and regulations. Legal expertise may be required in interpreting laws. Other inputs of lawyers may be in drafting regulations and enforcing protection of resources, infrastructure, etc. The water company may have to be protected against excessive claims from irate consumers, the public and subcontractors. Contracts for outsourcing of services, employment and supply will have to be drawn up and on occasions defended.

18.7.3 Inspectorates

Apart from a regulator, there may be government inspectors monitoring the water company. This may be to ensure adequate water quality and general standards of supply. There may also be building inspectors, safety inspectors, health inspectors and financial and stock auditors, required by laws and for good business management practice.

18.7.4 Regulation

If water supply is privatized or taken out of the hands of government, it needs to be controlled or audited in some way by a public body. In the UK, the office 'OFWAT' was established for this purpose and the duties of the Director General are as follows:

Set standards   -   Water quality

		Level of supply, assuredness
		Technical
Inspectorate	-	Monitor
		Report
Control of Core Business	-	Cannot be sold off
		Can alter areas
Limit charges		Increase limited to inflation plus a factor (K)
Maintain registers	-	Assets
		Performance
		Economy

The establishment of a regulatory authority is a wise move in respect of privatization, but to whom the authority should report and what its duty and source of money is, needs to be considered. The regulatory council could comprise one or more of the following:

- Government
- Local authority
- Public
- Water Board
- Consumers
- Technical consultants
- Auditors
- Shareholders
- Independent advisors

In the case of small, e.g. rural, water schemes, the duty of the regulators could be more of a technical advisory nature, whereas a large more capable water authority may only need a watchdog.

18.7.5 Competition

Although efficiency is generally improved by competition, in the case of bulk and public services, this may be impractical and the duplication costly. So competitive practices rarely appear. But within consumers, there may be competition for limited resources, pressure or service.

The use of a 'common pool' is inclined to waste resources. The competition to draw from an aquifer or network can result in overexploitation because the resultant risk is shared by other consumers.

Competition is in other forms such as subcontracts, bids for takeover. The day may yet come when alternative services are offered e.g. by container water, or low quality bulk supply.

18.8   SEWER MAINTENANCE DATA PROCESSING

Johannesburg has nearly 4000 kms of sewerage to operate and maintain on a continuous basis. Many of the areas are prone to abuse and blockage and the nature of the topography and climate make maintenance a high cost in the system. That is, intense storms often result in ingress into sewers and this may bring surface debris and other foreign matter which block the sewers. There is also unauthorized access in many places it is suspected, as articles, obviously not from the sanitation system, are found in sewers. Despite the high rate of growth in Johannesburg, many of the sewers are old and some are of poor quality requiring regular maintenance and replacement and repairs.

Overflows and lifting of manhole lids in certain areas may point to in adequate sewer capacities. Alternatively they may indicate corroded sewer linings or roots which block the sewers. Here again, identification of frequency and locality of such inadequacies indicates where maintenance is most urgently needed.

The human management side is also very complex. The outlying depots where such maintenance takes place employ some 600 people, which are generally organized into gangs at each depot. The supervisors report to managers to take messages and transmit the teams to problem points. Even managers and frequently supervisors are not highly trained and the type of logs they keep are often difficult to process. However, the computerization of the log keeping on an experimental basis at one of the depots has proved satisfactory and within the capabilities of the existing type of staff. Terminals connected to the municipality's main computer at head office are used and once basic keyboard skills have been picked up, the spreadsheet type of data logging has proved possible and in fact of great advantage to the engineers at head office concerned with planning and the engineers concerned with budgeting and maintenance and design.

The use of computers also enables micro graphics to be used to identify trouble areas. A screen map can highlight zones with frequent blockages. With the advent of computers, many fields in Civil Engineering have been opened up through the benefits which can accrue in both design and constructional areas as well as management and administration areas. Due to initial costs and a natural reluctance to adopt new methods, progress is sometimes slow but it can usually be said that while computers do not necessarily save money they can definitely give better results at the end of the day.

18.8.1 Application

The analysis of sewer systems to identify potential overloading by sewers has already been established and has been used to analyse townships for existing and future flows. In some cases, the effect of subdivision of stands has been assessed and accurate estimates of costs given for additional sewerage work (Stephenson and Hine, 1982 and 1985).

Sewer reticulations need regular planned cleansing if serious flooding and subsequent danger to health is to be avoided. If regular cleansing of public sewers is well organized, many of the blockages which occur can be avoided. Maintenance of privately owned sewers is not the responsibility of the sewerage authority, but in Johannesburg it is the official policy to unblock these sewers if asked to do so by the owner. In many cases, the owner is the local authority so that there is a vested interest to ensure that these are well maintained so as to reduce the number of blockages.

Conventional systems have been used to record the work carried out using cards, etc., which have been successful but time consuming. It was considered that records of cleansing work and the clearing of blockages could be more effectively done by computer and that retrieval of records and planning of work would be made easier.

Consequently the Maintenance Data System has been established and is being applied where Sewer Data has been established giving sizes and lengths of sewers together with a unique manhole numbering system.

Data is compiled by depot administrative staff on Forms which have a numerical format suitable for input to the computer. Details are abstracted from work reports daily.

Forms used in the field given township and street names which become numerical township codes and manhole numbers before being entered into the computer. Incorporated in the cleansing report is an inspection of each manhole and sewer length including the measurement of the depth of flow.

18.8.2 Processing of sewer maintenance data

The processing of sewer maintenance data has reached an advanced stage using the programs and techniques described below.

The workforce is divided into gangs which work on either cleaning of sewers or clearing of blockages.

The cleaning of sewers is recorded by the gang leaders in the field each day and manhole numbers are obtained from keyplans showing the sewer network.

On the following working day, information is abstracted by depot administrative staff and inserted in a numerical format (Table 18.1).

Program UPDATE is then used to provide details of sewer diameter, length and slope which is added to the data file. These details are obtained from the Sewer Data File.

Table 18.1. Cleansing of sewers

RECORD OF SYSTEMATIC CLEANING & SEWER/MANHOLE CONDITION

Table 18.2. Blockage data form B

NOTES:

Under "EFFECT" "Private": 1 = House Flooded, 2 = Yard flooded, 3 = House and yard flooded

Under "Referred to": 1 = Water Branch, 2 = Road and Works Branch

Program MAINTENANCE produces a report of the sewers cleaned between given dates as required and printed out according to each township.

Program GANGS produces a report of work carried out by each gang between given dates.

This report could form the basis for a bonus scheme.

Blockages are recorded as reported with the time and date recorded. The details of actual clearance giving the time started and completed appear on work report sheets and enable data to be completed. "Private" blockages which are within stand boundaries are denoted by a stand number and "Main" blockages which occur in public sewer are denoted by a manhole reference number.

Reports which are found to be problems in the water reticulation or storm water system are given a code which enables the computer to ignore that report apart from showing how many reports have been referred elsewhere for action (Table 18.2).

Program BLOCKMACRO produces a report of blockages in townships between given dates. The length of time taken to clear the blockage and possible cause is shown. The time which elapsed between the report and completion of clearance is also calculated to help identify administration problems, lack of staff, etc. The severity of a blockage is also shown by indicating the number of houses flooded as a result of a "main" blockage and for a "private" blockage if the house or yard is flooded.

A macro program produces a report of all the work each gang has done in unblocking sewers between given dates. Numbers of blockages and total time spent is shown (Table 18.3).

Table 18.3. Output file

```
SEWER BLOCKAGE RECORDS

REQUESTED START DATE:- 870105
REQUESTED END DATE :- 870107

GANG:    1     SIZE:    4

                         NO. OF    NO. OF    TOTAL
               TOWNSHIP  PRIVATE   MAIN      NO. OF    PRIVATE- MAIN-    TOTAL -
    TOWNCODE   NAME      BLOCKAGES BLOCKAGES BLOCKAGES JOB TIME JOB TIME JOB TIME
       159  ELDORADO PARK2    2         0         2      4.25    0.00    4.25
       177  ELDORADO PARK4    4         0         4      7.25    0.00    7.25
       342  KLIPSPRUIT WES    1         0         1      0.25    0.00    0.25

    TOTALS:-                  7         0         7     12.15    0.00   12.15
```

A program produces a report of all the stands and sewer lengths where there has been more than one blockage in a given time period. This information can be very useful in identifying possible defects and overloading of public sewers and also when answering queries about repeated blockages on private stands.

18.9   REFERENCES

Agthe, D.E., Billings, B. and Buras, N. (2003) *Managing Urban Water Supply*, Kluwer.

Dinar, A., Rosegrant, M.W. and Meinzen, Dick R (1997) *Water Allocation Mechanisms*. World Bank Policy Research Working Paper 1779.

Kalickman, S. and Nichols, N. (1996) *Water Distribution System Operation and Maintenance*, 3rd ed., California State Univ., Sacramento, in conjunction with National Environmental Training Association.

OFWAT (Office of Water Services) (1997).

South Africa (1998) Water Act, drafted by Dept. Water Affairs and Forestry, Pretoria.

Stephenson, D. and Hine, A.E. (1982) Computer analysis of Johannesburg sewers. *Proc. Instn. Munic. Engrs. SA, IMIESA*, 7(4), Apr., 13-23.

Stephenson, D. and Hine, A.E. (1985) Sewer flow modules for various types of development in Johannesburg. *Proc. Inst. Munic. Engrs. SA,* **10**(10), 31-41.

World Health Organization (WHO) (1994) *Operation and maintenance of urban water supply and sanitation systems, a guide for managers.* Geneva.

Index